Pest Management in Cotton

A Global Perspective

Edited by

Graham A. Matthews

Emeritus Professor, Imperial College London

and

Thomas A. Miller

Emeritus Professor, University of California, Riverside, California

CABI

CABI is a trading name of CAB International

CABI
Nosworthy Way
Wallingford
Oxfordshire OX10 8DE
UK

Tel: +44 (0)1491 832111
Fax: +44 (0)1491 833508
E-mail: info@cabi.org
Website: www.cabi.org

CABI
WeWork
One Lincoln Street
24th Floor
Boston, MA 02111
USA

Tel: +1 (617)682-9015
E-mail: cabi-nao@cabi.org

A catalogue record for this book is available from the British Library, London, UK.

Library of Congress Cataloging-in-Publication Data

Names: Matthews, G. A., editor. | Miller, Thomas A., (Entomologist), editor.
Title: Pest management in cotton : a global perspective / edited by G. A. Matthews, and T. Miller.
Description: Boston : CAB International, [2021] | Includes bibliographical references and index. | Summary: "This book describes the background and present state of pest management in cotton crops throughout the world. Cotton is one of the most economically important crops globally, and new, emerging pests driven by climate change, for example, are proving increasingly problematic. The book highlights new techniques in cotton crop protection"-- Provided by publisher.
Identifiers: LCCN 2021023836 (print) | LCCN 2021023837 (ebook) | ISBN 9781800620216 (hardback) | ISBN 9781800620223 (ebook) | ISBN 9781800620230 (epub)
Subjects: LCSH: Cotton--Diseases and pests--Control. | Agricultural pests--Control.
Classification: LCC SB608.C8 P462 2021 (print) | LCC SB608.C8 (ebook) | DDC 633.5/1--dc23
LC record available at https://lccn.loc.gov/2021023836
LC ebook record available at https://lccn.loc.gov/2021023837

References to internet websites (URLs) were accurate at the time of writing.

ISBN: 9781800620216 (hardback)
9781800620223 (ePDF)
9781800620230 (ePub)

Commissioning Editor: Ward Cooper
Editorial Assistant: Emma McCann
Production Editor: Tim Kapp

Typeset by SPi, Pondicherry, India
Printed and bound in the UK by Severn, Gloucester

Contents

Dedications

To my devoted wife Moira for her unending love and support.

Graham Matthews

I dedicate my portion of this volume to the late John Benson and the late Don Cox, two cotton growers in the Imperial Valley, California; the late Wally Shropshire, gin operator; Bob Roberson, branch chief of the California Department of Food and Agriculture, Integrated Pest Control branch; and Bob Staten, United States Department of Agriculture.

The people to whom this volume is dedicated taught me just how difficult farming is and how much even the smallest of research contributions are treasured and how the pink bollworm alone nearly destroyed the cotton industry in California and Arizona. All of this was done out of the public eye. In the Caribbean, Barbados Sea Island cotton is grown and they have the pink bollworm greatly restricting their crop, so there is a threat that it might return to the USA.

The five named above are heroes of mine because they went about their business unheralded; also they are examples of how the combination of university, industry and government can succeed when the right people are properly focused on goals and are unconcerned about who gets the credit.

This volume is dedicated to those who produce cotton in many parts of the world under such a wide range of pests and climates.

Thomas Miller

List of Contributors

Zeeshan Ahmed, Postdoc Fellow, Xinjiang Institute of Ecology and Geography, Chinese Academy of Sciences, Xinjiang, 830011, China

Yaşar Akişcan, Faculty of Agriculture, University of Hatay Mustafa Kemal, Hatay, Turkey

Abid Ali, Assistant Professor, Department of Entomology, University of Agriculture Faisalabad 38040, Punjab-Pakistan, and Adjunct Professor, College of Life Science, Shenyang Normal University, Shenyang 110034, Liaoning, China

Charles T. Allen, Professor and Extension Specialist and Associate Department Head Emeritus, Texas A&M AgriLife Extension Service, San Angelo, Texas 76904, USA

Silas Wintuma Avicor, Entomology Division, Cocoa Research Institute of Ghana, PO Box 8, New Tafo-Akim, Ghana

Steven M. Brown, Department of Crop, Soil and Environmental Sciences, Auburn University, Alabama 36849, USA

Charlie Cahoon, Crop and Soil Sciences, NC State University, Raleigh, NC 27695

Feza Can, Faculty of Agriculture, University of Hatay Mustafa Kemal, Hatay, Turkey

Keith Edmisten, Professor of Crop and Soil Sciences, North Carolina State University, Raleigh, North Carolina, USA

Pius Elobu, Research Officer, National Semi-Arid Resources Research Institute, Uganda

Bahodir Eshchanov, Associate Professor at the Department of Plants and Agricultural Products Quarantine, International Affairs at Scientific Research institute of Plant Protection, Michigan State University Co-ordinator in Uzbekistan

Gao Guizhen, College of Forestry and Horticulture, Xinjiang Agricultural University, Urumqi, China.

S.P. Gawande, ICAR-Central Institute for Cotton Research, Nagpur, Maharashtra, India

Paul Grundy, Principal Scientist, Crop and Food Science, Department of Agriculture and Fisheries, PO Box 102, Toowoomba, Queensland 4350, Australia

Joe C.B. Kabissa, Independent Consultant and Former Director General of the Tanzania Cotton Board, PO Box 36518, Kigamboni, Dar es Salaam, Tanzania

Billy R. Leonard, Louisiana State University, Baton Rouge, Louisiana, USA

Li Xueyuan, Cash Crop Research Institute, Xinjiang Academy of Agricultural Sciences, Urumqi, China.

Liang Fei, Xinjiang Academy of Agricultural Reclamation Sciences, China.

Lu Zhaozhi, Professor, College of Plant Health & Medicine, Qindao Agriculture University, China

Cafer Mart, Faculty of Agriculture, University of Hatay Mustafa Kemal, Hatay, Turkey

Graham A. Matthews, Emeritus Professor, Imperial College London, Silwood Park, Ascot, UK

Thomas Miller, Professor Emeritus, University of California, Riverside, California, USA

Anthony Muriithi, Director General, Agriculture and Food Authority, PO Box 37962-00100, Nairobi, Kenya

D.T. Nagrale, Scientist (Plant Pathology), Division of Crop Production, ICAR-Central Institute for Cotton Research, Nagpur, Maharashtra, India

V.S. Nagrare, Principal Scientist (Entomology), Division of Crop Production, Scientist (Plant Pathology), Division of Crop Production, ICAR-Central Institute for Cotton Research, Nagpur, Maharashtra, India

Shadmon E. Namazov, Uzbek Scientific Research Institute of Cotton Breeding and Seed Production, 702147 Tashkent Province, Kibray District, P.O. Salar, The Republic of Uzbekistan

Germain Ochou Ochou, Principal Scientist, Cotton Research Program and Scientific Coordinator of the National Center for Agricultural Research, 01 BP 1740 Abidjan 01, Côte d'Ivoire

Berkant Ödemiş, Faculty of Agriculture, University of Hatay Mustafa Kemal, Hatay, Turkey

Jane Pierce, Associate Professor, New Mexico State University, Las Cruces, New Mexico, USA

Dominic Reisig, Professor and Extension Specialist, Department of Entomology and Plant Pathology, North Carolina State University, USA.

Phillip M. Roberts, University of Georgia, Athens, Georgia, USA

Tian Jingshan, Key Laboratory of Oasis Eco-Agriculture, Xinjiang Production and Construction Corps, College of Agronomy, Shihezi University, China.

John Tunstall, Formerly Director, Cotton Pest Research Federation of Rhodesia and Nyasaland and then in Malawi (now retired)

Simone Silva Viera, Entomologist, Researcher and Consultant at PlantCare, Humberto Milanesi, 395, Botucatu, São Paulo, Brazil

M.V. Venugopalan, Principal Scientist (Agronomy), Division of Crop Production, ICAR-Central Institute for Cotton Research, Nagpur, Maharashtra, India

V.N. Waghmare, Head of Division of Crop Improvement, ICAR-Central Institute for Cotton Research, Nagpur, Maharashtra, India

Wang Juneduo, Cash Crop Research Institute, Xinjiang Academy of Agricultural Sciences, Urumqi, China

Yang Desong, College of Agriculture, Shihezi University, China

Zhang Wangfeng, Key Laboratory of Oasis Eco-Agriculture, Xinjiang Production and Construction Corps, College of Agronomy, Shihezi University, China

Zheng Guo, Professor, College of Life Science, Shenyang Normal University, Shenyang 110034, Liaoning, China

Zheng Juyun, Cash Crop Research Institute, Xinjiang Academy of Agricultural Sciences, Urumqi, China

Acknowledgements

We thank the many who agreed to participate in this project and contributed chapters from their country. Special thanks go to Abid Ali who has also assisted during the gathering of the chapters. We thank Peter Beeden who helped with the sections of the chapters covering Malawi and Nigeria. Information for other countries has been based on published articles and experience obtained during visits to 23 countries in relation to cotton growing. Many of the photographs in this book were taken during these visits or have previously been used in publications. I thank the authors of certain chapters for supplying the photographs for their chapter and to Fernando Carvalho for photographs of aerial and tractor spraying in Brazil. Data on production of cotton in the countries has been based on information from indexmundi.com, which sources data from the United States Department of Agriculture PSD database.

Prologue

Cotton is the world's most important natural fibre that has provided clothing for people throughout the world. Many books have provided information about the production of cotton and about the global trade in the fibre. Cotton is now grown in a vast number of countries and while some books cover the problems of controlling pests and diseases, this new book aims to examine how the crop is grown at a time when there is a major change in how the crop can be protected using modern technology.

The climatic conditions in the cotton growing areas vary enormously from the severe winters in the most northerly parts to the tropical regions, with similar contrasts of rainfall. Farm sizes have varied from small-scale family farms, relying on manual harvesting of the crop, to very large areas

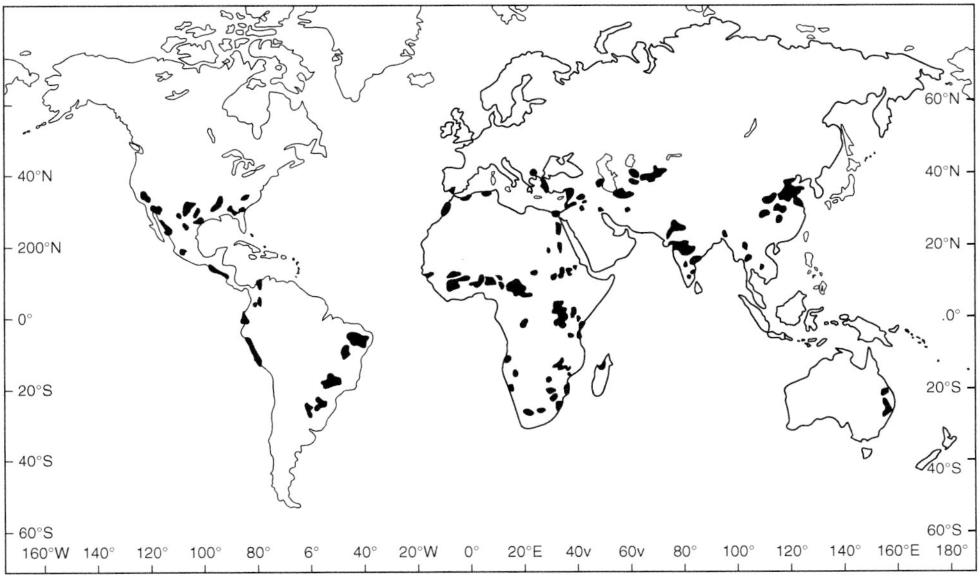

Fig. P1. Distribution of cotton growing areas around the world. (Author's figure, previously published by Longman in 1989)

with mechanization. While rain is adequate in some areas, irrigation systems have been developed successfully in many arid areas.

Over the centuries, cotton seed has been moved and plant breeders have vastly changed the important factors in relation to the length and strength of fibres and potential yield. Production was increased dramatically with the arrival of insecticides that significantly reduced the extent to which insect pests damaged the crop, while similar progress has been made with weed and disease management. New tools to help the plant breeder now have made a quantum change in the development of new genetically modified traits that in parts of the world have vastly improved yields.

This book has been prepared by many scientists in different parts of the world to reflect how the growing of cotton has changed. Emphasis has been put on the complex interdisciplinary aspect of how the cotton crops have been protected and especially in reducing the impact of insect pests. It will hopefully guide future developments as changes in climate affect how the plants grow.

The variation in temperature and rainfall throughout a year is shown in Fig. P2 wih data from eight areas where cotton is grown. While the cotton plant can grow as a perennial plant, the presence of a

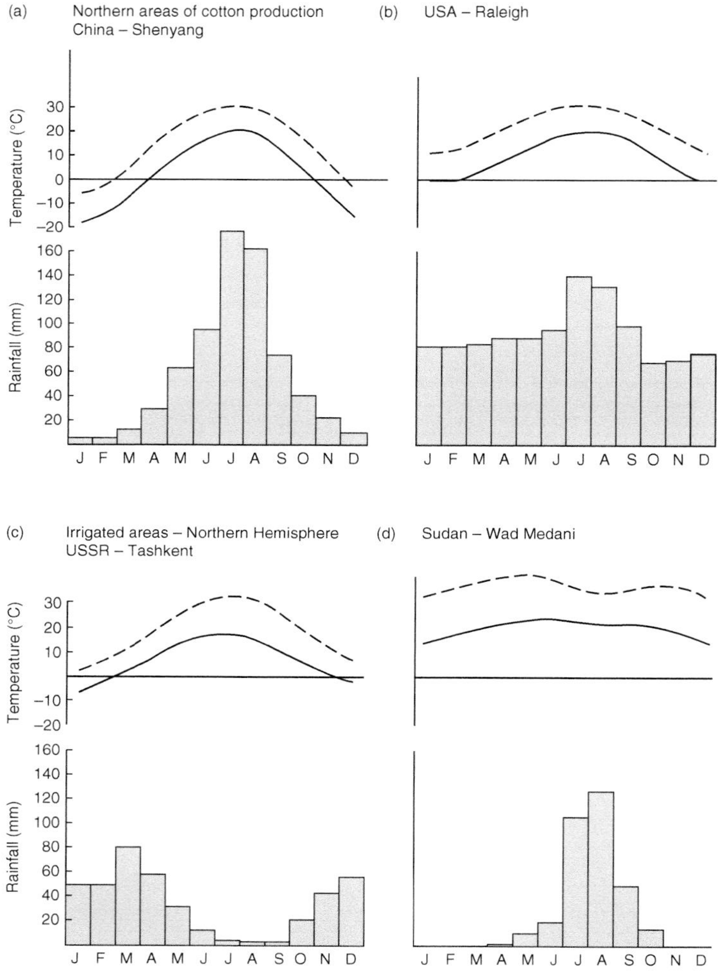

Fig. P2. Temperature and rainfall data for eight areas of the world where cotton is grown.

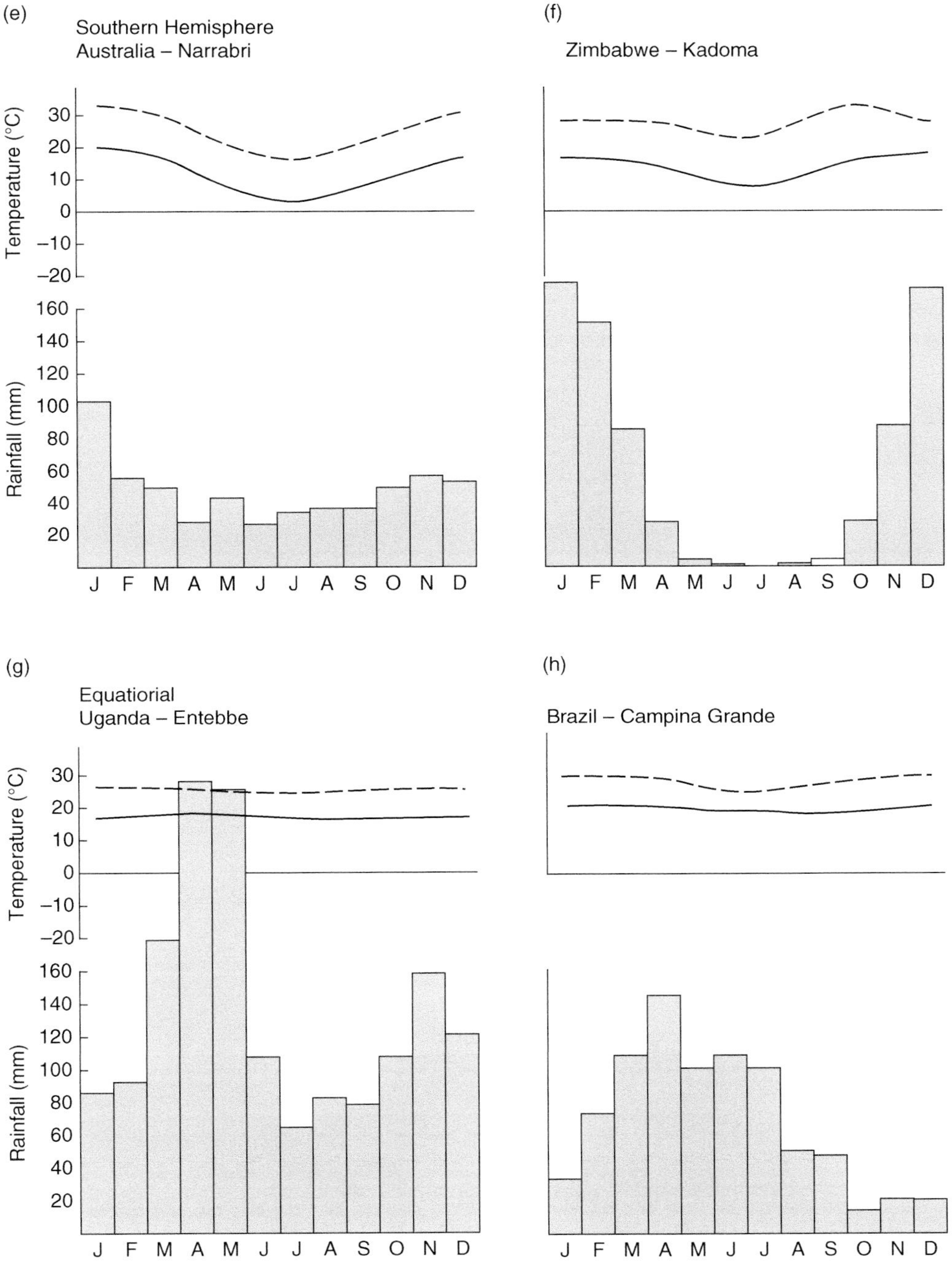

Fig. P2. Continued.

closed season is needed to minimize the impact of pests and diseases, so it is now grown as an annual crop, ideally in rotation with other crops. The growth of cotton plants and need for weeding during the early stages is illustrated in Figs P3 and P4. As cotton develops a deep root, this is important in relation to maintaining soils alongside legumes and other food crops.

Fig. P3. (a) Young cotton plants; (b) starting to develop baranches; (c) sympodial branch with buds and flower; (d) late weeding of a crop; and (e) a contrast between unsprayed and sprayed cotton plants showing a crop of seed cotton ready for harvest. (Photos: the author)

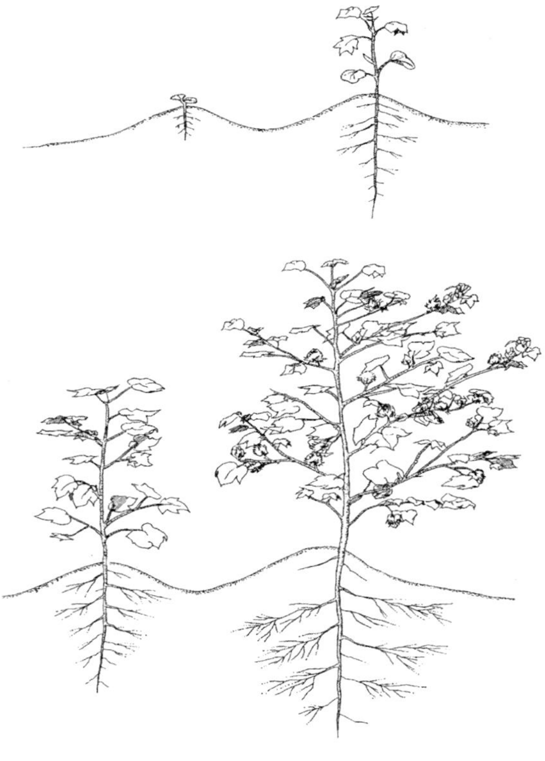

Fig. P4. Stages in plant growth from seedling to production of bolls ready for harvest.

The period when insects attack cotton plants during a season is indicated in Fig. P5. The early sucking pests, such as jassids, and the loss of buds and bolls by bollworms are illustrated in Figs P6 and P7.

In 2020, cotton was grown in 71 countries, as recorded by indexmundi.com. Production recorded ranged from over 28 million bales of 480 lb weight of cotton down to 1000 bales. Countries that produced the highest quantities of cotton in 2020 are shown in Fig. P8. Farmers have managed to get higher yields where there has been investment in research and extension to develop varieties suited to the national conditions and farmers have been trained to adopt tested techniques to control pests and diseases. The latest development of genetically modified cotton varieties has proved to be very effective in increasing yields and reducing pesticide use, but, nevertheless, in a changing climate, other pest problems have arisen and the quality of fibres has to be maintained.

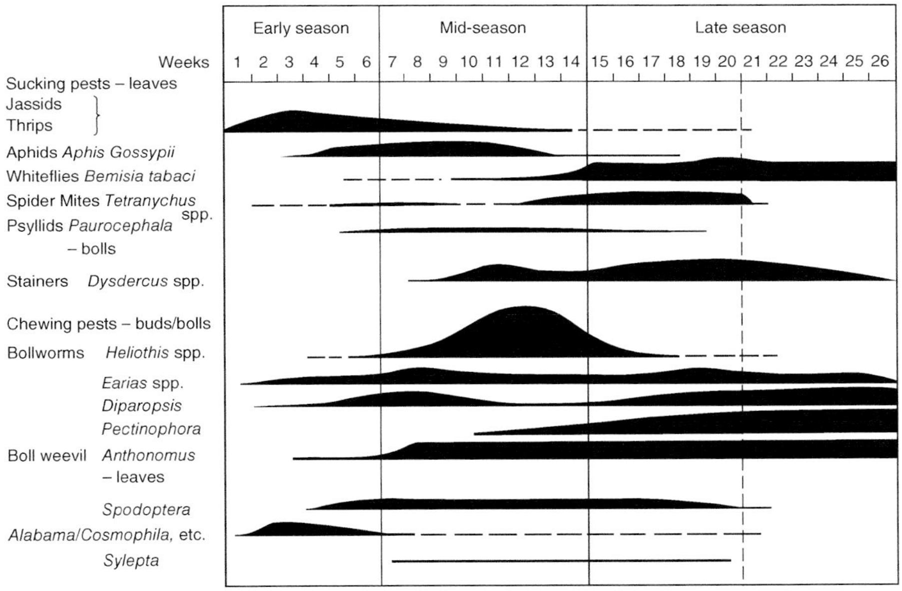

Fig. P5. The range of pests that can attack cotton crops in relation to the season.

Fig. P6. Cotton plants showing jassid damage alongside a plant (top left) with resistance to jassids. (Photo: the author)

Fig. P7. (a) A sympodial branch full of buds; (b) a similar branch that has lost all the buds and bolls due to bollworm infestation. (Photos: the author)

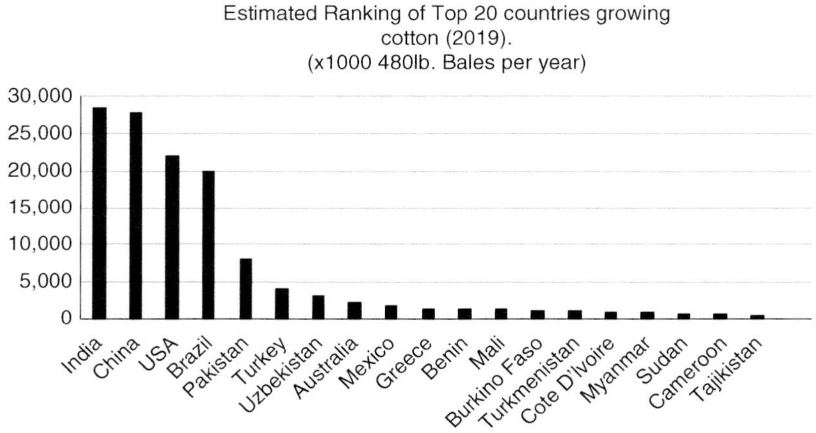

Fig. P8. The countries producing the most cotton.

1 Origins of Cotton

Graham Matthews*

The history of the domestication of cotton is very complex and is not known exactly. Cotton has been found during archaeological research at sites in both the old and new world, independently, indicating that the crop had been grown and domesticated with the fibres converted into fabric, using the earliest forms of combs, spindles and primitive looms. Samples of cotton fabric were detected in graves and ruins of ancient civilizations where the fabrics did not decay completely due to the dry conditions. The oldest fabric has been found in Peru (*c.*6000 BC).

Similarly, an archaeological examination at Mehrgarth (a neolithic site in Balochistan province of Pakistan), to the west of the Indus valley in Pakistan, has revealed evidence of domestication of cotton (Moulherat *et al.*, 2002; McIntosh, 2008). A close study of these relics indicated the coarse cotton from which fabrics were manufactured related to *G. arboreum* types (Sethi, 1960). Early observations by the Greek historian Herodotus described Indian cotton in the 5th century BC as 'trees, bearing as their fruit, fleeces which surpass those of sheep in beauty and excellence'. Sufficient evidence has been recorded by the Arabian travellers describing Indian fabrics and a flourishing export trade in cotton and cotton goods as early as 569–525 BC, and troops with Alexander the Great, during his sojourn in India, described cotton as 'a plant from which the natives plucked the vegetable wool which they spun to admirable clothing'. Marco Polo, who travelled to India in the 13th century, Chinese travellers to Buddhist pilgrim centres, Vasco Da Gama, who entered Calicut in 1498, and Tavernier in the 17th century have all praised the superiority of Indian fabrics.

Knowledge about cotton cultivation and manufacture of cotton textiles was introduced into Spain in the early centuries of the Christian era. Evidence indicates that India was the original habitat of cotton and an exporter of fine fabrics since ancient times. According to Hutchinson *et al.* (1947), trade between northwestern India and eastern Africa via the Arabian Peninsula in very ancient times may have introduced the first cotton plants into the Indus civilization, which differed little from their wild, lint-less plants, except they had a fluffy coat of lint fibres that were only gradually developed as a textile material. Subsequent early development occurred in the Sind (now the Sindh province, which ranked second after Punjab province among leading cotton-growing provinces of Pakistan) as the superiority of cotton over wool or flax for hot-weather clothing was soon apparent. Fibres from samples of

*g.matthews@imperial.ac.uk

 Pest Management in Cotton: A Global Perspective
(eds G.A. Matthews and T. Miller)
DOI: 10.1079/9781800620216.0001

fragments of cloth discovered in that area during excavations at Mohenjo-Daro at levels dated at approximately 3000 BC were shown to be similar to traditional Indian cottons in the 1930s.

Archaeological studies in Egypt raised some uncertainty about whether ancient Egyptians had imported domesticated cotton from the Indian subcontinent, as had happened with other crops, or whether they were growing a native African variety which had been domesticated locally. Studies suggested that seeds found at Qasr Ibrim (fort of Ibrim located in Egypt) were of the *Gossypium herbaceum*, native to Africa, rather than *G. arboreum*, which is native to the Indian subcontinent. Excavations in Egypt have discovered very fragile black fibres that were identified in one instance as *G. arboretum Sudanense* and another as *G. herbaceum*, suggesting that the ancient civilizations had achieved significant genomic reorganization when the ancient and modern varieties are compared.

Gossypium barbadense seeds were spread further north from Peru into Mexico, where discoveries have indicated the use of *G. barbadense* (*c.*3000 BC) and to Venezuela. It was spread into the Caribbean by the Arawak (Amerindians) so that by the 15th century 'Sea Island cotton' was well established in Barbados, when Christopher Colombus made his voyages to the Caribbean in 1492, 1493, 1498 and 1502. He was greeted by people wearing cotton ('the costliest and handsomest ... cotton mantles and sleeveless shirts embroidered and painted in different designs and colours'), a fact that may have contributed to his incorrect belief that he had landed on the coast of India. He also was presented with balls of cotton and found that the Arawaks could spin and weave the cotton and were sleeping on cotton hammocks. It is more than likely that he took seeds back to Spain. English colonists in the Caribbean established cotton as a commercial plantation crop with enslaved workers imported from West Africa. By the 1650s, Barbados was exporting cotton to England and Europe. Seemingly, the scientific name *G. barbadense* was chosen by Carl Linnaeus as his specimens came from Barbados.

Sea Island cotton was also introduced into the USA and grown in South Carolina and Georgia about 1790, although Hutchinson and Manning (1943) have suggested that the original introductions came from west-Andean Peruvian stocks but that it died out due to the boll weevil and the civil war, and one report indicated that it did not flower until late in the season, 'waiting for the autumn equinox'. The discovery of 4.56 kg of unginned cotton in excellent condition in two sealed prehistoric jars in the Pinaleno mountains in east-central Arizona in 1982 revealed the early domestication of *Gossypium hirsutum* in Mexico between around 3400 and 2300 BC (Huckell, 1993).

With the invention of the cotton gin at the end of the 18th century, *G. hirsutum*, known as upland cotton in the USA, could be grown successfully, so its shorter fibre became the prime commodity in the southern states. Growing cotton expanded very rapidly, principally due to the arrival of slaves from Africa, the availability of vast areas of land and a suitable climate. Much of the land was transferred from the native inhabitants to white settlers, whose slaves prepared the land, sowed the seeds and later hand-harvested the cotton. Much of the cotton was exported to England, where the development of spinning machines accelerated the processing of cotton fibres and manufacturing of cloth, in contrast to the traditional hand-spinning that had survived for centuries on a small scale in many parts of the world, but especially in India and other parts of Asia. The industrialization of cotton had begun.

Some of the *G. barbadense* seeds on arrival in Europe were taken to Africa, as in 1820 M. Jumel, who was a Frenchman, remarked, in a garden near Cairo, on certain cotton plants, of which the seed had been imported from the Soudan, and was quoted in a journal article in *Bulletin of Miscellaneous Information* (Royal Botanic Gardens, Kew), vol. 1987, pp. 102–104. His examination suggested that these plants had adapted to the soil and quickly became acclimatized, and, after many observations and experiments, he was convinced that the cultivation of the long staple cotton might be easily propagated throughout lower Egypt. M. Jumel's project experienced early difficulties, but ultimately established Egyptian cotton as a highly successful crop, initially often referred to as Jumel cotton.

The word 'cotton' is derived from the Arabic word قطن (*qutn* or *qutun*), which was the usual word for cotton in medieval Arabic, and was used in the mid-12th century and in English a century later. Cotton fabric was known to the

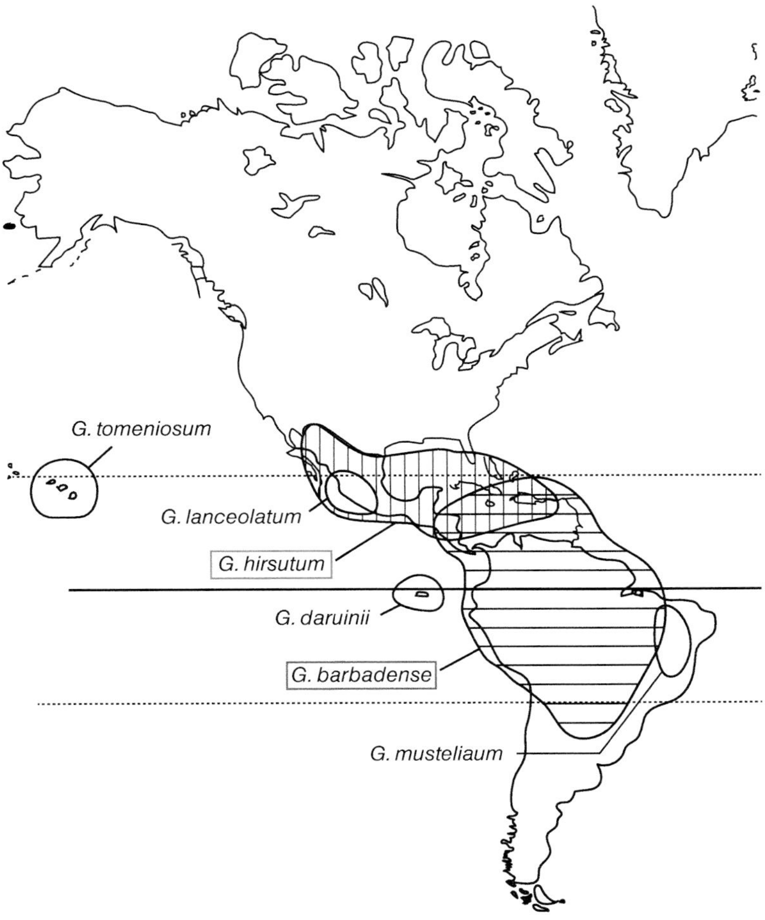

Fig. 1.1. Map showing *Gossypium* species in the New World. (From Fryxell, 1979)

ancient Romans as an import, but cotton was rare until less expensive imports were available from the Arabic-speaking lands in the later medieval era. Due to Marco Polo and other explorers, the Indian cotton fabric spread and the speed of cotton spinning improved when the spinning wheel was introduced to Europe around 1350. By the 15th century, Venice, Antwerp and Haarlem were important ports for cotton trade, and the sale and transportation of cotton fabrics had become very profitable. In 1730 cotton was first spun by machinery in England, and with the Industrial Revolution and the invention of the cotton gin ('gin' being a shortened version of 'engine'), patented in 1793 by Eli Whitney in the USA, resulted in a rapid expansion of the textile industry and paved the way for the important place cotton holds in the world today.

The first cotton exported from the USA was not initially accepted on arrival in Liverpool, as cotton had been imported from the Caribbean and Brazil as well as from India and other parts of Asia. However, it was easier to import from the USA, so the Lancashire cotton industry expanded as cotton imports rose steadily and the number of factories increased, employing more women, but also children as young as 8 working long periods to operate the spinning machines that were developed by Richard Arkwright and others. During the American Civil War, American cotton exports slumped due to a Union blockade on southern ports.

Large stocks of cotton, already shipped, meant the cotton mills were able to continue for much of the time when exports from the USA were not available. This prompted the main purchasers of cotton, Britain and France, to turn to Egyptian cotton, but after the war ended in 1865, British and French traders abandoned Egyptian cotton and returned to cheap American imports. Subsequently, in the USA, cotton became the major crop ('cotton is king') in the southern states, largely as result of emancipation and the end of the Civil War, so Africans who had previously been regarded as slaves were employed following the banning of slavery. Much of this cotton was trans-shipped to Europe. Across the south, share-cropping evolved, in which landless black and white farmers worked land owned by others in return for a share of the profits. Some farmers rented the land and bore the production costs themselves. Until mechanical cotton pickers were developed, cotton farmers needed the additional labour to hand-pick cotton. Picking cotton was a source of income for families, with rural and small-town school systems organizing vacations so that children could work in the fields during the cotton-picking season.

Cotton has many uses besides clothing, linens, draperies, upholstery and carpet. As early as 1813, nitrocellulose, or gun cotton, for explosives was made from raw cotton. In 1868 the combination of nitrocellulose and camphor made celluloid, an artificial plastic. Contemporary uses include fertilizer, paper, tyres, cake and meal for cattle feed, and cottonseed oil for cooking, paint and lubricants. Cotton was an ingredient in the first lightbulb, the telegraph, the Wright Brothers' plane, and the first automobile tyres.

The textile industry then started in the USA and increased its requirement for cotton, which resulted in less being exported to the UK and other European countries. In the UK, the British Cotton Growers Association (BCGA) was set up in 1902, following a severe shortage of cotton in Lancashire (Onyeiwu, 2000). The BCGA, located in Manchester, was funded by manufacturers and the textile workers' unions. The idea was that instead of relying on supplies of cotton from the USA, all the cotton required in Lancashire could be grown within the limits of the Empire and concentrated on the following African countries: Nigeria, Egypt, Uganda, Sudan, Nyasaland (now Malawi), Tanganyika (Tanzania) and South Africa. India was considered, as cotton growing was established there, but transport costs would be higher and there was no incentive to export the crop, so the BCGA invested mostly in Africa (Robins, 2016). Some African farmers were growing 'native' cotton to supply a local industry, but the BCGA offered a higher price if farmers grew a variety with a longer fibre and higher-quality cotton. Apart from the choice of variety, the area to which cottonseeds were distributed needed to be located to minimize problems of transport of fibre for export.

The BCGA was advised to establish model farms to undertake experimentation with different varieties of seed and to subsequently distribute the varieties found most suitable. Recognition of the importance of the identification and control of cotton pests, such as the bollworm, required assistance from the Imperial Bureau of Entomology.

The French had similar plans to improve the supply of cotton from Africa, and specifically in the French Soudan, a French colonial territory in the Federation of French West Africa, which became the independent state of Mali in 1960. A group of French industrial weavers and spinners, 'the Association Cotonniere Coloniale' (ACC), obliged farmers to grow cotton, but they had complete freedom to sell it, so most was sold to the spinners and weavers of the domestic and regional handicraft textile industry with very little being exported to France despite the activities of the ACC.

Later, the First World War again interrupted supplies of cotton from the USA to Europe. The Russians took over much of Central Asia as part of the Soviet Union and developed cotton mainly in Uzbekistan, but after independence, the area of cotton was reduced to grow more maize, wheat and other crops previously grown elsewhere within the Soviet Union. The UK and France then decided to actively encourage more cotton growing in many parts of Africa. By 1916, the BCGA was set up by the UK government to help the colonies prosper by encouraging as rapidly as possible the growing of cotton. While the government intended to get African farmers to grow cotton, the BCGA was more interested in getting settlers to establish farms, as they regarded

African agriculture as primitive and in need of improvement. BCGA was unable to secure a large and reliable supply of cotton to Lancashire during the First World War, although it did supply more than 90,000 bales from West Africa by the end of 1914.

In the 1920s, the agricultural policy of the British colonies, decided after the amateurish efforts of the BCGA, was that it had to support a state-building project to encourage economic development. This was needed because from the viewpoint of African farmers it was not worth the effort of growing cotton, as they were not paid to grow it. While continuing support for setting up facilities for ginning the seed cotton, the BCGA ceded its scientific responsibilities to the Empire Cotton Growing Corporation (ECGC), which obtained its charter in November 1921 and worked closely with colonial agricultural departments to improve African agriculture. The Cotton Industry Act in July 1923 enabled the ECGC to collect a levy of 6d per 500 lbs of raw cotton supplied to the spinners for a five-year period. The ECGC continued to support research in Africa with staff working in Uganda, Sudan, Nigeria, Kenya, Tanzania and Malawi on plant breeding, agronomy and crop protection until 1975. By 1924, the ECGC had established a journal – *The Empire Cotton Growing Review* – to assist information transfer between the scientists working in different parts of Africa. The journal continued until 1975 when the ECGC was closed, as the UK government decided to withdraw its support.

In West Africa, the francophone countries were supported by the Institute for Research in Cotton and Exotic Textiles (IRCT) (Bassett, 2001).

The Species *Gossypium*

Scientifically, we call cotton *Gossypium*, which is derived from the Arabic word *goz*, which means a delicate/fine material (Baffes and Ruh, 2005).

In more recent times, further taxonomic study of *Gossypium* has centred on the four domesticated species: the New World allopolyploids *G. hirsutum* and *G. barbadense* (2n = 52) and the Old World diploids *G. arboretum* and *G. herbaceum* (2n = 26). However, Wendel *et al.* (2009) point out that, despite the considerable diversity of these species, they are dwarfed when compared with the whole genus. While the wild cottons are perennial, *G. hirsutum*, cultivated by the Pueblo Indians of the south-western USA, may have achieved a day-neutral response and the annual habit as early as the first century AD, whereas the same pattern in *G. barbadense* was not achieved until much later in the 18th century with the development of Sea Island cotton (Fryxell, 1979).

The specialized form of *G. barbadense*, var. *Brasiliense*, originally from the Amazon basin, is also of interest. It is generally referred to as 'kidney cotton' as its seeds are fused in a solid mass that is somewhat kidney-shaped. The lump of seeds was probably an advantage in primitive farming as it was easier to sow, and it was also easier to remove fibres from the seed. These advantages no longer applied with the development of machinery to remove the fibres and sow individual seeds (Fryxell, 1979). Seeds were certainly distributed to Asia and Africa, where the kidney cotton can be found sporadically along the old trading tracks from the coast.

In the Indian subcontinent, under the Mughal Empire, which ruled from the early 16th century to the early 18th century, Indian cotton production increased in terms of both raw cotton and cotton textiles. The Mughals introduced agrarian reforms such as a new revenue system that was biased in favour of high-value cash crops, such as cotton and indigo, providing state incentives to grow cash crops, in addition to rising market demand (Richards, 1995). The largest textile manufacturing industry in the Mughal Empire produced piece goods, calicos and muslins, available unbleached and in a variety of colours, and was responsible for a large part of the empire's International trade. India had a 25% share of the global textile trade in the early 18th century, with exports in the 18th century across the world from the Americas to Japan. The most important centre of cotton production was the Bengal province (Subah), particularly around its capital city of Dhaka. After many experiments with cottonseeds from Malta and Mauritius, upland cotton was introduced to the Indian subcontinent with successive efforts started from 1790s to 1914 when it spread widely over millions of acres.

Much later in the 1950s the cotton industry in the USA changed with reliable harvesting

Table 1.1. Listing of cotton species derived from Hutchinson *et al.* (1947).

Chromosome number	*Gossypium* species	Races/variety	Distribution
13	arboreum	burmananicum	Myanmar*, Bengal, Malaysia
		indicum	Peninsular India
		bengalense	Northern and central India
		cernuum	Assam, east Bengal, India
		soudanense	Sudan, south to Mozambique
		sinense	China, Korea, Japan, Formosa
	herbaceum		Asia minor, Central Asia, SE Europe, Mediterranean
		africanum	Bushveld South Africa, Botswana, Mozambique
		acerifolium	Gujerat, east India, parts of Africa – Senegal, Sudan, Zambesia, Shire, Malawi
26	tomentosum		Hawaiian islands
	hirsutum		Central America, southern states of USA
		punctatum	C. America, Caribbean Gulf coast, USA. Also in Eritrea, Zanzibar, Tanzania, Sudan, Egypt, West Africa
		Marie-galante	Antilles, Spanish main, Ecuador, eastern Brazil
	barbadense		South America, from NW Argentina northwards, occasionally due to introduction in Caribbean and C. America.
		brasiliense	S. America, sporadic in Africa and India
		darwinii	Galapagos

*known as Burma in 1947

machinery that reduced the demand for labour and the arrival of DDT and other pesticides to provide more effective control of insect pests. This has enabled cotton to remain a major export of the USA, although growing cotton has become a key factor in many African countries and has expanded in Australia. The latest change has been the ability to genetically modify cotton plants. This has so far been directed at introducing a toxin to control bollworm larvae, namely the Bt cotton, using genes from *Bacillus thuringiensis*, and genetically making the plants tolerant of certain herbicides, initially allowing glyphosate to be applied without affecting cotton plants. More recently, the textile industry in Europe has declined, as more manufacturing of textiles has developed extensively in Asia.

References

Baffes, J. and Ruh, P.A. (2005) Part 1: The cotton trade: history and background. In: Townsend, T. (ed.) *Cotton Trading Manual*. Woodhead Publishing, Cambridge, UK.

Bassett, T.J. (2001) *The Peasant Cotton Revolution in West Africa – Côte d'Ivoire, 1880–1995*. Cambridge University Press, Cambridge, UK.

Fryxell, P.A. (1979) *The Natural History of the Cotton Tribe*. Texas A & M University Press, College Station, Texas.

Huckell, L.W. (1993) Plant remains from the Pinaleno cotton cache, Arizona. *Kiva* 59, 147–203.

Hutchinson, J.B. and Manning, H.L. (1943) The efficiency of progeny row breeding in cotton improvement. *Empire Journal of Experimental Agriculture* 11, 140.

Hutchinson, J.B., Silow, R.A. and Stephens, S.G. (1947) *The Evolution of Gossypium.* Oxford University Press, Oxford, UK.

McIntosh, J. (2008) *The Ancient Indus Valley – New Perspectives*. ABC-CLIO, Santa Barbara, California.

Moulherat, C., Tengberg, M., Haquet, J.-F. and Mille, B. (2002) First evidence of cotton at Neolithic Mehrgarh, Pakistan: analysis of mineralized fibres from a copper bead. *Journal of Archaeological Science* 29, 1393–1401.

Onyeiwu, S. (2000) Deceived by African cotton: the British Cotton Growing Association and the demise of the Lancashire textile industry. *African Economic History* 28, 89–121.
Richards, J.F. (1995) The Mughal Empire. Cambridge University Press, Cambridge, UK.
Robins, J.E. (2016) *Cotton and Race Across the Atlantic: Britain, Africa, and America, 1900–1920*. University of Rochester Press, New York.
Sethi, B.L. (1960) History of Cotton. In: Sethi, B.L., Sikka, S.M., Dastur, R.H., Gadkari, P.D., Balasubrahmanyan, R. *et al*. (eds) *Cotton in India – A Monograph*. Indian Central Cotton Committee, Bombay.
Wendel, J.F., Brubaker, C.L. and Seelanan, T. (2009) The origin and evolution of *Gossypium*. In: Stewart, J.M., Oosterhuis, D., Heitholt, J.J. and Mauney, J.R. (eds) *Physiology of Cotton*. Springer, The Netherlands.

2 Cotton in the United States of America and Mexico

C.T. Allen, Steven M. Brown, Charlie Cahoon, Keith Edmisten, Rogers Leonard, T. Miller*, Jane Pierce, Dominic Reisig and Phillip Roberts

Introduction

Cotton became a major crop in the Sunbelt States of the USA (Table 2.1) after the Civil War ended in 1865, utilizing the labour freed from slavery. Landless farmers were known as sharecroppers.

Gossypium hirsutum, upland cotton from Mexico, was grown with a range of varieties selected by commercial plant breeders. Popular varieties including Deltapine, Coker Wilds and Stoneville were sown with selection for glabrous leaves when machine harvesting was introduced to minimize the trash when ginning the crop. Seeds of the Deltapine brand were the most popular in 2020 with Americot varieties second and American Pima varieties third.

Transgenic cotton varieties were initially developed to include the toxins of *Bacillus thuringiensis*. These included Bollgard I and II and were referred to as Bt cottons. Later attention was on creating herbicide-tolerant varieties, notably to allow glyphosate (Roundup Ready) to be used in weed management, and later to allow the use of Dicamba. WideStrike and Liberty Link varieties have become popular.

Cotton-growing States

Note

In this chapter, the area of fields is in acres. To convert acres to hectares, multiply by 0.405.

Information in some sections has been obtained from publications issued by the Extension Service that links the outcomes of research and guides farmers with recommendations to follow for pest management. In many countries the farming industry is not supported by a similar independent advisory service.

Major Cotton Pests in the USA

Cotton crops are attacked by a range of insect pests, some having a limited host range, for example the boll weevil, while others can infest a wide range of plants. The pests are either sucking pests, including cotton aphids, plant bugs, stink bugs, thrips and whiteflies, or the larvae of Lepidopteran insects, notably the bollworms, but also some leaf eaters. Losses caused by these pests were described by Luttrell *et al.* (2015).

*Corresponding author: chmeliar@ucr.edu

 Pest Management in Cotton: A Global Perspective
(eds G.A. Matthews and T. Miller)

DOI: 10.1079/9781800620216.0002

The boll weevil

The boll weevil, *Anthonomus grandis*, a native of central Mexico, entered the lower south east of the cotton belt in 1892 and from Texas spread east, causing billions of dollars of damage. Early attempts were made to stop its spread, but in the 1950s many insecticide sprays were applied to control it as well as bollworms.

The programme implementing integrated pest management initially involved late-season insecticide treatments as soon as diapause was detected and continued until the cotton plants were destroyed. Pheromone traps were used to monitor populations to determine if in-season sprays were needed, with diflubenzuron used at pinhead-square (young-bud) stage. Release of sterile boll weevils in the early fruiting period was also tried. Defoliant was applied prior to harvesting and stalks destroyed after harvest (Smith and Harris, 1990). Ultimately, the Boll Weevil Eradication Program was sponsored by the United States Department of Agriculture (USDA), which sought to eradicate the boll weevil from all cotton-growing areas. As a result of these efforts, boll weevil was eradicated from the United States, except in areas of Texas bordering Mexico.

Table 2.1. Sunbelt States of the USA

Region	States
South-East	Virginia, North and South Carolina, Georgia, Alabama, Florida
Mid-South	Missouri, Tennessee, Mississippi, Arkansas, Louisiana
South-West	Texas, Oklahoma
West	New Mexico, Arizona, California

Bollworm

Prior to the development of transgenic Bt cotton varieties, the tobacco budworm (*Heliothis virescens*)

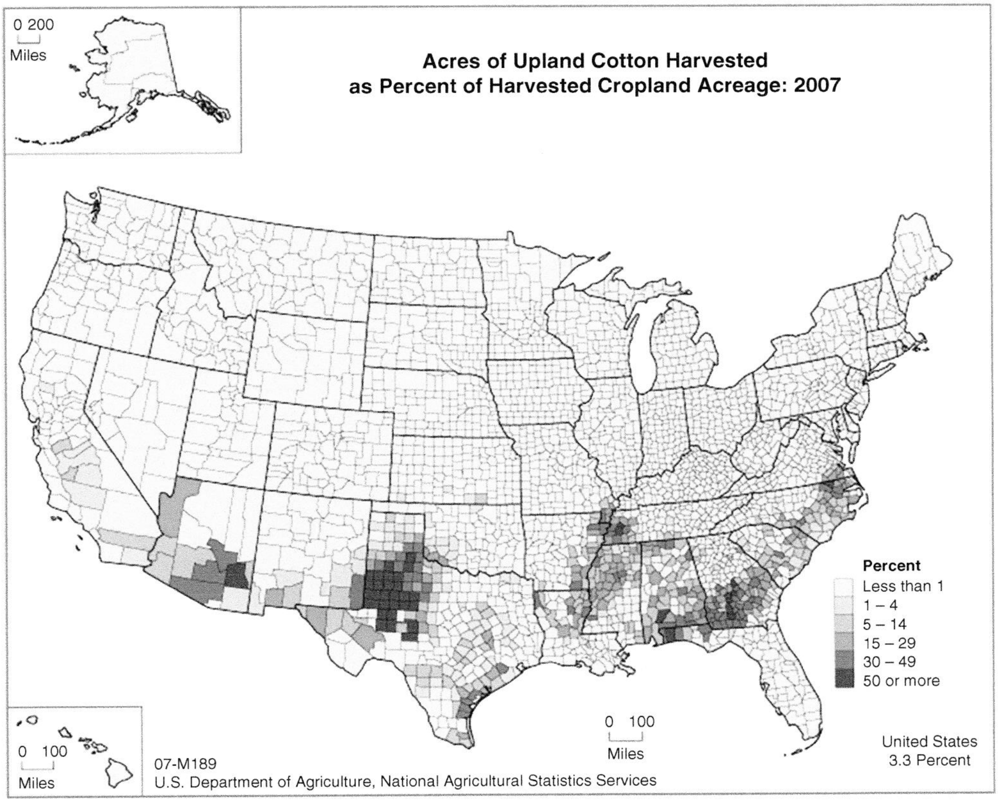

Fig. 2.1. The distribution of cotton-growing areas across the USA. (From US Department of Agriculture, National Agricultural Statistics Service)

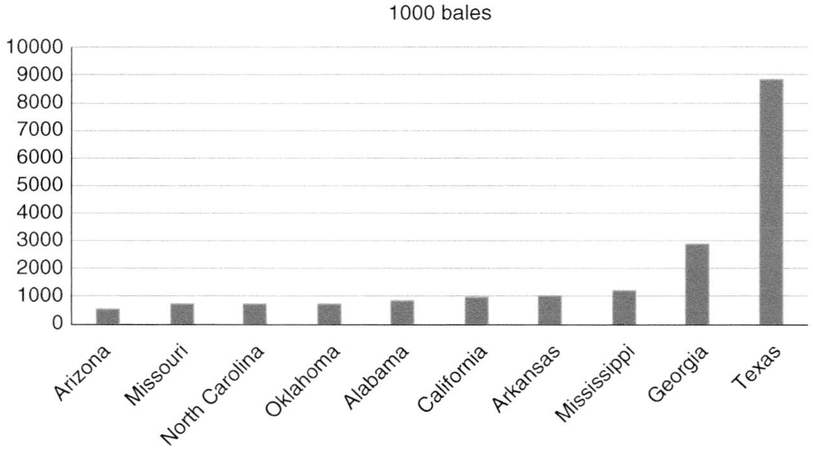

Fig. 2.2. The leading cotton-producing US states.

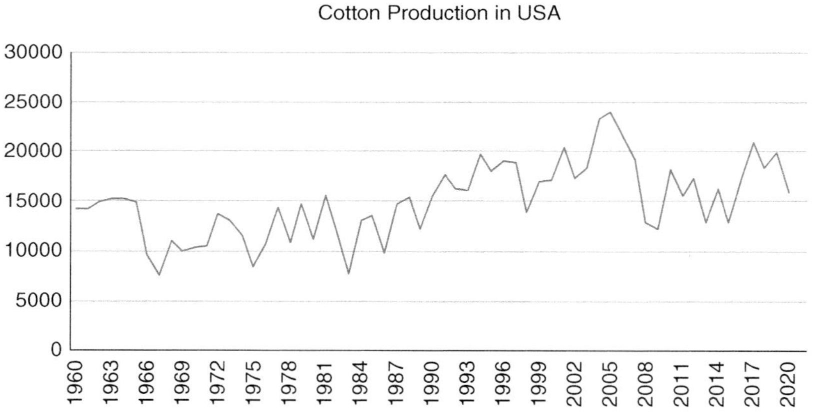

Fig. 2.3. Changes in production of cotton, 1960–2020.

and bollworm (*Helicoverpa zea*) were major pests, which resulted in farmers applying DDT sprays, often mixed with toxaphene (Brown *et al.*, 1962). To control boll weevil, resistant to chlorinated insecticides (Smith, 1998), low doses of methyl parathion were also added to the spray (Hardee and Harris, 2003). When DDT was banned, farmers started to use higher concentrations of organophosphate and other insecticides to control these pests (Blanco, 2012).

While tractor sprayers were used with a wide boom, many farmers adopted aerial spraying yet control was still inadequate. Part of the reason was that although both methods of application gave a good deposit on the exposed upper part of plants, distribution within the crop canopy was much less effective.

The pink bollworm

Pectinophora gossypiella (pink bollworm) was first detected in the USA in 1917, possibly on seeds from Mexico. The pest originated in Australia but was detected in India prior to 1910 and then spread to Egypt. It was the key pest in south-western US cotton-growing areas. The eggs were laid on or near the developing cotton bolls and the larvae spent their entire development inside the bolls.

Fig. 2.4. Helicoverpa bollworm. (Photos: Graham Matthews)

This made it impossible to control under normal circumstances. Pheromone trapping was done to monitor its presence (Haynes *et al.*, 1987). Efforts were then made to eradicate it but not until the advent of Bt (GMO) cotton was the boll able to withstand the feeding of the pink bollworm larvae. However, as some soon learned, Bt cotton plants exert 100% selection pressure for resistance. Towards the end of the season the toxin expression begins to drop, leading to selection for resistance. Nevertheless, growing Bt cotton suppressed the pink bollworm population in the first half of the season and this enabled use of sterile insect technique (SIT) on a regional level to be effective.

The US Department of Agriculture issued a proclamation on 19 October 2018 announcing the eradication of pink bollworm, *Pectinophora gossypiella*, from the USA. This effort was made more complicated by requiring that the eradication also be accomplished in neighbouring Mexico to establish a 'buffer zone' to prevent reinvasion from infested fields further into the interior of Mexico (Staten and Walters, 2020; Tabashnik *et al.*, 2021).

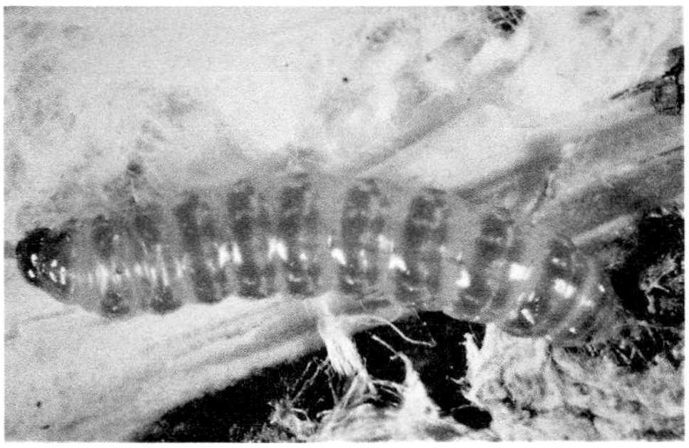

Fig. 2.5. Pink bollworm. (Photo: Graham Matthews). A recent issue of the *ICAC Recorder* provides an extensive overview of the present global management of this pest (Kranthi *et al.*, 2021).

Alabama

Cotton was an economic force in creating Alabama, some of the first settlers were attracted into the state's river valleys in search of the broad fertile plains for growing cotton. Two dominant labour systems in the early days were slavery in the Old South and share-cropping in the New South. It was estimated that as much as 90% of the farmers were engaged in the production of cotton and corn.

The presence of the boll weevil compelled cotton farmers in the Coastal Plain region in the 1920s to consider and embrace an alternative cash crop, peanut. However, the town of Enterprise, in 1919, erected a monument, a female Greek statue with a larger-than-life weevil hoisted above her head, in tribute to the boll weevil. An adjacent historical marker reads, 'In profound appreciation of the Boll Weevil and what it has done as the Herald of Prosperity…'. Still the two crops, cotton and peanut, thrive in that region of Alabama.

Alabama produces around half a million acres of cotton annually. The state is divided into six production areas with the highest percentages of cotton planted in the north-west (TN Valley) and south-east (Wiregrass) corners. Many producers began a shift to reduced- or no-till practices in the 1980s and today many growers also plant cover crops during the winter to help with soil erosion, weed control and increased soil organic matter.

Cotton production in Alabama can be traced as far back as the late 1700s. Throughout the early and mid-1800s, production was focused on the Black Belt soils of west-central Alabama. During this time, cotton was the primary driver of the state's economy. Until the Civil War, Alabama cotton farmers faced few challenges they could not overcome. The Civil War devastated the State's economy and cotton production for a variety of reasons. However, after the war, cotton made a comeback and acres harvested began to rise. Not long after this came the event that would change cotton production in Alabama for most of the 20th century.

Insect pest management

Alabama's first boll weevil was found in a Mobile County cotton field in 1910. For the next 80 years the boll weevil dominated cotton production, causing an average loss of $30 million annually. No economic losses have been incurred since 1995 and the last 'hitchhiking' weevil was captured in Mobile County in 2003. In the era following the Boll Weevil Eradication Program (BWEP), cotton insect pest management has been driven by plant-incorporated protectant (PIP) technology for tobacco budworm and bollworm control. The rapid adoption of Bt cotton in 1996 led to a low-spray environment which has caused the bug complex of stink bugs and tarnished plant bugs to become key pests. The shift to reduced- and no-till production practices decades ago caused secondary pests, such as grasshoppers, slugs, snails, cutworms and spider mites, to become a more consistent threat in recent years.

The future of cotton insect pest management is likely to be in the form PIP technologies and more selective insecticides. A new trait for thrips and tarnished plant bug management is on the horizon and the most recent insecticides that have been developed generally provide control of just a few species or insect complexes. Moving forward, managing insecticide resistance, and developing strategies to monitor and manage sporadic and secondary pests will be critical for economic cotton production.

Nematodes, including root-knot and reniform, are important pests of cotton across Alabama. In south-east Alabama, populations are reduced by using peanut as a rotational crop while corn has been used as a rotational crop to manage reniform nematodes in the north. The development of herbicide-tolerant cotton varieties in the early 1990s caused a major shift from reliance on tillage and pre-emergent herbicides to the use of over-the-top broad-spectrum herbicides that controlled most species in the field. However, the development of herbicide-resistant weeds, including Palmer amaranth, common ragweed and ryegrass, has caused problems for growers across the state.

Conservation tillage systems

With the advent of certain herbicides and improvements in planting equipment, reduced-tillage systems have become common in the state since the 1980s. These systems often include cover crops such as fall-planted small grains or cool-season legumes as well as minimum in-row tillage or no-till. The overall results include soil and water conservation, increased surface soil organic matter, reduced erosion, improved soil health, and often increased yields.

Cotton pests

The boll weevil had a huge impact on growing cotton in Alabama, which was recognized by the citizens of Enterprise placing a statue to commemorate it in 1919 (Fig. 2.6a).

The introduction of Bt-cotton, and its effective, 'built-in' worm control, is a critical ongoing component of the sustainability of cotton

Fig. 2.6. (a) Boll weevil statue in Alabama. (Photo: Martin Lewison, used under a Creative Commons Attribution-ShareAlike licence); (b) boll weevil on a boll.

in the state, at least from the standpoint of insect management. Other pests include tarnished plant bugs, which generally move to cotton from surrounding hosts in early to mid-June to begin laying eggs and feeding. Without control, several generations can occur in a cotton field during the growing season.

Drought early in the season dried out host plants so the plant bugs moved into cotton where they will feed on pinhead squares. There are also stink bugs waiting on the first bloom to drop so they can feed on the tiny bolls. Plant bugs feed on pinhead squares just as they form. This causes them to dry up, turn dark brown and finally fall off the plant. With frequent and continued feeding by large numbers of plant bug adults, a tall, spindly plant has no fruit.

Farmers are advised to begin by scouting the earliest mid-May planted cotton. The cotton is lush, with better shade and more fruit. Using sweep nets is the best way to catch and identify these insects (https://alabama.growingamerica.com/news/2019/06/plant-bugs-put-heavy-pressure-cotton – accessed 18 July 2021).

Weed pests

Historically, control was accomplished by hand weeding and tillage, including 'chopping', which means the use of a hoe to remove weed pests. Later, plowing with a mule, and still later cultivating with a tractor, became integral to weed management. The late 1950s and early 1960s brought the introduction of herbicides as a common tool, and the 21st-century farmers employ tillage, cover crops, herbicides and even hand weeding to deal with weeds. Herbicide-tolerant crops were introduced in the 1990s and are widely used today. Herbicide-resistant weeds have become more common in recent years, requiring greater timeliness and persistent effort.

Common, troublesome weeds in Alabama include Palmer amaranth, yellow nutsedge, purple nutsedge, morning glories, annual grasses, horseweed and bermudagrass.

Arizona

'Arizona' comes from a Uto-Aztecan Indian word. It means 'little spring' in the Tohono O'odham language. The descendants of the original inhabitants from 21 tribal groups are still present. Pima cotton, prized for its long threads, was developed in Arizona and is associated with the ancient Hohokam people and the Maricopa area of the state.

The southern half of the state is mainly desert and is good for year-round crop growth in irrigated areas. The northern half of the state is mountainous and cattle and sheep are the main agricultural commodities in this northern region. Most of the average rainfall of 12.7 inches is during the late-summer period, but several irrigation systems divert water from the Colorado River.

The Central Arizona Project

Early in the 20th century, Arizona's leaders knew that the state's future depended on a water supply that was secure, stable and renewable, and a 336-mile system was developed to deliver the state's largest renewable water supply which serves 80% of the population.

Growing techniques, pest control, fertilizers and harvesting of cotton

Both upland and Pima cotton are planted between February and April and harvested in the fall. It can be planted wet or dry – in a field with the soil already wet or dry, to be irrigated later.

Before its eradication was officially declared in 2018, pink bollworm was the key pest in south-western USA cotton-growing areas. As the egg is laid near or on the developing cotton boll, and the larvae immediately chewed in and remained inside the boll, it was impossible to control with the usual sprays. The advent of Bt cotton meant that a pink bollworm larva was exposed to the toxin as soon as it started to feed on the outer part of the boll. However, as some soon learned, a GM cotton plant expressing an insecticidal factor represents 100% selection pressure for resistance. Toxin expression decreases late in the season. This would naturally lead to a selecting dose rather than remaining a killing dose.

Along with Bt cotton, whitefly-specific insect growth regulators (Courier & Knack) ushered in a new era of 'selective' pest control beginning in 1996. Thus, the target was controlled while most or all other non-target organisms (e.g. predators, parasitoids, pollinators) remained unaffected. In 2006, Arizona gained access to a fully selective, Lygus-specific feeding inhibitor (flonicamid) (Ellsworth *et al.*, 2011).

Despite rising costs of growing cotton with fluctuating prices for cotton fibre, Arizona farmers continue to grow cotton, including having loans that require them to plant cotton. Their viewpoint is that while cotton might not be profitable one year, that could change the next year and it is difficult to shift to another crop.

Arkansas

The University of Arkansas System Division of Agriculture crop IPM extension faculty members partner with research scientists and county agents to develop and deliver needed information to growers, consultants and industry representatives. IPM is an essential part of row-crop production, helping producers farm more efficiently and reduce reliance on pesticides. Input costs for items such as fertilizer and weed control have continued to rise even as prices remain in the doldrums. China has warehoused millions of tons of cotton, which has driven world commodity markets lower. Weeds and other wild grasses remain a continuous problem.

According to the Farm Bureau of Arkansas, the state is currently ranked fourth in cotton and cottonseed production, and with exports valued at $463 million it is the fifth-largest exporter of cotton. Cotton is grown in the Delta region with a concentration in the far north-east of Arkansas. In 2014, 335,000 acres of cotton were planted in the state resulting in 820,000 bales of lint being produced. The average yield for Arkansas was 1193 pounds of lint per acre. Average production has dropped from more than a million acres to around 250,000–300,000 in the past decade. This is part of a larger trend. In 2015, US cotton production declined by some 3.4 million bales from the previous year. Cotton prices hit a six-year low point in 2015 and, according to data from the US Department of Agriculture, farmers planted the fewest acres of cotton since 1983.

Cotton-breeding research was initiated at the University of Arkansas (UA) in the early 1900s. Early work focused on evaluating cultivars and on making plant selections out of established cultivars. From 1948 until 1986, UA maintained two cotton-breeding programmes, from which several Rex cultivars, two stripper cultivars, and Arkot 518 were released. Others focused on early maturity, seedling vigour, host-plant resistance (including the Frego bract trait) and naked and tufted seed. In 1988, the two traditional breeding programmes were merged into one campus-based programme. The programme was subsequently moved to the Northeast Research and Extension Center, Keiser, AR, in 1997. Bourland (2018) has led the programme since 1988 and has been responsible for almost 100 germplasm and cultivar releases and has established methods for evaluating and selecting several cotton traits. The cotton-breeding programme at UA continues to develop well-adapted lines and concepts that promote profitable cotton production in Arkansas.

In 2014, 335,000 acres of cotton were planted in the state resulting in 820,000 bales of lint being produced. The average yield for Arkansas was 1193 pounds of lint per acre. Average production has dropped from more than a million acres to around 250,000–300,000 in the past decade. This is part of a larger trend. In 2015, U.S. cotton production declined by some 3.4 million bales from the previous year. Cotton prices hit a six-year low point in 2015 and, according to data from the U.S. Department of Agriculture, farmers planted the fewest acres of cotton since 1983.

California

History of cotton in California

Cotton production in California started with the Spanish missionaries. After the arrival of Europeans brought diseases that decimated the local native American tribes, surviving local American Indians were recruited to work at Spanish missions to grow crops. The missionaries wanted cotton to make clothing, partly to clothe the native Indians who wore modest or brief covering. This stable mission culture lasted until the Gold Rush of 1849, which completely disrupted the state and brought in waves of new immigrants from the east.

The first cotton grown in California in 1888 was on 30 acres and produced five full bales of cotton. This was considered a successful venture, but the first commercial planting of cotton in the San Joaquin Valley was not until 1919 – the end of the First World War. Subsequently, California became a major cotton-producing state. Cotton is also grown in the Palos Verde Valley and, more recently, has returned to the Sacramento Valley.

Geisseler and Horwath (2013) illustrated the expansion of upland cotton (Fig. 2.7) and reported the decline in California cotton acreage from a peak of 1.6 million acres harvested around 1980 to below 0.2 million acres in 2009. The decline is attributed to several things. Drought conditions led to water shortages (climate change, although California does have historic drought cycles). This drought accompanied increased severity of pests, particularly pink bollworm (that arrived in 1968) and sweet potato whitefly.

The cause and effect of drought on pest abundance is not clear. Outside of cotton, other commodities became more attractive. In Fresno

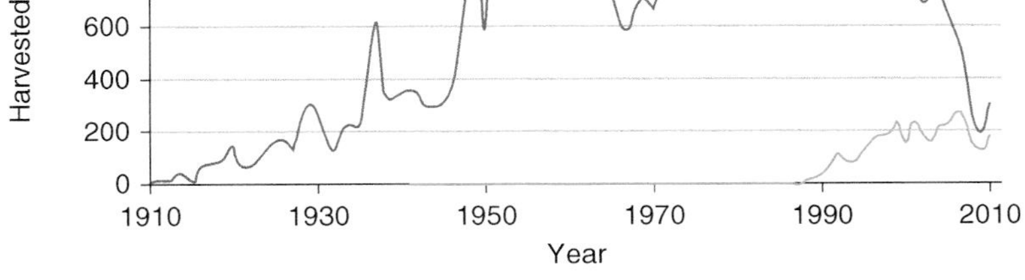

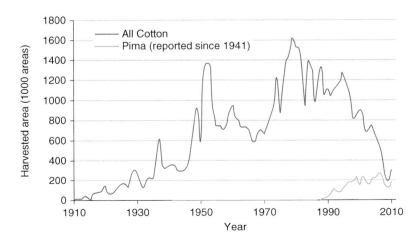

Fig. 2.7. Harvested cotton acreage in California since 1911. (From Geisseler and Horwath, 2013)

County, while cotton acreage started to decline, those of almonds, grapes and processing tomatoes increased, beginning in the 1970s. Since cotton is an annual crop, the growers have options and decide what to grow in any given year based on market projections and past experiences. When insecticides have needed to be applied, the choice in California is limited as several, including dicrotophos, aldicarb, chlorpyrifos and clothianidin used elsewhere in USA, are prohibited in California.

The institutional and environmental settings associated with cotton cultivation in California differed markedly from other states. Factors such as the size of farms and the region's dry weather during the harvest season help explain the more rapid mechanization of picking in California (Constantine *et al.*, 1994). The present California Cotton Ginners and Growers Association (CCGGA), formed in 2016, evolved from the original Western Cotton Growers Association, founded in 1949, that also covered Arizona and New Mexico until 1990 and became the CCGA in 1991.

As the short staple cotton *Gossypium hirsutum* declined in acreage, the area of long staple Sea Island-type Pima cotton began to increase at the end of the 1980s, as the softer, longer and stronger lint was more valuable. The main type of upland cotton was the Acala variety.

Georgia

Introduction

Growing cotton in Georgia began near Savannah in 1734. The crop inspired a schoolteacher, Eli Whitney, to invent and patent the cotton gin in 1793. The first major textile mill was built in 1811 near Washington, Georia. Cotton production in Georgia in 1860 was 584,000 bales, but the Civil War interrupted commercial production. However, by 1866 production had climbed back to 300,000 bales (Howard, 1867). Today, Georgia ranks third, nationally, in cotton production and acres planted, and UGA Extension continues to aid in the growth of cotton production. Cotton-related professions provide 53,000 jobs in the state of Georgia, and cotton's overall impact exceeds $3 billion.

Georgia, where peanut and a host of other crops can compete with cotton for acreage, respondents projected a decrease of 14% from 2019. Nevertheless, with Georgia's farmers planting 1.4 million acres of cotton, which is the third-highest planted acreage for the past decade, down 30,000 acres from 2018, it means the state is once again second in the nation for cotton acreage, trailing only Texas. The average yield is forecast at 932 lbs of seed cotton per acre. Production is forecast at 2.7 million bales, which would be the second highest on record.

Farmers and agricultural producers have been constantly under pressure to achieve higher yields yet keep the price of cotton low. Extension researchers have therefore aimed to develop precision agriculture methods to make planting more efficient for farmers while protecting profits.

Georgia currently ranks second in the USA in bale production and planted cotton acreage (five-year averages are 1.9 million bales produced on 1.1 million acres). Increased management and the use of IPM programmes have replaced numerous insecticide applications resulting in increased profits for growers, a more sustainable production system and an improved environmental profile.

Cotton pest management

Elimination of the boll weevil and adoption of Bt transgenic varieties have changed the situation significantly and enabled farmers to utilize natural controls more effectively and implement much improved IPM programmes (Haney *et al.*, 2012). The number of insecticide applications has been reduced from more than 15 during the early 1980s to less than five (five-year average is 2.6). However, the complex of key pests has changed in the reduced insecticide use environment.

Now, stink bugs are a primary pest of cotton. With more knowledge of basic stink bug biology, the ecology in cotton and the farmscape has improved in recent years. This information will be used to refine thresholds, improve sampling and develop innovative management programmes such as in-field border applications of insecticides where only a portion of the field is treated. The extension programme is very important as cotton continues to be an intensively managed crop, so effective IPM programmes are critical to sustain profitability and promote environmental stewardship in cotton-production systems in Georgia and the south-east.

Louisiana

Introduction

Cotton has been the most important crop grown in north-east Louisiana, supported by the research conducted by the Louisiana Agricultural Experiment Station, to increase yields, reduce production costs and minimize losses from insect pests, weeds, nematodes and plant diseases. Much of the initial research in cotton was done on the Calhoun Research Station in Ouachita Parish. As cotton was one of the most expensive crops to grow in Louisiana (Fig. 2.8) a key factor was the cost of controlling insect pests, notably the boll weevil, which plagued farmers in the state since the early 1900s, causing them to rely heavily on pesticides, which created other pest issues and environmental problems.

The 1950s–1970s records are replete with references from L.D. Newsom, James R. Brazzel, Jack Jones, John Roussel, Dan Clower and others who reported on their efforts at boll weevil control (for example, Brazzel and Newsom, 1959; Earle and Newsom, 1964; Brazzel *et al.*, 1961). When the boll weevil eradication effort began in the 1990s, these historically important entomologists had mostly retired, but their work provided critical information to eradicate the boll weevil from Louisiana in 2011.

As in other states, farmers were introduced into growing Bt cotton in the early 1990s in which genes from a naturally occurring soil bacterium, *Bacillus thuringiensis*, toxic to certain insects including the tobacco budworm and the bollworm, were inserted into cotton plants. This enabled a further reduction in applying insecticides, but with Louisiana's subtropical climate and the diversity of cotton insect pests, some pesticides were still required to maintain optimum crop productivity. A technique for limiting sprays to those areas of a field which actually needed the pesticide was subsequently developed.

Cotton breeding

Cotton breeding from the 1950s to the 1990s focused on the plant disease fusarium wilt, nematodes, key insect pests and plant canopy characteristics. Late-season rotting of bolls before harvesting was a major problem in south Louisiana, so three new varieties with unique narrow leaf shapes similar to okra leaves were introduced to open up the canopy to enable plants to dry out after rainfall. A variety with resistance to root-knot nematode and fusarium wilt was also produced. One variety, Louisiana 887, was later crossed with other varieties, including transgenic varieties that have the Bollgard gene for insect resistance and the Roundup Ready gene for herbicide resistance. However, over the last ten years, the acreage of cotton has fallen as

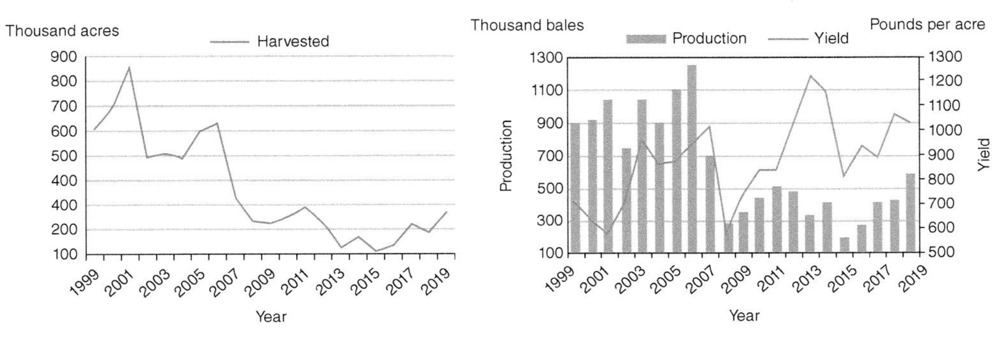

Fig. 2.8. Harvested cotton acreage. (a) production and yield; (b) in Louisiana during 1999–2019.

farmers have switched to planting more corn as markets shifted and corn varieties improved.

Mississippi

Cotton was first grown in what is now Mississippi in 1795 in the Spanish-ruled Natchez district as an alternative to tobacco and indigo. Cotton cultivation in the Mississippi Valley previously had been either unsuccessful or unproductive. Despite Eli Whitney inventing the modern cotton gin in 1794, cotton remained a marginal crop in the early 1800s, largely because the strains initially used – Creole seed and then Tennessee green seed – provided only mediocre yields. Mississippi growers began using the more productive Mexican seed around 1820 and within a decade Dr Rush Nutt had developed Petit Gulf cotton, a Mexican–Tennessee green hybrid, on his Rodney plantation. This seed produced the white gold that helped to make Mississippi the nation's top cotton-producing state.

During the second half of the 20th century, many Mississippi planters and farmers moved away from cotton production and towards other row crops such as soybeans and corn as well as highly commercialized catfish and poultry operations. Cotton has recently ranked third, behind poultry and forestry, among Mississippi's leading forms of agriculture. Today, approximately 1.1 million acres are planted to cotton in Mississippi, depending on various factors, including weather, price and commodity markets.

The success of the BWEP and technological advances in transgenic cotton varieties have facilitated management of the cotton crop. The tarnished plant bug has now become the key cotton pest in Mississippi. Farmers have also realized the benefits of reduced tillage programmes to increase yields and profit margins, so these advances continue to be major reasons that cotton yields have continued to increase over the past few years.

New Mexico

Archeological digs in New Mexico have uncovered evidence of cotton being grown as early as 300 BC (Smith and Cothren, 1986). Findings from the sites have yielded roasted seed hulls and cotton cloth, indicating the plant was used as a food source as well as for fibre. Although cotton had been grown earlier, the first commercial production began in the Pecos Valley around 1910. Cotton was established as a major crop by 1922. A cotton field station was opened in Las Cruces in 1926, and G.N. Stroman began breeding for local varieties there in 1928. An important development was New Mexico's most recognizable variety, Acala 1517. Advanced strains resistant to verticillium wilt and bacterial blight were developed. The programme also increased the quality and yield of New Mexico cotton. Advanced strains resistant to verticillium wilt and bacterial blight were developed.

As New Mexico was considered to be subject to fewer pests than other areas, some research examined the possibility of growing glandless cotton as the seeds could be used more widely as an animal feed. However, previous studies had shown that severe yield losses were caused by piercing and sucking insect pests such

Fig. 2.9. Mechanical harvesting of cotton in New Mexico. (Photo: Graham Matthews)

as lygus, boll weevil, and vertebrate pests (Jenkins *et al.*, 1966; Benedict *et al.*, 1977).

Glandless cotton (cotton plants without gossypol) can potentially widen cottonseed use beyond ruminant animal feed and increase the value of cottonseeds due to absence of gossypol. However, since gossypol is a natural defence mechanism against pests, glandless cotton is subject to a higher pest pressure, which may affect lint and seed yields (Idowu *et al.*, 2012). Bollworm and beet armyworm survival was 2–6 times higher at pupation on glandless cotton (Pierce *et al.*, 2014).

Using conventional breeding techniques, three new glandless cultivars were released in New Mexico (Zhang *et al.*, 2016; Zhang *et al.*, 2017; Zhang *et al.*, 2020) despite being more susceptible to insect injury, as the variety could be manageable as a valuable niche crop.

North Carolina

Introduction

The history of cotton in North Carolina is similar to that in Georgia and Mississippi. Most of the cotton is grown in the Coastal Plain region in the eastern portion of the state. About 3% of the acreage is located in the southern Piedmont.

Weed control in cotton is critical. As a perennial, cotton competes poorly with weeds. Growers rely primarily on herbicides for weed control. Mechanical cultivation to control weeds is rare due to the adoption of reduced tillage systems. Growers rely on the use of herbicides with multiple modes of action to combat weed resistance to herbicides.

North Carolina producers selected varieties with at least two Bt gene constructs, but with bollworm resistance to two gene varieties, growers are now looking for varieties with three gene constructs. Plant bugs, once considered to be a rare problem in North Carolina, have emerged as a significant pest and the pressure has been spreading from the north-east portion of the state into other cotton-producing areas. A new variety that contains the Bt protein mCry51Aa2 is expected to target two types of plant bug.

Thrips pressure in North Carolina is generally high during the early season leading growers to adopt prophylactic treatment at planting, as well as foliar sprays. Resistance to neonicotinoids for both thrips and cotton aphids is scattered through the state and growers need to be vigilant in terms of the possibility of additional treatment once the cotton emerges. Researchers at North Carolina State University have developed a model to predict thrips pressure for individual fields and approval for a new Bt trait to manage both thrips and plant bugs (mCry51Aa2, or ThryvOn) is expected to change insect pest management in the system.

Essentially, all cotton in the state is defoliated to allow for harvesting without fibre-quality loss. Cotton harvested without defoliation is very likely to be of low quality because green leaves can stain the lint. Defoliation also allows the harvested material to be low enough in moisture to not heat and degrade in modules or bales, a key consideration because most cotton in North Carolina is stored in modules before ginning. Defoliation further causes a harsh environment for aphids, reducing the need for chemical treatments in opening cotton.

Insect and mite pests

Thrips (various species, but primarily *Frankliniella fusca*) have the potential to cause significant yield losses and maturity delays, so this pest group must be controlled annually. Foliar insecticides applied to manage thrips include acephate, dicrotophos and spinetoram. In most years, most of the cotton is treated with a foliar insecticide for thrips control. Nearly all acres are treated with an at-plant insecticide, either as a seed treatment or in-furrow application.

The importance of early planting for agronomic reasons (i.e. the short growing season) and in minimizing the impact of late-season insects overshadows other cultural practices that would help in lowering thrips damage.

Cotton aphids (*Aphis gossypii*)

Cotton aphids are resistant to organophosphates, as well as the pyrethroid bifenthrin and other pyrethroids and neonicotinoids, which include imidacloprid and thiamethoxam.

Fortunately, aphid populations are low due to a combination of predators, parasites and fungi, along with ineffective insecticides, so the general recommendation is not to treat cotton aphids, especially in early to mid-season, except under dry, stressed conditions. Very high aphid levels occur with little evidence of mummies or the fungus. Prior to harvesting, aphid-caused sooty mould or sticky cotton (from the heavy presence of honeydew) may become a problem. After the defoliant has been applied, however, cotton aphids are typically found only at very low levels.

Spider mites

In a typical year, approximately 1–2% of North Carolina's cotton acreage is treated for spider mites. Because spider mites occur as a result of factors largely outside the control of producers (dry weather and mowing of highway rights of way), little in the way of cultural practices can be effectively followed in non-irrigated cotton. Growers can affect mite populations somewhat by avoiding mowing field borders in cases where mites are present.

Plant bugs (*Lygus lineolaris*)

Plant bug damage has become more widespread in recent years, particularly in eastern counties. However, treatable populations are now widespread throughout the Coastal Plain. Early season monitoring of plant bug populations is recommended. Early planting, adequate plant spacing (three or fewer plants per foot) and holding down excessive plant growth have been shown to decrease plant susceptibility to plant bug damage.

Stink bugs (*Chinavia hilare, Euschistus servus*, and others)

The green stink bug and the brown stink bug can be very damaging pests of cotton. Stink bugs often invade cotton fields in early to mid-July and may reach damaging levels from this time through late August and sometimes into September. They may introduce boll-rot pathogens, resulting in partially or entirely destroyed locks, hard-lock and a lower grade of harvested cotton.

Stink bug damage is more prevalent in fields where bollworm treatments have been minimal (that is, none or one), although significant stink bug damage may occasionally occur prior to applications for bollworms. Where the bollworm population is high enough that the field has been treated twice or more (as is often the case with conventional cotton), stink bug numbers will usually, but not always, be reduced enough to limit damage to low levels.

Bollworm and tobacco budworm (*Helicoverpa zea* and *Chloridea virescens*)

Bollworm, also called corn earworm and soybean podworm, can be a significant bud and boll-damaging pest of cotton, especially following foliar insecticide sprays for other insects. This species emerges from the soil from diapause in early to mid-May and completes at least two generations, primarily in wild hosts and field corn, before flying to blooming cotton and soybeans.

Cotton lines that have been genetically altered to express a toxin of Bt have been available to North Carolina producers since 1996. Bt cotton varieties, referred to as Bollgard II, Bollgard 3, TwinLink, TwinLink Plus, WideStrike, and WideStrike 3 have been available for North Carolina farmers since 1996. Some of these express up to three Bt toxins. Bt varieties will not control insect pests other than caterpillars. Also, different caterpillar pests are not controlled to the same degree. For example, tobacco budworms attempting to feed on Bt varieties have shown zero survival in the field while other caterpillar pests, such as beet and fall armyworms and cabbage and soybean loopers that may be present at low levels, are no longer economic pests. In contrast, bollworms can become established, especially if a prior 'disruptive' spray has been used that reduces or eliminates beneficial arthropods. Finally, only limited control of cutworms is provided by the Bt toxins, at least in part because cutworms are often partially to fully grown when cotton seedlings are available in the spring.

Since 2016, bollworm resistance has been widespread in Bollgard II, TwinLink, and WideStrike varieties, but not in Bollgard 3, TwinLink Plus and WideStrike 3 varieties.

Diseases

Seedling diseases cause an estimated average annual yield loss of 5% and are usually the major disease problems in cotton production. Several soil-borne fungi are responsible; however, cultural and environmental factors that delay seed germination and seedling growth make the problem more severe.

Several species of fungi can cause seedling disease, but the primary agents are *Pythium* spp., *Rhizoctonia solani*, *Phoma exigua* (Ascochyta) and *Fusarium* spp. These disease-causing organisms can attack the seed before or at germination. In some years, replanting is necessary. Poor stand establishment causes problems with the management of other pests and may reduce yields.

Pythium spp.

Several species of fungi in the genus *Pythium* can cause seedling disease in cotton as well as several other crops. *Pythium* spp. are generally classified as water moulds, producing spores that move actively in soil water. In general, *Pythium* is commonly the culprit if the soil has remained saturated for several days or is poorly drained. Mefenoxam (Ridomil Gold) or etridiazole (ETMT, Terrazol) is necessary to control *Pythium* spp. seedling disease.

Rhizoctonia solani

This fungus typically causes sore shin and is more common on sandy, well-drained soils. Plants injured by sand blasting are particularly susceptible to this pathogen. Fungicides containing PCNB (Terrachlor), iprodione (Rovral) or azoxystrobin (Quadris) are generally effective against *Rhizoctonia solani*.

Phoma exigua (Ascochyta gossypii)

This fungus can cause postemergence damping-off. This disease is characterized by premature dying of cotyledons, which turn brown and shrivel; and it has been observed when night temperatures fall into the 50s and are accompanied by foggy or misty conditions. Fungicide effectiveness against *P. exigua* has not been evaluated.

All cottonseed offered for sale in North Carolina is treated with fungicides. Seed treatments are categorized as protectants and systemics. Protectant fungicides, such as captan or thiram, provide surface protection from disease organisms carried on the seed and from organisms found in nearby soil that cause seed rot.

In most years, seed treatment fungicides are sufficient for controlling seedling disease, unless the quality of the seed is low or weather conditions are unfavourable for germination. If additional fungicide is desired, it is best to use an in-furrow treatment. Hopper-box seed treatments are also available, but coverage and effectiveness are much better with in-furrow sprays or granules.

Fungicides are not a substitute for high-quality seeds and good planting conditions. In-furrow fungicides will not be profitable in most years; however, if conditions are less than optimal, they can result in better and more uniform stands.

Crop rotation, good seed quality, optimum soil temperature, and destruction and incorporation of cotton residue are beneficial in suppressing most diseases. Seed treatments and in-furrow fungicides may become more important in no-tillage cotton-production systems.

Nematodes

Cotton nematode control is accomplished through crop rotation, resistance and nematicides. Resistant varieties are available only for root-knot nematodes. Some varieties have shown extreme susceptibility to Columbia lance nematode and should be avoided in heavily infested fields. Long-season cotton varieties generally perform better than short-season ones when Columbia lance nematode is present.

Weeds

Cotton requires better weed control than either corn or soybean. Because cotton does not compete well with weeds, especially early in the season, a given number of weeds will reduce cotton yield more than corn or soybean yield. Weeds also may interfere more with harvesting of cotton, and they can reduce lint quality because of trash or possibly stain.

The first step in a weed management programme is to identify the problem, which is accomplished by growers through weed mapping and in-season monitoring of weeds in their cotton fields. Both of these activities are dependent on proper weed identification. The state extension service provides advice on controlling a range of weeds which can affect cotton (https://content.ces.ncsu.edu/cottoninformation/weed-management-in-cotton, https://ipmdata.ipmcenters.org/).

Crop rotation aids in the management of nematodes and diseases. Additionally, it can be a significant component of a weed management programme. Crop rotation allows the use of different herbicides on the same field in different years. By rotating cotton with other crops and selecting a herbicide programme for the rotational crop that effectively controls the weeds that are difficult to control in cotton, one can reduce or prevent the build-up of problem weeds and help keep the overall weed population at lower levels. Crop rotation and properly planned herbicide rotation also prevent evolution of herbicide-resistant biotypes of weeds.

Transgenic cotton

Very little non-transgenic cotton is now being grown in North Carolina. Roundup Ready cotton is any variety of transgenic cotton containing the gene that imparts resistance to the herbicide glyphosate. Roundup Ready varieties were grown on 95% of North Carolina's acreage in 2004.

Liberty Link refers to transgenic cotton resistant to the herbicide glufosinate, which can be applied over the top of this variety from emergence until the early bloom stage without concern over injury or fruit shed. On plants larger than about 10 inches, a semi-directed application may be preferred to obtain better coverage on weeds under the cotton canopy.

The latest transgenic cotton cultivars include multiple herbicide traits. XtendFlex cotton varieties are resistant to dicamba, glyphosate, and glufosinate whereas Enlist cotton varieties are resistant to 2,4-D, glyphosate, and glufosinate. Dicamba, 2,4-D, and glufosinate are necessary for management of troublesome glyphosate-resistant weeds like Palmer amaranth.

South Carolina

Introduction

South Carolina, one of the original 13 colonies, is one of the states that originally grew Sea Island cotton. This was an original long staple from the *Gossypium barbadense* plant, as opposed to the *G. hirsutum* plant. As implied, the Sea Island cotton plant was presumed to have originated in the Caribbean region and with silky and extra-long fibres, the variety is particularly desirable. William Elliott first imported *G. barbadense* to grow at Myrtle Bank on Hilton Head Island in 1790.

The Civil War and its aftermath destroyed the Sea Island economic and social structure, and agriculture was neglected. The postbellum period saw some adjustments, but the industry deteriorated and was no longer economically viable by the 1910s. The final blow was the invasion of the boll weevil in the 1910s and 1920s. A few surviving barbadense seeds went to an agricultural research centre in Texas, where they were crossbred with another cotton variety.

While upland cotton planters knew of various conservation practices, few were practised. They normally followed a pattern of clearing forest, planting cotton until yields declined, planting corn, abandoning the field, and then clearing new land. The major investment was in labour (slaves), and yield was measured in labour units. Land was inexpensive relative to labour costs. When soil nutrients were depleted and the yield per slave unit declined, fresh land was cleared.

Cotton production totalled about 280,000 bales in 1860 but declined to less than 180,000 bales in 1870. By 1911, however, production reached its peak at 1.6 million bales. The upper Piedmont and inner coastal plain were the chief production areas in the 1920s and 1930s. After World War II, the Piedmont counties of Anderson, Spartanburg and Greenville, for example, harvested cotton on more than 150,000 acres in 1945, but cotton slowly disappeared from the Piedmont and became concentrated in the inner coastal plain. Marginal land was taken out of production. Land in farms was just over 11 million acres in 1945, only about 5.6 million acres in 1982, and 4.8 million in 2000. Cotton production declined dramatically, dropping to a low of 53,000 bales in 1983. Strong demand, good prices, more effective boll weevil control, and decreased demand for soybeans all contributed to a revival of cotton production in the 1990s. By the year 2000, 379,000 bales were produced. Cotton remained concentrated in the inner coastal plain, but some was grown in the lower Piedmont.

Cotton-growing conditions

Historically, South Carolina was one of the first colonies, and then states, to grow cotton. Since revolutionary times to the present day, it has been an important cash crop. Although the acreage in the state does fluctuate with the price for cotton, currently almost 300,000 acres are in cotton production in the state. Cotton can tolerate the heat and still produce consistent yields and is a strong complement in rotation with the peanut crop as peanuts decrease the nematode population, which creates a favourable environment for the cotton crop the next year. As South Carolina seldom gets an early frost, the cotton crop can do well if the temperatures stay in the 85–95° F. range. Cotton is harvested starting in October and ends around Thanksgiving (26 November) (Lockman, 2013).

Texas

Introduction

Cotton was first grown commercially in Texas by Spanish missionaries. A report of the missions at San Antonio in 1745 indicates that several thousand pounds of cotton were produced annually, then spun and woven by mission craftsmen. Cotton cultivation by Anglo-American colonists was begun in 1821. A census of the cotton production of the state reported 58,073 bales (500 pounds each). In 1849 and by 1852 Texas was in eighth place among the top ten cotton-producing states of the nation. After the Civil War, production of cotton increased to 350,628 bales in 1869 and then by 1879 some 2,178,435 acres produced 805,284 bales. Forty years after the loss of the Civil War and Texas' recovery from a disruptive period of reconstruction, Texas emerged in the early 20th century as the largest cotton-producing state in the USA – a position it has maintained for nearly 120 years. In 2019, Texas produced 33% of the US cotton crop (Hundl, 2020).

Cotton acreage increased strongly in West Texas between 1889 and 1919. This increased cotton acreage was the result of farmers leaving central Texas because of the crop losses caused by boll weevil (Britton *et al.* 1994; Stavinoha and Woodward 2001). The early 1950s marked another period during which cotton plantings in West Texas increased. This time, mechanized agriculture – tractors and cotton strippers and irrigation – made it profitable for farmers in West Texas to enlarge farms and increase cotton production (McArthur *et al.* 1980; Fite, 1984).

Cotton is the leading cash crop in Texas, generating about $2.2 billion in crop value each year – adding in the value of cottonseed and cottonseed oil, the Texas cotton crop value is $2.8 billion. The overall economic impact of cotton in Texas is estimated to be as high as $24 billion annually (Hawkes, 2017). The quality of cotton grown in Texas has improved significantly over the last 30 years and has closed the gap relative to the high-quality cotton grown in California - which has demanded a premium price for years.

Cultural practices

Upland cotton (*G. hirsutum*) is the dominant type throughout Texas, with planting usually beginning in February and March in the lower Rio Grande Valley and later into June in the south High Plains. Pima cotton is grown in the far west of Texas, because of premium prices for the extra-long fibre, as it requires an extended growing season and grows well in a desert environment. Certain genetic variations of upland cotton, resulting in coloured cotton fibre ranging from green to brown, are currently produced in small quantities for novelty or special fashion markets (Kohel and Lewis, 1984) and some white and coloured lint cottons are produced organically.

There are eight areas of Texas in which cotton is grown (see Table 2.2 and Fig. 2.10).

Acid-delinted seed is treated with a fungicide for planting in all regions of Texas as this

Table 2.2. Cropping systems in different parts of Texas.

Location	Cropping Cotton is planted in April–May and harvested mostly from October–December.	Annual inches	Percentage irrigated
Panhandle	Cotton is a rotational crop with corn in areas of dryland wheat production.	20	50
South Plains	Largest cotton area can exceed 3 million acres. Alongside peanuts, corn, grain sorghum and wheat	18	60
Permian Basin	Rangeland with cotton and grain sorghum	15	52
Trans-Pecos	Rangeland with vegetables, alfalfa and pecans grown as well as cotton	10	100
Rolling Plains	Rangeland with cotton and wheat	21	15
Buckland Prairies	Pasture and hardwood trees with cotton, wheat, corn and grain sorghum	33–39	20
Winter Garden	Rangeland and notable corn, wheat, grain sorghum, cotton, vegetable and pecan	30	78
Coastal Bend	Pasture, rice, corn, grain sorghum, cotton and soybean	27–48	6
Lower Rio Grande Valley	Extensive agricultural production with a variety of vegetables, citrus, corn, grain sorghum, soybean and cotton	27	40

improves germination, planting uniformity and the threat of seedling diseases. Virtually no fungicides are applied after planting. Herbicides are commonly applied on all cotton land to reduce annual and perennial weed problems. Insect pests are scouted on 83% of cotton farms, and insecticides are generally applied based on economic thresholds or criteria for the Boll Weevil Eradication Program.

Row spacing will vary depending on rainfall to improve use of soil moisture and type of irrigation adopted, but where narrow rows are used, insect pest control is more challenging and harvesting is with stripper equipment. Picker harvesting is used in areas with long growing seasons and longer staple cotton as these harvesters remove only the seed cotton (fibre and seeds), which results in less trash, but higher harvest costs. Desiccants and defoliants are commonly applied prior to mechanical harvesting.

Farmers can get advice on managing cotton pests by consulting a booklet (Vyavhare *et al.*, 2018) as frequent and careful scouting for insect pests and beneficial insects is critical for successful cotton production. Information on the various pests and varieties available are provided with methods of sampling pest populations. Approaches to cotton production vary from one region to another because of differences in climate, harvest techniques, irrigation availability, pest pressure, soil types and variety planted. As cotton is vulnerable to attack throughout the crop season, frequent and careful scouting for insect pests and beneficial insects is a necessary component of successful cotton production.

Using multiple pest suppression tactics allows growers to control pests at a low cost, preserve natural enemies and slow the development of pest resistance to insecticides and insect control traits. Applying insecticides at the proper rates and only when necessary, as determined by frequent field inspections and economic thresholds, helps prevent economic losses that pests can cause.

Recent developments have significantly impacted managing pests in cotton. The availability of transgenic, insect-resistant traits in cotton has reduced the incidence of damaging populations of caterpillar pests. Eradication programmes have eliminated the boll weevil from everywhere but the far south of Texas and the pink bollworm throughout the state. Neonicotinoid seed treatments have greatly decreased the pest status of thrips and other early-season cotton pests.

Cotton varieties with third-generation Bt technologies have excellent activity against cotton leaf perforators, loopers, pink bollworm and tobacco budworm, and good activity against beet armyworm, cotton bollworm, fall armyworm

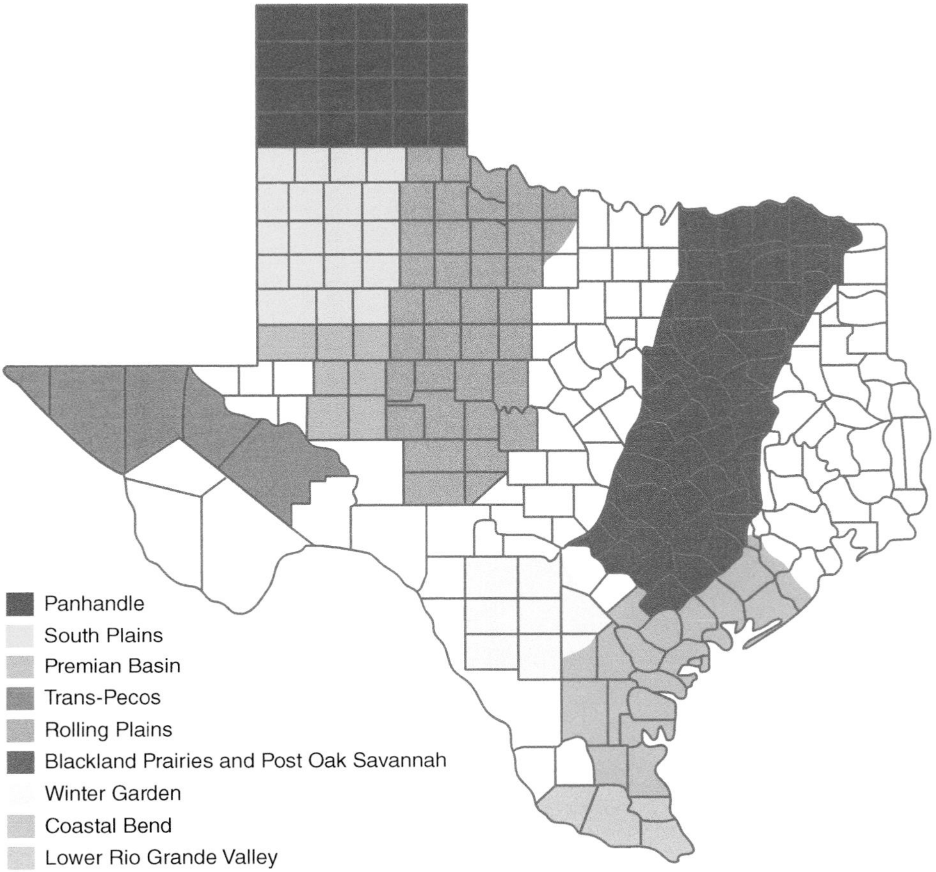

Fig. 2.10. Major cotton production regions of Texas (From: https://cottonbugs.tamu.edu/cotton-production-regions-of-texas – accessed 19 July 2021).

and saltmarsh caterpillar. Some situations may require supplemental insecticide treatment for bollworm and fall armyworm. Recommended economic thresholds used to trigger insecticide applications on Bt cotton are the same as those used for non-Bt cotton but should be based on larvae larger than a quarter of an inch.

Without effective weed control, cotton cannot be grown successfully in Texas. Historically, the necessity for 8–12 weeding operations per season limited the amount of land growers could plant in cotton (Reeves, 1975). Soil erosion due to wind and water compelled Texas cotton growers to develop conservation tillage systems (Bryson and Keeley, 1992). The development and use of herbicides revolutionized cotton production and greatly increased farmer efficiency. In 1997, the release of glyphosate-tolerant cotton varieties further improved the efficiency of Texas cotton growers. But herbicide-resistant weeds threatened the progress they had gained. Other herbicide-tolerance traits were developed and used, thus maintaining the efficiencies farmers had gained from over-the-top herbicide-application technology.

Cotton farming in Texas – progress and challenges

Texas cotton producers have made remarkable progress in the last 40 years. Growers typically fertilize their soil based on soil testing. Years ago, farmers on rolling land adopted terracing, and more recently they and flat land farmers have embraced the use of herbicide resistant cotton varieties and reduced/minimum tillage practices, greatly reducing soil erosion on their farms. They carefully study and consider their needs and varieties, traits and seed treatments to suppress diseases and pests and increase yields. Growers are contentious about optimizing cotton plant health and maintaining rapid fruiting through vigilant monitoring of planting soil and weather conditions, seeding rates and planting depths. They are vigilant about crop scouting for pest and beneficial insects and diseases. They have adopted integrated pest management practices and utilized agricultural research as they adopt best management practices on their farms. Between 1995 and 2019, foliar insecticide use in Texas has declined by 84% – largely due to boll weevil and pink bollworm eradication, Bt cotton and Texas farmers long-held confidence in cultural control as the foundation of their integrated pest management programme for insect pests (also for nematodes and diseases). Adoption of reduced tillage has kept tons of soil in place on Texas farms. Plant growth regulators, boll openers, defoliants and/or desiccants are used to condition the crop and help ensure that it is harvested as early and as cleanly as possible. Modern, efficient cotton harvesting equipment and technology have been adopted by Texas cotton growers. In stripper harvested areas of Texas, farmers quickly began utilizing strippers with burr extractors when they became available some years ago. Farmers in Texas quickly adopted modern post-harvest field/gin yard cotton storage systems – using module system which was developed in Texas. And, they are increasingly utilizing fast, modern gins to process their cotton crop cleanly and efficiently. Texas farmers have been diligent about early defoliation, early harvest, early stalk destruction and control of regrowth cotton to inhibit pests. They have maintained enforced early stalk destruction programmes since mechanical means to accomplish the job became available over 70 years ago.

The continuing upward trend in Texas cotton yields and planted acreage support the conclusion that the Texas cotton industry is healthy. Stresses on the industry will continue. But as a review of the history of cotton production in Texas demonstrates, many serious challenges have been faced and overcome. Optimism ingenuity, hard work and willingness to adopt better methods are qualities Texas cotton farmers have consistently maintained for nearly 200 years.

Cotton Growing in Mexico

As indicated earlier, domestication of *Gossypium hirsutum* occurred in Mexico between *c.*3400 and 2300 BC, and cotton clothes were worn there long before Europeans arrived in the Americas. Cotton spinning began before the Aztecs. An early explorer, Cortez, found cotton fabrics of the Aztecs in Mexico in 1519 and, impressed by their excellence, sent some to the King of Spain (Garloch, 1944). In modern times, the earliest cotton mills began in *c.*1840.

With the arrival of insecticides after the Second World War, up to 18 sprays were applied in a single crop season. The sprays were aimed at lepidopteran pests, namely the pink bollworm (*Pectinophora gossypiella*), which had been brought into Mexico in seeds from Egypt, bollworm (*Helicoverpa zea*), tobacco budworm (*Chloridea virescens*) as well as the boll weevil (*Anthonomus grandis*). Despite intensive spraying, often with a mixture of DDT, toxaphene and methyl parathion, as in another part of Central America, Nicaragua, growers lost 30–50% of their crop. With the pests becoming resistant to the sprays, production costs soared.

Cotton is mainly grown in Chihuahua which represents about 63% of total production, followed by Baja California (18%), Coahuila (11%), Sonora (5%) and Durango (2%).

The situation has changed as they embarked on an integrated programme endeavouring to enforce a close season of at least two months, initially treating the last cotton plants prior to stalk destruction with methyl parathion to reduce the overwintering boll weevils.

Now, with the availability of Bt cotton over the last 20 years, cotton pest management in the country has changed substantially, as growers can now use significantly less insecticide. Although the boll weevil has not been controlled with the initial Bt cotton cultivars, a bi-national eradication programme implemented in different regions of Mexico and the USA has reduced the boll weevil, which remains only in one small region of Mexico (Nava-Camberos *et al.*, 2019).

However, the current Bt cotton cultivars do not control sucking insects such as *Bemisia tabaci*, *Nezara viridula*, *Lygus* spp. and *Chlorochroa ligata*. The production of cotton has increased again since 2002.

The Secretariat of Environment and Natural Resources (SEMARNAT) has prohibited the use of transgenic seed and in 2020 warned the National Committee on Cotton Product System, which meant that Mexico would not be able to grow cotton the following year. The urgency for new genetically modified seeds of better quality for the next planting cycle led the organization to request a hearing from the presidency to expose the problem, pointing out that to go back to planting conventional cotton was not possible as no seeds were available to sow and it was not economically profitable to grow cotton in 2021 without GM seeds. It was pointed out that the vast majority of the more than 7000 cotton producers would have to grow other crops such as sorghum or corn.

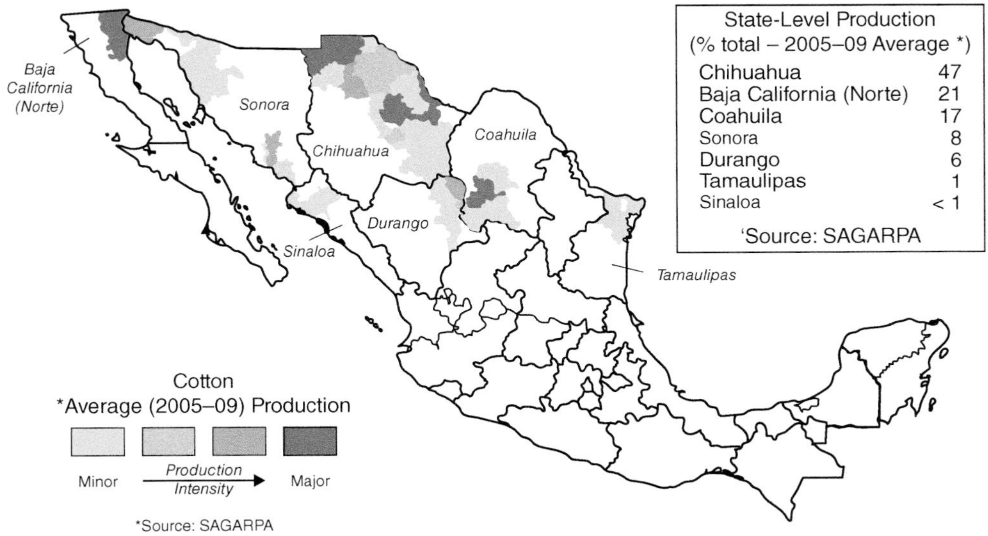

Fig. 2.11. Map showing the areas in which cotton is grown in Mexico. (From SAGARPA)

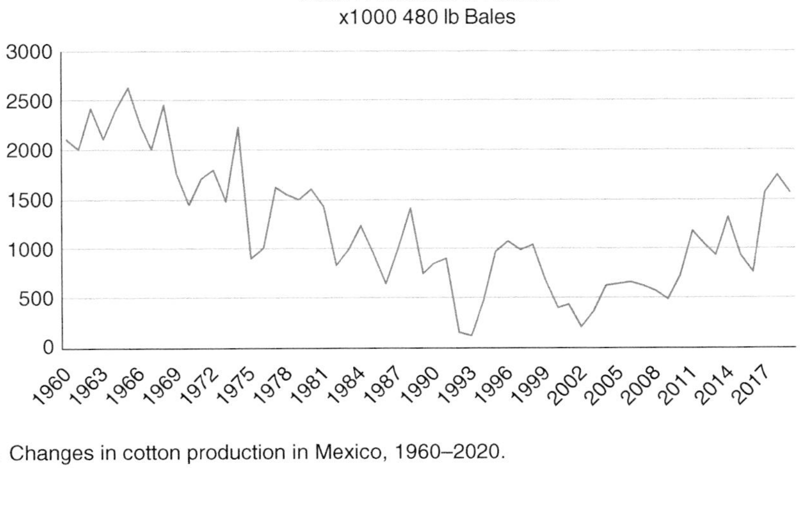

Fig. 2.12. Changes in cotton production in Mexico, 1960–2020.

References

Benedict, J.H., Leigh, T.F., Tingey, W. and Hyer, A.H. (1977) Glandless Acala cotton: more susceptible to insects. *California Agriculture* 31, 14–15.

Benedict, J.H., Treacy, M.F. and Kinard, D.H. (1994) Vegetable oils and agrichemicals. *The Cotton Foundation Reference Book Series (USA)* 4.

Blanco, C. (2012) *Heliothis virescens* and Bt cotton in the United States. *GM Crops & Food: Biotechnology in Agriculture and the Food Chain* 3, 201–212.

Bourland, F.M. (2018) History of cotton breeding and genetics at the University of Arkansas. *Journal of Cotton Science* 22, 171–182.

Brazzel, J.R. and Newsom, L.D. (1959) Diapause in *Anthonomus grandis* Boh. *Journal of Economic Entomology* 52, *603–611.*

Brazzel, J.R., Davich, T.B. and Harris L.D. (1961) A new approach to boll weevil control. *Journal of Economic Entomology* 54, 723–730.

Britton, K.G., Elliott, F.C. and Miller, E.A. (1994) Cotton Culture Handbook of Texas. Texas State Historical Association. Available at: https:www.tshanoline.org/handbook/entries/cotton-culture (accessed 14 August 2021).

Brown, L.C., Cathey, T.G. and Lincoln, C. (1962) Growth and development of cotton as affected by toxaphene-DDT, methyl parathion, and calcium arsenate. *Journal of Economic Entomology* 55, 298–301.

Bryson, C.T. and Keeley, P.E. (1992) Reduced tillage systems. Chapter 9. In: McWhorter, C.G. and Abernathy, J.R. (eds) *Weeds of Cotton: Characterization and Control*. The Cotton Foundation, Memphis, Tennessee, pp. 323–363.

Constantine, J.H., Alston, J.M. and Smith, V.H. (1994) Economic impacts of the California One-Variety Cotton Law. *The Journal of Political Economy* 102, 951–974.

Earle, N.W. and Newsom, L.D. (1964) Initiation of diapause in the boll weevil. *Journal of Insect Physiology* 10, 131–139.

Ellsworth, P.C., Brown, L., Fournier, A., Li, X., Palumbo, J. and Naranjo, S. (2011) *Keeping Cotton Green!* University of Arizona Co-operative Extension.

Fite, G.C. (1984) *Cotton Fields No More: Southern Agriculture, 1865–1980*. University of Kentucky.

Garloch, L.A. (1944) Cotton in the economy of Mexico. *Economic Geography* 20, 70–77.

Geisseler, D. and Horwath, W.R. (2013) *Cotton Production in California*. Fertilizer Research and Education Program. Available at: http://apps.cdfa.ca.gov/frep/docs/Cotton_Production_CA.pdf (accessed 16 July 2021).

Haney, P.B., Lewis, W.J. and Lambert, W.R. (2012) Cotton production and the boll weevil in Georgia: history, cost of control, and benefits of eradication. *Georgia Agriculture Experiment Station Research Bulletin* 428.

Hardee, D.D. and Harris, F.A. (2003) Eradicating the boll weevil. *American Entomologist* 49, 82–97.

Hawkes, L. (2017) *Cotton's economic benefit to Texas tops $24 billion*. Farm Progress. https://www.farmprogress.com/cotton/cotton-s-economic-benefit-texas-tops-24-billion

Haynes, K.F., Miller, T.A., Staten, R.T., Li, W.G. and Baker, T.C. (1987) Pheromone trap for monitoring insecticide resistance in the pink bollworm moth (Lepidoptera: Gelechiidae): new tool for resistance management. *Environmental Entomology* 16, 84–89.

Howard, C.W. (1867) Condition and resources of Georgia. *Report of the Commissioner of Agriculture for the Year 1866*. Government Printing Office, Washington, DC, pp. 567–580.

Hundl, W., Jr (2020) *Annual Cotton Review*. USDA, National Agricultural Statistics Service, Southern Plains Region. Available at: https://www.nass.usda.gov/Statistics_by_State/Texas/Publications/Current_News_Release/2020_Rls/tx-cotton-review-2020.pdf (accessed 14 August 2021)

Idowu, O.J., Pierce, J.B., Bundy, C.S., Zhang, J., Flynn, R.P., Carrillo, T. and Wedegaertner, T.C. (2012) Evaluation of glandless cotton cultivars in New Mexico. *Proceedings of the Beltwide Cotton Production Conference*, pp. 90–94.

Jenkins, J.N., Maxwell, F.G. and Lafever, H.N. (1966) The comparative preference of insects for glanded and glandless cottons. *Journal of Economic Entomology* 59, 352–356.

Kohel, , R.J. and Lewis, C.F. (1984) *Cotton*. American Society of Agronomy. Madison, Wisconsin.

Kranthi, K.R. (2021) Pink bollworm. *ICAC Recorder* 34, 1–56. [17 authors contribute to a special issue from the International Cotton Advisory Committee].

Lockman, C. (2013) Cotton remains a staple crop for South Carolina. *Farm Flavor*. Available at: https://www.farmflavor.com/south-carolina/south-carolina-ag-products/cotton-remains-staple-crop-for-south-carolina/ (accessed 16 July 2021).
Luttrell, R.G., Teague, T.G. and Brewer, M.J. (2015) Cotton insect pest management. In: Fang, D.D. and Percy, R.G. (eds) *Agronomy Monograph* 57, 2nd edn. American Society of Agronomy, Inc., Crop Science Society of America, Inc. and Soil Science Society of America, Inc., Madison, Wisconsin, pp. 509–546.
McArthur, W.C., Bolton, B., Ethridge, D., Heagler, A.M., Ghetti, J.L., Shaw, D.L., Cooke, F.T. Jr and Lawler, J. (1980) The Cotton Industry in the United States – Farm to Consumer. USDA National Economics Division and Texas Tech University. College of Agricultural Sciences Publication No. T-1-186.
Nava-Camberos, U., Terán-Vargas, A.P., Aguilar-Medel, S., Martínez-Carrillo, S.L., Rodríguez, V.A. *et al.* (2019) Agronomic and environmental impacts of Bt cotton in Mexico. *Journal of Integrated Pest Management* 10, 15, 1–7.
Pierce, J.B., Monk, P. and Garnett, A. (2014) Glandless cotton in New Mexico beet armyworm and cotton bollworm development and field damage. In: *68th Proceedings Beltwide Cotton Conferences.* National Cotton Council, New Orleans, Louisiana, pp. 688–692.
Reeves, B.G. (1975) Minimum tillage: summary of cotton belt results. *Proceedings of Beltwide Cotton Producers Mechanization Conference*, pp. 33–34.
Smith, C.W. and Cothren, J.T. (eds) (1986) *Cotton, Origin, Technology and Production*, John Wiley and Sons, Inc.
Smith, J.W. (1998) Boll weevil eradication: area-wide pest management. *Annals of the Entomological Society of America* 91, 239–147.
Smith, J.W. and Harris, F.A. (1990) *Anthonomus* (Coleoptera: Cuculionidae). In: Matthews, G.A. and Tunstall, J.P. (eds) *Insect Pests of Cotton.* CAB International, Wallingford, UK.
Staten, R.T. and Walters, M.L. (2020) Technology used by field managers for pink bollworm eradication with its successful outcome in the United States and Mexico. In: Hendrichs, J., Pereira, R. and Vreysen, M.J.B. (eds) *Area-wide Integrated Pest Management: Development and Field Application.* CRC Press, Boca Raton, Florida, pp. 51–92.
Stavinoha, K. D. and Woodward, L.A. (2001) Texas boll weevil history. In: Dickerson, W.A., Brashear, A.L., Brumley, J.T., Carter, F.L., Grefenstette, W.J. and Harris, F.A. (eds) *Boll Weevil Eradication in the United States Through 1999*. Number 6. Cotton Foundation Reference Book Series. National Cotton Council. Memphis, Tennessee.
Tabashnik, B.E., Liesner, R., Ellsworth, P.C., Unnithan, G.C., Fabrick, J.A. *et al.* (2021) Transgenic cotton and sterile insect releases synergize eradication of pink bollworm a century after it invaded the United States. *Proceedings of the National Academy of Sciences* 118.
Vyavhare, S.S., Kerns, D., Allen, C., Bowling, R., Brewer, M. and Parajulee, M. (2018) *Managing Cotton Insects in Texas.* Available at: https://agrilifecdn.tamu.edu/texaslocalproduce-2/files/2018/07/Managing-Cotton-Insects-in-Texas.pdf (accessed 16 July 2021).
Zhang, J., Wedegaertner, T., Idowu, O.J., Flynn, R., Hughs, S.E. and Jones, D.C. (2016) Registration of glandless 'NuMex COT 15 GLS' cotton. *Journal of Plant Registrations* 10, 223–227.
Zhang, J., Wedegaertner, T., Idowu, O.J., Sanogo, S., Flynn, R., Hughs, S.E. and Jones, D.C. (2017) Registration of a glandless 'Acala 1517–18 GLS' cotton. *Journal of Plant Registrations* 13, 12–18.
Zhang, J., Idowu, O.J. and Wedegaertner, T. (2020) Registration of glandless 'NuMex COT 17 GLS' upland cotton cultivar with Fusarium wilt race 4 resistance. *Journal of Plant Registrations* 14, 1–9.

3 Cotton Growing in India

V.N. Waghmare*, M.V. Venugopalan, V.S. Nagrare, S.P. Gawande and D.T. Nagrale

Cotton has been an economically important commercial crop for India from the earliest times. Cotton is grown for its fibre (lint), seed oil and as feed for animals. India is the only country where all four cultivated species of cotton are being grown. India has pioneered the hybrid cotton technology and become the sole country where more than 90% of its acreage is under hybrids. India also pioneered cotton-cultivation skills, ginning, spinning, weaving and excellence in cotton fabrics, famed as 'webs of woven wind'.

The history of cotton in India can be traced to its domestication, which is very complex. The latest archaeological discovery in Mehrgarh (now in Pakistan) puts the dating of early cotton cultivation and its use to 5000 BC (Menon and Uzramma, 2017). The ancient Indus Valley Civilization discovered through Mohen-jo-daro relics makes the time of cotton cultivation and manufacture of cotton fabrics about 3000 BC. A close study of these relics at the Technological Laboratory of the Indian Central Cotton Committee (now CIRCOT) indicates the coarse cotton from which fabrics were manufactured related to *G. arboreum* types (Sethi, 1960). Alexander the Great, during his sojourn in India, described cotton as 'a plant from which the natives plucked the vegetable wool which they spun to admirable clothing'. Herodotus, an ancient Greek historian, described Indian cotton in the fifth century BC as 'tree, bearing as their fruit, fleeces which surpass those of sheep in beauty and excellence'. Sufficient evidence has been recorded by the Arabian travellers describing Indian fabrics and the flourishing export trade in cotton and cotton goods as early as 569–525 BC. Knowledge about cotton cultivation and manufacture of cotton textiles was introduced into Spain in the early centuries of the Christian era. In fact, available evidence proves that India was the original habitat of cotton and an exporter of fine fabrics since the ancient times. Marco Polo, who travelled to India in the 13th century; Chinese travellers to Buddhist pilgrimage centres; Vasco Da Gama, who entered Calicut in 1498; and Tavernier in the 17th century all have praised the superiority of Indian fabrics.

In the 13th–14th centuries, during the early Delhi Sultanate era, the roller cotton gin was invented and is still used in India, but processing was further advanced with the spinning wheel across the country which greatly lowered the cost of textile production and facilitated expanding textile trade from India. The agricultural reforms in favour of higher-value cash crops, such as cotton, during the 16th to the early 18th centuries provided the much-needed incentives to grow cash crops. This increased the

*Corresponding author: vijayvnw@yahoo.com

DOI: 10.1079/9781800620216.0003

production of cotton, helped to expand textile manufacturing and export of calicos, muslins and other manufactured cotton products capturing 25% share of the global textile industry by early 18th century.

Introduction of cotton fabrics by the East India Company into Britain was initially regarded as a novel addition to the trade in spices, but the colourful printed fabrics became more popular and surpassed the spice trade. The inflow of Indian cheaper calicoes and cotton fabrics threatened British interests. As a result, the British parliament, in 1708, prohibited the import of Indian printed calicoes for domestic use either as apparel or furniture, under penalty of £200. To protect the textile industry in Britain, import of cotton printed fabrics was completely forbidden and only raw cotton was imported. The Industrial Revolution in England brought about the development of mechanized spinning and weaving machines and increased Britain's requirement of raw cotton enormously, which rose from 4 million pounds in the 1770s to about 56 million pounds by 1800.

While Indian cotton textiles, particularly those from Bengal, continued to maintain a competitive advantage until the 19th century, British colonization of India opened the large Indian market to British goods, which could be sold without tariffs or duties, in contrast to the heavily taxed Indian goods. So, Britain eventually surpassed India as the world's leading cotton textile manufacturer in the 19th century. The East India Company made special efforts for production of better and cleaner cotton and facilitated extensive trials with exotic cotton (*G. hirsutum*) to compete with cotton from America.

The history of the introduction of American cotton in India dates back to the last quarter of the 18th century. The first attempt was made in 1790 when the seeds of Bourbon cotton from Malta and Mauritius were distributed in Bombay state, but failed. It was only in the Hubli-Dharwar area; good results were not obtained until the 1840s where seeds from New Orleans were grown successfully. Dharwar-American cotton soon became popular and registered an area of 178,682 acres in Hubli Taluka by 1861–62. The acclimatized American upland Georgian cotton variety Buri was released for the first time from Nagpur Farm in 1903–04.

Scientific studies on different problems of cotton cultivation started only after the establishment of the Agricultural Departments in various provinces and princely states in 1904, but these lacked proper co-operation. To ensure extending cultivation of long-staple cotton, the Indian Cotton Committee was established in 1917 and later the Indian Central Cotton Committee (ICCC) was established in 1921. The main functions of the committee were to act as an advisory body to the government in all matters pertaining to cotton and to assist the Agricultural Departments to develop improved cotton varieties.

Until 1947, when India became independent, the old-world diploid Asiatic cottons, namely *G. arboreum* and *G. herbaceum*, were grown covering 97% of the acreage under cotton. Concerted efforts to improve tetraploid cotton *G. hirsutum* and *G. barbadense* yielded good results and several improved varieties were developed. The cotton improvement efforts got further fillips with the abolition of the ICCC in 1966 and the establishment of the All India Coordinated Cotton Improvement Project (AICCIP) at Coimbatore in 1967 and the Central Institute for Cotton Research (ICAR-CICR) at Nagpur in 1976, under the overall control of the Indian Council of Agricultural Research (ICAR). Since then, more than 350 improved varieties and hybrids have been released by the public sector for commercial cultivation in different cotton-growing states (Tables 3.1 and 3.2). Some of the prominent landmark events in cotton improvement efforts in India include: the release of the world's first cotton hybrid, 'H4', in 1970; first interspecific hybrid between *G. hirsutum* and *G. barbadense* – 'Varalaxmi' – in 1972; *G. barbadense* Sea Island cotton variety Suvin in 1978; LRA5166, a *G. hirsutum* variety in 1983 with wide adaptability in all three cotton-growing zones, which occupied more than 30% of the area under cotton for about a decade before the Bt era; introduction of Bt cotton (one gene event MON 531 containing Cry 1Ac gene Bollgard) in 2002; and approval of Bollgard II (Cry 1Ac and Cry 2Ab) for commercial cultivation in 2006. Since the introduction of GM cotton in 2002, private sector seed companies developed >2000 BGII cotton hybrids and made them available for planting in all cotton-growing states.

Table 3.1. Cotton varieties released for different states of India.

Name of state	Tetraploid cotton	Diploid cotton
Punjab	F 1378, LH 1556, LH 900, F 846, F 1054, LH 1134, F 505, F 1861	LD 327, LD 491, LD694
Haryana	H 1098, H 777, HS 6, H 974, HS 182, H 1117	HD 107, HD 123
Rajasthan	RST 9, RST 875, G. Ageti, RS 810	RG 8, RD 18
Uttar Pradesh	Vikas	Lohit, CAD 4
MP	Khandwa 2, Khandwa 3, Vikram, JK 4	Maljari, Jawahar Tapti, Sarvottam
Gujarat	G.Cot 12, G.Cot 16, G.Cot 18	G.Cot 15, G.Cot 19, G.Cot 13*, G.Cot 17*, G.Cot 21, G.Cot 23*
Maharashtra	DHY 286, Rajat, LRA 5166	AKH 4, AKA 5, AKA 8401, PA 183, PA 255, AKA 7, Y1, PA 402, CNA 1028, CNA 1032
Andhra Pradesh	L 389, L 603, Kanchana, LK 861	Srisailam, Mahanandi, Raghvendra*, Arvinda
Telangana	L 389, L 603, Kanchana, LK 861	Srisailam, Mahanandi, Raghvendra*, Arvinda
Karnataka	Sharda, Abadhita, Sahana	DB 3-12*, Raichur-51*, DLSA 17
Tamil Nadu	MCU 7, MCU 5 VT, LRA 5166, Surabhi, Sumangala, MCU 12, SVPR 2, Suvin	K 10, K 11, CNA 1003 (Roja)

**G. herbaceum*

Cotton Production and Consumption

At the time of Independence (1947–48), India produced only 0.39 million tonnes of cotton from 4.4 m ha with a productivity of 88 kg lint/ha. The production steadily increased as a consequence of a series of technological breakthroughs to reach an all-time high of 6.77 million tonnes in 2013–14 from 11.96 m ha. Currently, India has the largest area under cotton (Fig. 3.1), is the largest producer (Fig. 3.2) and the second largest consumer of cotton. The average productivity during the last decade was 512 kg lint/ha. There was a 5.2-fold increase in the domestic consumption in the last six decades. From 2005–06, India became a net exporter of cotton. It imported around 0.09 million tonnes of extra-long staple cotton from the USA, Egypt, Sudan and Australia and another 0.09–0.12 million tonnes of contamination-free cotton. India's major export destinations in 2018–19 were Bangladesh, China and Pakistan.

Cotton is cultivated in 11 states across three cotton zones – north, central and south (Fig. 3.3). The average national cotton production during the last five years was 6.06 million tonnes. The states of the north zone (Punjab, Haryana and Rajasthan) together contributed 0.854 million tonnes (14%) to the national cotton pool from an area of 1.45 million ha (12%). Similarly, the states of the central zone (Gujarat, Maharashtra and Madhya Pradesh) produced 3.417 million tonnes (56%) from an area of 7.35 million ha (60.6%). The states of the south zone, Telangana, Andhra Pradesh, Karnataka and Tamil Nadu, produced 1.73 million tonnes (28.4%) from 3.14 million ha (26%). Odisha produced around 0.058 million tonnes from an area of 0.14 million ha.

Cotton is an important monsoon season cash crop which competes with soybean, castor, maize, sorghum, sunflower and pigeon pea, but in recent years the area under cotton has increased (Venugopalan *et al.*, 2017). Comparative domestic price advantage, internal support price and better export prospects favour cotton over the competing crops. Only synthetic fibres offer fierce competition to cotton.

Cotton cultivation provides direct livelihood to 5.8 million farmers and another 40–50 million persons earn their living through cotton processing and trade. To make cotton production more sustainable, global brands, retailers and NGOs have introduced voluntary sustainability standards and codes, among them Better Cotton

Table 3.2. Popular non-GM hybrids in India.

Name of state	Tetraploid cotton	Diploid cotton
Punjab	FHH 209, F 2276, FATEH, LHH 144	DDH 11, Moti (LMDH 8), PAU 626 H (FMDH-3), FMDH-8, FMDH-9
Haryana	DHANLAXMI , OM SHANKAR,	AAH1, CICR-2, AAH 32
Rajasthan	MARU VIKAS	RAJH-9
Uttar Pradesh	–	–
Madhya	LAHH 4 and JKHy-1 and JKHy-2, JKHY 11	–
Gujarat	H 4 , H 6, H 8, H 10	DH 7, DH9
Maharashtra	PKV Hy 2 and NHH 44, NHH 250, SAVITRI, RHH 195, NHH 302, CICR HH 1	AKDH-7, AKDH-5, PhA 46
Andhra Pradesh	LAHH 1, LAHH 4, NHB 80	–
Telangana	LAHH 1, LAHH 4, NHB 80	–
Karnataka	VARALAXMI, DCH 32, DHB 105 and DHH 11, RAHH 455	DDH 2
Tamil Nadu	Savita, TCHB 213, Surya and Sruthi, TSHH 0629, CBS 156, Suguna	

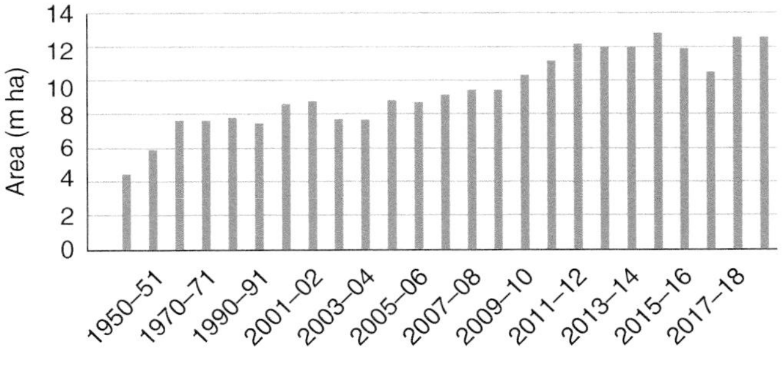

Fig. 3.1. Changes in area of cotton from 1950 to 2018.

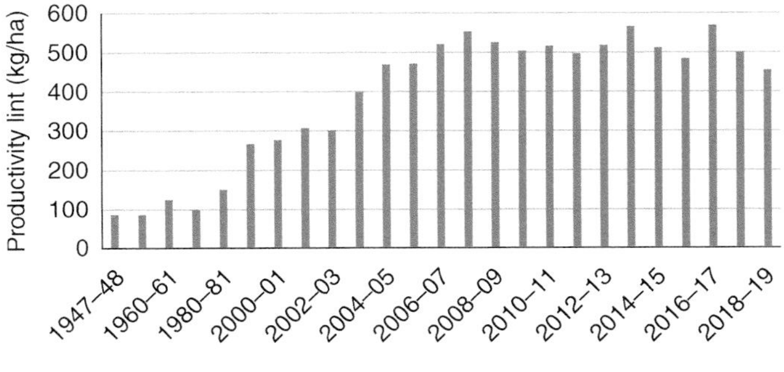

Fig. 3.2. Increase in cotton productivity since 1947.

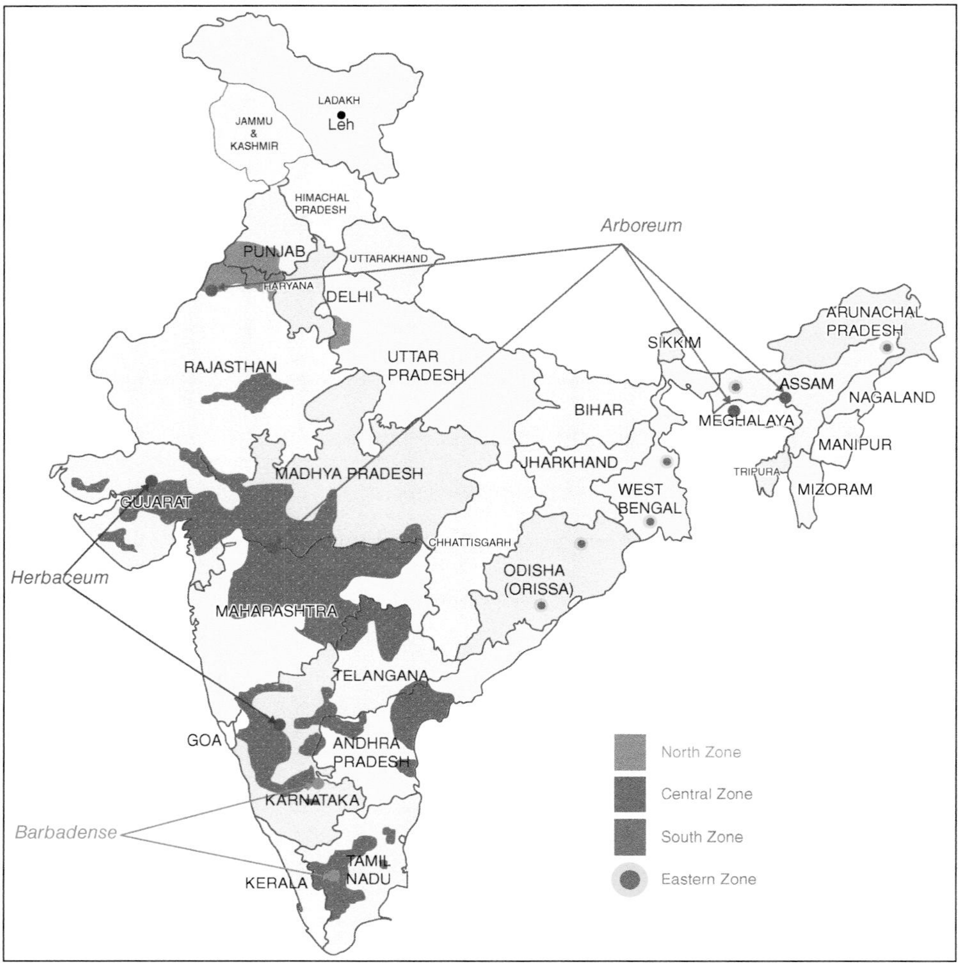

Fig. 3.3. Map showing areas of India where different cottons were grown.

Initiative (BCI), Fairtrade, Organic (Global Organic Textile Standard and Organic Cotton Standard) and REEL Cotton operate in India (Ward and Mishra, 2019). During 2018/19, India produced 123,072 (51% of global) metric tonnes of organic cotton, from 211,863 ha through 166,767 farmers (Textile exchange, 2020). India has the largest number of farmers participating in the BCI Programme. During 2018/19, 684,274 BCI farmers produced 652,000 tonnes of Better Cotton on 785,000 ha (https://bettercotton.org/where-is-better-cotton-grown/india/ – accessed 20 July 2021).

Table 3.3 provides a list of crops that compete with cotton for area. In recent years, the area under cotton is increasing at the expense of pulses, oilseeds and coarse cereals (Venugopalan *et al.*, 2017).

Government Support Policy for Cotton

Every year, before the commencement of the cotton season (October–September), the government of India, on the recommendation of the Commission for Agricultural Costs and Prices (CACP), fixes the Minimum Support Price (MSP) for two basic varieties of cotton: medium staple

Table 3.3. State-wise list of crops competing with cotton for area.

State	Competing crops
Punjab, Haryana and Rajasthan	Paddy, maize and cluster bean
Gujarat	Groundnut, castor and soybean
Maharashtra	Soybean, pigeon pea and sorghum
Madhya Pradesh	Soybean, maize
Odisha	Soybean, mung bean and urid bean
Telangana	Soybean, castor, corn and pigeon pea
Andhra Pradesh	Paddy, chillies and sunflower
Karnataka	Soybean, finger millet, sunflower
Tamil Nadu	Maize, vegetables, mung bean and urid bean

(24.5–25.5 mm with micronaire of 4.3–5.1) and long staple (29.5–30.5 mm with micronaire of 3.5–4.3). Based on these two basic varieties and taking into account the quality differential, normal price differential and other relevant factors, the MSP for other classes of seed cotton of Fair Average Quality (FAQ) is fixed by the Ministry of Textiles. The Cotton Corporation of India (CCI), Ministry of Textiles procures cotton from farmers whenever the market price of seed cotton falls below the MSP (Reddy, 2020).

Currently, the Department of Agriculture, Cooperation and Farmers Welfare is implementing the Cotton Development Programme through the National Food Security Mission (NFSM). Under NFSM, Front Line Demonstrations (FLD) on Integrated Crop Management (ICM), intercropping, desi/extra-long staple cotton and pink bollworm management are conducted. These programmes disseminate and upscale cotton technologies developed by ICAR-CICR and State Agricultural Universities (SAUs). The Committee on Cotton Production and Consumption (COCPC), Ministry of Textile, assesses the production, consumption, imports and exports of cotton at regular intervals and ensures adequate availability of cotton for the domestic spinning industry.

Applied Cotton-growing Methods

Cotton is widely grown on Vertisols and Vertic Intergrades, Entisols, Inceptisols and Alfisols. Double-cropping of cotton-wheat is dominant in the north zone. In the central zone, rain-fed cotton is either monocropped or strip intercropped with pigeon pea or occasionally intercropped with soybean, urid bean, mung bean or groundnut. In the south zone, cotton is monocropped or occasionally intercropped with vegetables, urid bean, mung bean. Under irrigated conditions, double-cropping of cotton-paddy or cotton-corn is also followed.

Land preparation is done by tractors using disc plough/harrows (Jalota *et al.*, 2008) or bullock-drawn blade harrows. In the central zone, deep ploughing using mould-board plough is done once every three years during summer (March–May). The benefits of reduced tillage (Blaise, 2011) and conservation tillage have been established (Choudhary *et al.*, 2016), but these systems are yet not widely adopted.

Cotton is sown using tractor- or bullock-drawn planters or is manually dibbled. Hand-dibbling of single seeds at predetermined intervals is practised in rain-fed areas, particularly for Bt hybrids, to save seeds and optimize plant stand and geometry. A spacing of 67.5 x 60 cm is uniformly adopted for Bt cotton in north India. In central India, various geometries ranging from 150 x 30 cm to 90 or 75 x 30, 45 or 60 cm are adopted to plant Bt hybrids. In South India, wide spacing (90 x 60 cm^2) is recommended for Bt hybrids in deep black soils and closer (60 x 60 cm^2) spacing in red soils (Malavath *et al.*, 2014). A spacing of 60 x 30 cm^2 is recommended for *G. hirsutum* varieties. For *desi* cotton, a spacing of 60–75 cm x 15 cm is followed and sowing is done by bullock- or tractor-drawn seed drill (Venugopalan *et al.*, 2013). Cotton is sown or harvested all year round in one or other regions of India (Table 3.4).

Nutrient management in cotton is complex due to the simultaneous production of vegetative and reproductive structures during the active growth phase. Sawhney and Sikka (1960) summarized the results of numerous trials with

Table 3.4. Crop-growing seasons in different regions of India.

Cotton growing states	Ecosystem	Cotton species under cultivation	Sowing period	Harvesting period
Punjab, Haryana and north & central Rajasthan	Irrigated – north zone	Intra-*hirsutum* hybrids, *hirsutum* varieties and *G. arboreum*	Mid-April – mid-May	October–mid-November
South Rajasthan	Rain-fed – north zone	Intra-*hirsutum* hybrids, varieties of *G. hirsutum* and *G. arboreum*	July	November–January
Gujarat	Irrigated	Intra-*hirsutum* hybrids	June (last week) – July (first week)	November–February
	Rainfed	*G. herbaceum*	First week of June	November–March
Madhya Pradesh	Irrigated	Intra-*hirsutum* and HxB hybrids	Mid-May	October–Feb/March
	Rainfed	Intra-*hirsutum* hybrids, *hirsutum* varieties and *G. arboreum*	End of June	February
Maharashtra	Irrigated western region	Intra-*hirsutum* hybrids	March–April	September–November
	Khandesh, Marathwada and Vidarbha	Intra-*hirsutum* hybrids, varieties of *hirsutum* and *G. arboreum*	Mid-June to first week of July	November–January
Karnataka	Rainfed and irrigated	Intra-*hirsutum* hybrids, *hirsutum* varieties, HxB hybrids	May–mid-July	January–February
	Rainfed	*G. herbaceum*	July–September	February
	Irrigated southern region	Intra-*hirsutum* hybrids and HxB hybrids	March–April	September–October
Telangana	Rainfed	Intra-*hirsutum* hybrids, *hirsutum* varieties	End of June–mid-July	November–January
Andhra Pradesh	Rainfed	Intra-*hirsutum* hybrids, *hirsutum* varieties	July	January–February
Odisha	Rainfed	Intra-*hirsutum* hybrids, *hirsutum* varieties	June	January
Tamil Nadu	Winter irrigated	Intra-*hirsutum* hybrids, *hirsutum* varieties, HxB hybrids and *G. barbadense*	First week of August	February–March
	Summer-irrigated	*G. hirsutum* varieties	February	July
	Rainfed	Intra-*hirsutum* hybrids, varieties of *G. arboreum* and *G. hirsutum*	September	February–March

manure conducted before 1960 and recommended the application of nitrogen to cotton and this was popularized during the 'Grow More Cotton' campaigns in the 1950s. Prior to 1980, the single-nutrient concept, with focus on the three primary nutrients (N, P and K), was the philosophy of nutrient management. With the introduction of hybrid cotton in the 1970s, the emphasis shifted to balanced application of nutrients. During the 1990s, the concept of integrated nutrient management (INM) picked up (Venugopalan *et al.*, 2013). Soil test-based nutrient application using the 4R concept (right rate, right placement, right source and right timing) is currently advocated. Grid-based site-specific nutrient management using nutrient prescription maps based on the target yield and customized fertilizer recommendations using a decision support system (DSS) (Majumdar and Prakash, 2018) are also being attempted.

For the Bt hybrids currently grown, 100–120 kg N/ha was found to be optimum in the north zone. In the central cotton and southern zone, the recommendation is 80–100 kg N/ha under rain-fed and 90–120 kg N/ha under irrigated conditions. Higher N dose (240 kg N/ha) is recommended for hybrids in Gujarat (Kairon and Venugopalan, 1999). To improve the efficiency of fertilizers, split application of N and foliar nutrient spray, use of slow-release fertilizers and nitrification inhibitors are recommended. Recently, the government of India mandated all the fertilizer manufacturers to coat urea with neem seed oil and fortify urea to improve its use efficiency. Foliar applications of 2% DAP, 2% urea and 2% KNO_3 have been found to improve yields at several locations (Venugopalan *et al.*, 2016). Studies conducted during the 1980s showed responses to P application (Venugopalan and Pundarikakshudu, 1999). Although soil K status in the cotton-growing regions is high, K is recommended as a prophylactic measure (Blaise and Prasad, 2005). On farmers' fields in Punjab, irrigated Bt cotton hybrids responded to soil application of K up to 41.6 kg K/ha on soils testing low or medium in available K and there was no response on soils with high available K (Brar *et al.*, 2008). Currently, 45–50 kg/ha each of P_2O_5 and K_2O is recommended for Bt hybrids.

Among the secondary and micronutrients, deficiency of S, Zn and B is widespread. In the Zn-deficient areas, soil application of 25 kg $ZnSO_4$/ha is advised. On B-deficient Vertisols, foliar spray of borax at peak flowering or initial boll-formation stage increased retention of bolls and improve yield (Blaise *et al.*, 2016). Systematic experimental evidence indicated a 25% higher nutrient requirement for Bt hybrids over their non-Bt counterparts (Blaise *et al.*, 2014). However, farm-level data indicated that within a decade of introduction of Bt hybrids, there was a three-fold increase in fertilizer consumption in Punjab, 2.5–3.0-fold in Maharashtra, Karnataka and Gujarat, and 2.0–2.5-fold in Tamil Nadu and Haryana (Venugopalan *et al.*, 2017). This indicated that the assurance of protection against bollworms encouraged farmers to invest more in fertilizers. As per the *ICAC Databook 2020*, the current average fertilizer consumption is 183.7 kg/ha (128 kg N+38kg P_2O_5 and17.7 K_2O/ha).

Water Management

About 35% of the cotton is raised with assured irrigation. Cotton in the north zone is grown under irrigated conditions (>97%), while the irrigated area in the central zone is 23% and south zone is 40% (Kranthi, 2020). In the rain-fed regions, the annual rainfall ranges from 700 to 1220 mm. The consumptive use of water ranges from 650 to 1100 mm depending on the crop duration, soil and climatic conditions. In the north of India, it is around 700–750 mm, in Gujarat 900–1100 mm, in Madhya Pradesh 660–685 mm, in Karnataka 800–900 mm and in Tamil Nadu 650–750 mm (Sivanappan, 2004). The water requirement of cotton in Nagpur (Maharashtra) is about 939.8 mm (Ghadekar and Patil, 1990).

The water requirement is low from emergence to squaring (<2.5mm/day), gradually increases from 2.5 mm to 5.0 mm/day from squaring to first white flower, and rapidly rises to 5–7 mm/day during peak bloom and early boll development, gradually tapering thereafter (Datta *et al.*, 2019). Figure 3.4 illustrates the typical rainfall and crop water requirement pattern for currently grown cultivars in central India. The crop experiences excess rainfall during the pre-flowering period and, later, moisture deficit during the peak bloom and boll development phases when the requirement is high.

The success of rainfed farming depends on the post-bloom rain and the ability of the soil to hold rainwater in its profile and release it later. Rainfed farmers adopt a variety of *in situ* mechanical/physical measures – contour bunding, land shaping (ridge furrow, dead furrow, broad bed furrow), sub-soiling and bench terracing – and agronomic measures – tillage, mulching (dust, straw, synthetic), cover crops and intercrops

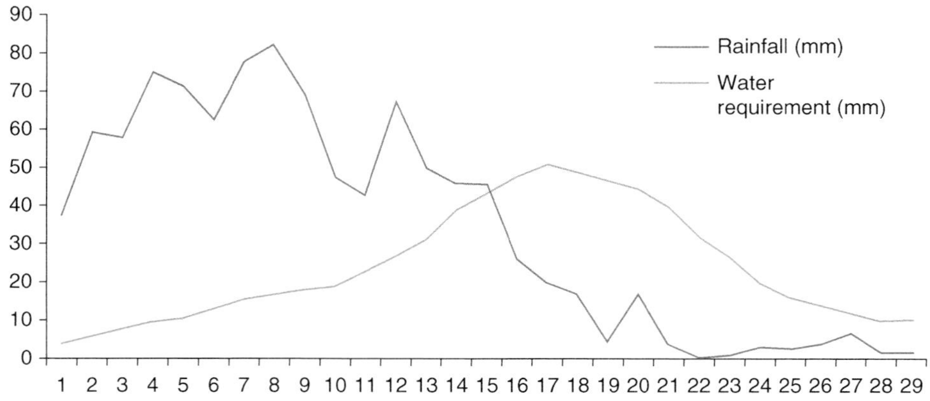

Fig. 3.4. Typical pattern of weekly rainfall and water use by cotton during the season in central India.

to conserve soil moisture. Excess run-off rainwater is collected in water-harvesting ponds and recycled to irrigate during the critical boll development phase.

Canals followed by lift irrigation are the main source of irrigation. In the north zone, a pre-sowing irrigation is given after the harvest of wheat to facilitate the seed bed preparation, lower soil temperature and create a favourable environment for germination and early seedling growth of cotton (Venugopalan *et al.*, 2013). Cotton is irrigated 2–10 times depending on the soil, season, climate and crop duration etc. (Venugopalan *et al.*, 2009). Consequently, the irrigation requirements vary between 140 and 330 mm in the north zone, 200 to 730 mm in the central zone and 500 to 700 mm in the south zone. Flood and furrow irrigation are widely practised in north India, whereas furrow irrigation and alternate furrow irrigation is adopted in the central and south zones. On sandy loam soils of the north zone, 3–5 irrigations are given. On red sandy loam soils of Tamil Nadu, with low water-retention capacity, more than ten light irrigations are done regularly. Drip irrigation is becoming popular in the states of Maharashtra, Gujarat, Tamil Nadu, Andhra Pradesh and Telangana.

Weed Control

Weeds compete with the cotton crop for light, water and nutrients. The extent of economic loss due to weeds ranges from 40 to 60% (Rao *et al.*, 2014) depending on the composition of weed species, intensity of weeds, duration of competition, and cotton plant density. The composition and abundance of weed species vary considerably across cotton-growing locations due to differences in growing season, cropping systems, edaphic and climatic conditions. For instance, in the northern cotton-wheat regions the major weeds are *Trianthema* sp. and *Echinochloa* sp. and *Digera arvensis* and *Cyperus* sp. (Balyan *et al.*, 1983) whereas water-loving species like *Portulaca oleracea* and *E. crusgalli* are major weeds in the rice fallows (Solaippan *et al.*, 1992).

The most common weeds occurring in the cotton ecosystem include *Cenchrus catharticus*, *Chloris barbata*, *Cynodon dactylon* (Pers.), *Dactyloctenium aegyptium* (Beavu.), *Dicanthium annulatum*, *Digitaria* spp., *Dinebra retroflexa*, *Echinochloa* spp., *Eleusine indica* (L.), Gaerth, *Eragrostis* spp., *Panicum repens*, and *Tragus bifloris* (Schult) among monocots. *Abutilon indicum* (L.), *Ageratum conyzoides* L., *Alysicarpus rugosus*, *Amaranthus* spp., *Aristolochia bracteata*, *Boerhaavia diffusa*, *Calotropis gigantea* (R. Br.), *Cleome viscosa*, *Coccinia indica*, *Commelina benghalensis*, *Corchorus olitorius*, *Cynaotis dactylon*, *Datura fastuosa*, *Desmodium* sp., *Digera arvensis* (Forsk.), *Eclipta* spp., *Euphorbia* spp., *Flaveria australasica* (Hook), *Merremia emarginata* (L.), Cufod., *Mimosa pudica*, *Parthenium hysterophorus*, *Phyllanthus* spp., *Portulaca oleracea*, *Sesbania aculeata* (Pers.), *Trianthema portulacastrum*, *T. terrestris* and *Xanthium strumarium* are the dominant dicotyledonous weeds. Among sedges,

Cyperus rotundus and *Cyperus iria* are the most abundant (Blaise and Kranthi, 2019).

The critical period of weed competition extends to 60–90 days after planting (Siddagangamma and Channabasavanna, 2018). Hand weeding and intercultural operations using bullock- or tractor-drawn implements are largely employed for managing weeds. However, the increasing labour costs and shortage of labour are driving farmers to use herbicides (Table 3.5). Pre-emergence herbicides of pendimethalin, fluchloralin and trifluralin combined with two mechanical hoeings and one weeding by hand are commonly recommended. However, farmers cultivating cotton under rain-fed conditions seldom adopt pre-emergence herbicides. Hence one hand-weeding combined with two hoeings and post-emergence application of pyrithiobac-sodium + quizalofop-ethyl is also recommended.

The biosafety related tests for herbicide-tolerant (HT) Roundup Ready Flex (MON88913) cotton were completed in 2013 but HT cotton is not yet approved for commercial cultivation. Some farmers resort to directed spray of glyphosate by covering the cotton plants since an over-the-top application would destroy the cotton crop.

Harvesting and Ginning

Fully mature cotton bolls are picked almost exclusively by farm women and the picking continues for about two to three months or more in a staggered manner depending on the fruiting pattern of the cultivars, labour availability, the crop management, the crop duration and market prices. Manual picking of cotton is labour-intensive, requires around 465 labour-hours/ha (Vithal, 2018) and is expensive (Rs 800–1200 [11–16 US$]/100 kg). To circumvent this, both small hand-operated cotton-picking machines were evaluated (Singh *et al.*, 2014), but these have not become popular because their performance was significantly inferior to the conventional manual picking. Cotton-picking machines, successful in other countries, are not suitable for India due to small holding size and scattered fields, bushy plant habit, asynchronous and staggered boll opening, non-uniform plant row spacings and perceived low return on asset investment (Bautista *et al.*, 2017). Self-propelled and tractor-mounted spindle-type cotton harvesters developed by M/S John Deere and M/S New Holland Tractors were tried at several locations on Indian farms but the results were not encouraging and the trash content (15–20%) and present set-up of ginning machinery available is not equipped to process it.

Manually picked cotton is clean, with a trash content of less than 2.5% while picking. However, post-picking, the trash content increases to 5–6% and extraneous materials, including cloth pieces, fabric sheets and hessian strings, are other sources of contamination.

Seed cotton yields range from as low as 0.5 tonnes/ha to 5 tonnes/ha in some farms. High yields are realized under irrigated, better managed and input-intensive farms. The average seed cotton yield in the majority of farms under rain-fed conditions is 1–1.5 tonnes/ha. The cotton

Table 3.5. CIB&RC approved herbicides for use on cotton in India.

Herbicide name	Dosage/ha (a.i.) (g/kg)
Alachlor 10% GR	2.0–2.5
Diuron 80% WP	0.75–1.5
Fenoxaprop-p-ethyl 9.3% w/w EC (9% w/v)	67.5 g
Fluchloralin 45% EC	0.9–1.2
Glufosinate Ammonium 13.5% SL (15% w/v)	375–450
Pendimethalin 30% EC	0.75–1.25
Pendimethalin 38.7% CS	677.27
Paraquat dichloride 24% SL (post-emergence directed inter-row application at 2–3 leaf stage of weeds)	0.3–0.5
Pyrithiobac Sodium 10% EC	62.5–75
Quizalofop-ethyl 5% EC	50.5

stalk after harvest is either consumed as domestic fuel or stacked on farm bunds, and less than 10% is ploughed back into the soil.

Raw cotton comprises about 34% of cotton, 63% of cottonseed and about 3% wastage. The Indian ginning industry is the second largest in the world. At present, there are about 4300 cotton-ginning and pressing factories, mainly situated in rural and semi-urban areas. The majority of them have double-roller cotton-ginning machines while in the north, i.e. Punjab, Haryana and Rajasthan, there are saw-gins at some ginning factories. A massive programme was undertaken under the Technology Mission on Cotton from the year 2000 to modernize the ginning and pressing sector of India. Quality of cotton after ginning depends on the quality of harvested cotton as well as the type and amount of cleaning performed. Increased productivity of ginning machines, reduction of energy use, reduction in contamination and improved cotton quality are benefits of these developments, which resulted in increased export of cotton from India. A recent report by Patil (2020) highlighted that modernized automated ginning and pressing units provide quality output at 40% less operational cost.

Insect Pest Management in Cotton

During the last two decades, the cotton crop has experienced frequent invasions by new insect pests as well as resurgence of existing insect pests. Cotton plants have succulent leaves, attractive flowers as well as bolls at reproductive phase; thus, several arthropod pests get attracted for food and shelter and, in due course, damage the crop throughout the crop season till harvesting. In a cotton crop, the presence of 251 arthropod pest species (including insects and mites), and on them, about 174 species of predators and 194 species of parasitoids/parasites in cotton agro-ecosystems have been documented (Nagrare *et al.*, unpublished data). These arthropod pests cause losses in quality and quantity by feeding on plant parts as well as transmitting viral diseases. However, about a dozen species of insects are identified as key/major pests causing significant losses to cotton crop, while the remaining species are occasional, sporadic, localized or minor in nature.

The major pests of cotton are: the bollworm complex – American bollworm, *Helicoverpa armigera* (Hubner); spotted bollworm, *Earias insulana* (Boisduvel), *E. vittella* (Fab.); and pink bollworm, *Pectinophora gossypiella* (Saunders); sucking insect pests – leafhopper, *Amrasca biguttula biguttula* (Ishida); aphid *Aphis gossypii* Glover; thrips, *Thrips tabaci* Lind.; whitefly, *Bemisia tabaci* (Gennadius); cotton mealybug, *Phenacoccus solenopsis* Tinsley; and papaya mealybug, *Paracoccus marginatus* Williams and Granara de Willink. Other pests, such as the Indian cotton mirid bug, *Creontiades biseratense* (Distant); stem weevil, *Pempherulus affinis* Faust and tobacco caterpillar, *Spodoptera litura* Fabricius are also categorized as major/key pests especially in south India. In the pre-Bt era, overall, all these pests were reported attacking cotton crop at different stages of growth, causing losses ranging between 50–60% (Puri *et al.*, 1999), but currently, losses are estimated at 20–30%.

Prior to the introduction of Bt cotton, farmers were endeavouring to control pests by applying insecticides. Many farmers complained that the insecticides were not very effective. To a large extent, this was due to use of the less-expensive spraying equipment, so coverage of the crop was poor (Fig. 3.5). Some farmers tried to obtain different insecticide, but they were sold with a different trade name and the active ingredient was not different to the one they complained about. Subsequently, more attention was given to studying resistance of bollworms to the pyrethroids.

The Indian cotton ecosystem experienced a phenomenal change in its pest profile since the introduction of cotton containing *Bacillus thuringiensis* called 'Bt cotton' with single-gene Cry 1Ac in 2002 (Bollgard) and two-gene Cry 1Ac+Cry 2Ab (Bollgard II) during 2006. At present, 95% area is now sown with Bt cotton hybrids. Bt cotton provides protection against bollworm and is more effective against *H. armigera*. It also offers protection against spotted and pink bollworm.

However, unusual survival of pink bollworm on Bt cotton producing Cry1Ac during 2008 confirmed pink bollworm resistance to Cry1Ac in Gujarat (Mohan *et al.*, 2015). Resistance development in pink bollworm to Bt cotton was confirmed in subsequent research findings (Bagla, 2010; Dhurua and Gujar, 2011; Fabrick

Fig. 3.5. A syringe-type sprayer spraying cotton near Guntur, Andhra Pradesh. (Photo: Graham Matthews)

Fig. 3.6. Late-season spraying of Bt cotton due to pink bollworm. The spray operator is walking directly into the spray at head height, which resulted in the death of farmers in 2017 using a highly hazardous insecticide. (Photo: Indian Express)

et al., 2014; Ojha *et al.*, 2014). From 2014 onwards, high incidences of pink bollworm were recorded on Bollgard II, substantiating resistance development to both the genes employed in Bollgard II. Naik *et al.* (2018) confirmed the resistance development by pink bollworm against Bt cotton, expressing Cry1Ac and Cry1Ac+Cry2Ab in a study conducted during 2010–17 in 38 districts of the ten major cotton-growing states.

Widespread infestation of pink bollworm on Bt cotton was reported from major cotton-growing Indian states, i.e., Gujarat, Maharashtra, Andhra Pradesh, Telangana, Karnataka and Madhya Pradesh, from 2015 onwards (Kranthi, 2015a; Naik *et al.*, 2018; Fand *et al.*, 2019; Naik *et al.*, 2020). Until 2017/18, BGII cotton in north-Indian states (Punjab, Haryana and Rajasthan) was free from pink bollworm infestation. However, the pest had been seen infesting Bt cotton during 2018/19 and 2019/20, especially in the Jind district of Haryana in the vicinity of oil-extraction mills (Kumar *et al.*, 2020). Over the last 5–6 years, this pest has become a serious menace on Bt cotton in India, causing widespread damage and approximate yield losses to the tune of 20–30% (Fand *et al.*, 2019) (Fig. 3.6). Studies of the DNA have shown low genetic diversity among pink bollworm populations across India (Sridhar *et al.*, 2017) and close genetical similarity in populations occurring in the early and late season have been verified with respect to the partial cytochrome oxidase (CO1) region of the DNA (Naik *et al.*, 2020). Developmental thresholds of 13.4°C/35.5°C (lower and higher temperatures, respectively) have been reported (Hemant *et al.*, 2020) with thermal requirements of 504.05 ± 4.84 degree days from the historical data across ten locations with diverse geographical and bioclimatic features (Fand *et al.*, 2020). Seven generations of pink bollworm in a cropping season, the length of which varied between 35 and 73 days in response to temperature have been clearly estimated (Fand *et al.*, 2020).

Bt cotton still provides complete protection against *H. armigera*; however, the pest can cause damage to non-Bt cotton that results in yield loss up to 40%. *H. armigera* infestation reduced significantly and in the last two decades it hardly exceeded economic threshold levels in the majority of the cotton-growing regions. The case is similar for *Earias* spp. The reasons behind the two bollworms under control are availability of multiple alternative hosts in the nature and efficacy of genes employed in Bollgard II, and also little or no development of resistance against *Cry* Bt proteins.

In cotton, there have been recurrent resurgences of existing pests as well as infestation of exotic pests which were hitherto not reported to be pests of cotton. During 2007/08, outbreak of invasive mealybug, *Phenacoccus solenopsis* Tinsley, recorded on cotton in all three cotton-growing zones of India, caused economic losses of up to 50% in affected cotton fields (Nagrare *et al.*, 2009). The infestation of *P. solenopsis* was high during 2007 and 2008 but subsequently went down and now it is found in traces.

A high population of another mealybug species, Papaya mealybug, *Paracoccus marginatus* Williams and Granara de Willink, was recorded on cotton in Coimbatore in 2008/09 (Dhara Jothi *et al.*, 2009). The mealybug was also found in other districts of Tamil Nadu (Tanwar *et al.*, 2010). Over the years, the pest has been seen in traces. Besides these two species of mealybug, five more mealybug species, namely spherical mealybug, *Nipaecoccus viridis;* striped mealybug, *Ferrisia virgata*; pink hibiscus mealybug, *Maconellicoccus hirsutus*; mango mealybug, *Rastrococcus iceryoides* (*Pseudococcidae*); and ber mealybug, *Perissopneumon tamarindus* (*Monophlebidae*) have been recorded as minor pests of cotton (Nagrare *et al.*, 2014).

Leafhopper, also called Indian Jassid, *Amrasca biguttula biguttula* (Ishida), is dominant and a regular major pest of cotton. Existence of a single species of leafhopper was recorded across India (Kranthi *et al.*, 2017). Leafhopper infestation significantly reduces chlorophyll and relative water content in cotton (Prabhakar *et al.*, 2011) and causes 'hopperburn' symptoms leading to loss of crop vitality and cotton yield up to 30%. Monoculture of Bt cotton hybrids that has saturated almost every cotton area, subsequent reduction of broad-spectrum pesticide sprays on Bt cotton, and the insects' ability to develop resistance are seen to be the reasons for higher incidence of leafhopper, especially on upland cotton. The indiscriminate use of insecticides against the leafhopper resulted in the development of resistance in leafhoppers against organophosphates (Singh and Jaglan, 2005; Sagar *et al.*, 2013; Sandhu and Kang, 2015) and neonicotinoids (CICR, 2011; Chaudhari *et al.*, 2015; Halappa and Patil, 2016).

Cultivated cotton is one of the important commercial crops ravaged by aphid *Aphis gossypii* Glover (Hegde *et al.*, 2011). Prevailing climatic conditions, especially seasonal changes in temperature and relative humidity, are highly conducive to the proliferation and the extent of damage caused by *A. gossypii* in cotton. The pest is also reported as a vector of polerovirus infecting cotton in India (Mukherjee *et al.*, 2012). Potential risk of establishment and survival of *A. gossypii* in India is based on the simulation of a temperature-dependent phenology model, temperatures between 22 and 27°C favoured its optimum development and the lower and upper thresholds of 6.24°C and 34°C, respectively, have been reported for its development (Nagrare *et al.*, 2019a).

Cotton thrips, *Thrips tabaci* Lind, is a regular pest of cotton and generally appears early in the season (Sharma and Sharan, 2016). The insect is extremely polyphagous and attacks several hundred plant species. Its presence has been found throughout India but during the last 4–5 years, unprecedented increase in the thrips population was recorded in central India as well as in north India. Increasing infestation in north India has now become a serious concern. Up to 94.20 thrips per three leaves during the 30th standard meteorological week (23–29 July) have been recorded in north India (CICR, 2018). *T. tabaci* is also a vector transmitting tobacco streak virus (TSV) disease and is reported to be an emerging threat to cotton cultivation in India (Rageshwari *et al.*, 2017; Vinodkumar *et al.*, 2017).

Whitefly, *Bemisia tabaci* (Gennadius), is a major pest occurring throughout cotton-growing zones of India, but it is the most prominent sucking pest in north-Indian cotton-growing states (Punjab, Haryana and Rajasthan) and also by virtue of its being a vector to transmit cotton leaf curl virus disease (CLCuD), especially in upland cotton. Several outbreaks of whitefly have been reported (Jayaraj *et al.*, 1986) but the recent one witnessed during 2015 in north India caused havoc to cotton crop, inflicting huge economic losses to the extent of 40–50% (Kranthi, 2015b). *Bemisia tabaci* is a vector of begomoviruses (family *Geminiviridae*), reported to transmit 111 viruses (Tiwari *et al.*, 2013).

Three species of mirid bug, namely *Creontiades biseratense* (Distant), *Campylomma livida* Reuter and *Hyalopeplus lineifer* Walker were reported, out of which *C. biseratense* inflicts damage in south India (Nagrare *et al.*, 2013). Incidence of *C. biseratense* was observed from Karnataka and Tamil Nadu during 2006 (Patil *et al.*, 2006; Rohinim *et al.*, 2009) that resulted in significant reduction in seed cotton yield of Bt cotton. As of now, this *C. biseratense* is a major pest but is restricted to Tamil Nadu and Karnataka states. The remaining two mirid species are minor pests of cotton.

Tobacco caterpillar, *Spodoptera litura* Fabricius, was one of the economically important polyphagous pests of cotton which exhibited high resistance levels when pyrethroids were first introduced in India in 1982 (Ramakrishnan *et al.*, 1984; Kranthi *et al.*, 2002). The pest was severe in most parts of Andhra Pradesh (Armes *et al.*, 1997). During these periods, the pest exhibited high levels of resistance between 61- and 148-fold to cypermethrin, and 45–129-fold to chlorpyriphos in south India (Kranthi *et al.*, 2002). During the 2020 cropping season, sporadic incidences of *S. litura* were reported from Andhra Pradesh (Nagrare, personal communication).

Stem weevil, *Pempherulus affinis* (Faust), is an endemic pest to some parts of south India, particularly in Tamil Nadu, especially in irrigated crops where high moisture exists (Krishna Ayyar and Margabandhu, 1941). The pest caused 65.8% plant mortality, 72.0% reduction in boll production and 78.9% reduction in seed-cotton yield (Parameswaran and Chelliah, 1984; Dhara Jothi *et al.*, 2011).

In the cotton ecosystem, hemipteran insects such as stink bug, red cotton bug and dusky cotton bug are reported as minor pests of cotton in all the three cotton-growing zones; however, tea mosquito bugs, *Helopeltis bryadi* (Waterhouse) and *H. theivora* Waterhouse, are reported as emerging pests of cotton in the south zone, especially in Karnataka and Tamil Nadu (Udikeri *et al.*, 2011; Dhara Jothi *et al.*, 2018).

Currently, various methods of insect pest management are being employed to prevent and reduce pest populations during the season, as well as off-season, which include cultural, mechanical, behavioural and botanical pesticides, biocontrol agents and need-based application of chemical pesticides based on economic threshold level (ETL).

Pink bollworm management strategies

- Mass-awareness among cotton production stakeholders
- Timely crop termination and avoid ratooning
- Destroy residual stalks and partially opened bolls
- Crop rotation in the hot-spot areas

- Selection of sucking pests tolerant to early to medium duration varieties/hybrids
- Timely sowing and avoidance of pre-monsoon sowing
- Regular crop inspection/monitoring at squaring, flowering and boll-development stage. Deployment of pheromone traps @ 5/ha at 45 days after sowing
- Need-based use of insecticides with emphasis on use of organic/plant-origin insecticides at early growth stages (up to 60 days), and thereafter chemical insecticides
- Install pheromone traps in ginneries, cotton mills, market yards, etc.
- Gin sanitation

Insecticide resistance management (IRM)/ integrated pest management (IPM) for major insect pests of cotton

Crop growth stage: 0–60 days after sowing (DAS)

- Tobacco caterpillar: collect and destroy egg masses/gregarious larvae/solitary larvae by hand-picking
- Tobacco caterpillar, American and spotted bollworm: avoid chemical insecticide spray in the initial stage of crop; however, in emergency situation, spray Clorantraniliprole 18.5 SC
- Sucking pests: avoid sprays of neonicotinic group of insecticides against sucking pests; if necessary, spray Flonicamid 50 WG or Dinotefuran 20 SG
- Pink bollworm: install pheromone traps, 5/ha at 45 days after sowing for monitoring of pink bollworm; spray crop with Neem oil + NSKE at 50–60 days after sowing

Crop growth stage: 60–90 DAS

- Bollworms (American and spotted) in non-Bt: spray Chlorantraniliprole 18.5 SC
- Pink bollworm: initiate pink bollworm monitoring starting from flowering stage, observe for rosette flowers and destroy them. Assess ETL by picking 20 green bolls/acre (50/ha). On crossing ETL (10% infested flowers and/or 10% infested bolls and/or 8 male moth catches/trap/night for 3 consecutive nights), spray Chlorpyrifos 20 EC or Thiodicarb 75 WP or Quinalphos 20 AF or Profenofos 50 EC or Emamectin benzoate 5 SG.
- Leafhopper: spray Flonicamid 50 WG or Dinotefuran 20 SG
- Whitefly: install yellow sticky traps for monitoring and management of whitefly. Spray Spiromesifen 22.9 EC or Pyroproxifen 10 EC
- Thrips: Spinetoram 1.7 SC

Crop growth stage: 90–120 DAS

- American and spotted bollworm: spray Flubendiamide 39.35 SC or Indoxacarb 14.5 SC or Spinosad 45 SC
- Leafhopper, thrips: spray Thiomethoxam 25 WG
- Whitefly: spray Dinotefuran 20 SG
- Pink bollworm: release parasitoid *Trichogramma bactrae* @60,000/acre, if available, after a gap of 10 days. Spray Chlorpyrifos 20 EC or Thiodicarb 75 WP or Quinalphos 20 AF or Profenofos 50 EC or Emamectin benzoate 5 SG.

Crop growth stage: >120 DAS

- Pink bollworm: spray Fenvalerate 20% EC or Cypermethrin 10 EC or Lambda cyhalothrin 5% EC

Cotton Diseases and Their Management

In India, during the transgenic Bt cotton era, the disease-management strategies also changed according to changing climate and etiological factors specific to cotton-growing zones of the country. Among the diseases, grey mildew (*Ramularia areola*), Alternaria blight (*Alternaria* spp.) and bacterial blight (*Xanthomonas citri* pv. *malvacearum*) cause 26–30% yield losses (Chidambaram and Kannan, 1989; Chattannavar *et al.*, 2006; Bhattiprolu, 2012). Losses due to important diseases like cotton leaf curl virus (53.6%), bacterial leaf blight (20.6%), *Alternaria* leaf spot (26.6%), grey mildew (29.2%) and *Myrothecium* leaf spot (29.1%) have also been recorded (Monga *et al.*, 2011). Here we are emphasizing the important diseases of cotton crop, their causal agents, disease-causing factors, washing tests and integrated disease-management strategies.

Fungal diseases

Root rot

Causal organisms: *Rhizoctonia solani, R. bataticola* and *Sclerotium rolfsii*

Sudden wilting and drooping of plants occur; they can be easily pulled out of the ground and shredded bark of roots gives yellowish appearance. *R. solani* is reported as one of the most important pathogens of seedling complex of cotton (Rothrock, 1996); infected root becomes brown and wet with sunken lesions on stems known as 'sore shin' (Atkinson, 1892). *R. bataticola* causes black and dry root, *S. rolfsii* infection noticed with white mycelial growth on the collar region of plants and may produce small, brown, spherical sclerotial bodies leading to the rotting of roots and drying of seedlings. The disease is prevalent in all cotton-growing zones of India.

Grey mildew

Causal organism: *Ramularia areola*

Pale, irregular and angular spots initially appear on older leaves delimited by veinlets (Chohan *et al.*, 2020). Dirty white powdery growth spreads on lower and upper surface of the leaves. As the disease advances, leaves turn yellow, necrotic and dark brown (Sharma and Bambawale, 2008) as well as dry leading to premature defoliation and boll opening. Grey mildew disease is more prevalent in central and south zones of India.

Alternaria leaf spot

Causal organisms: *Alternaria macrospora, Alternaria alternata*

Alternaria cause brown or tan spots on cotyledons, leaves, bracts and bolls. Concentric rings develop within the spots, mostly on the upper surface. Later, spots coalesce and cause blighting of the leaves. Favourable conditions lead to severe defoliation. The disease is prevalent in major cotton growing tracts of the country but seriously affects the cotton crop in Gujarat, Karnataka and Andhra Pradesh (Rane and Patel, 1956).

Corynespora *leaf spot*

Causal organisms: *Corynespora cassiicola, Corynespora torulasa*

Initially, leaves show circular to irregular dark red spots which turn to brown lesions surrounded by a dark border. In the later stages of the disease, alternate light- and dark-brown rings may develop on the lesions with 'shot hole' appearance. Under severe conditions defoliation may occur. The disease is emerging in central India (Salunkhe *et al.*, 2019).

Bacterial diseases

Bacterial leaf blight (BLB)

Causal organism: *Xanthomonas citri* pv. *malvaceraum*

Appears as water-soaked, light- to dark-green, small spots measuring 1–5 mm on cotyledons and lower surface of leaves. Lesions darken, veins also become black with age. Leaves are shed prematurely resulting in extensive defoliation. Known as bacterial blight, angular leaf spot, black arm and boll rot (Hillocks, 1992). BLB disease is prevalent in all cotton-growing zones of India.

Inner boll rot

Causal organism: *Pantoea* spp.

Apparently green, healthy bolls when cross-sectioned, the developing seeds and fibres or lint observed as discoloured yellowish-orange to reddish in colour. Seeds swell and rotten in one or two locules (Hudson, 2000), occasionally the complete bolls. The disease is emerging and currently prevalent in Maharashtra and central India (Nagrale *et al.*, 2020).

Viral diseases

Tobacco streak virus (TSV) disease

Causal organism: *Ilarvirus*

Chlorotic appearance of growing tip in young leaves. Bronzing and curling with necrosis of leaves and plants become stunted (Gawande *et al.*, 2019). It is transmitted by thrips (*Thrips tabaci*) and usually prevalent in southern states of India, but recently reported from Maharashtra.

Cotton leaf curl disease (CLCuD)

Causal organism: Genus: *Begomovirus* (family: *Geminiviridae*)

The prominent symptoms of CLCuD include yellowing and small veins thickening (SVT) on the lower surface of young leaves. Downward or upward curling of leaves with stunted plant growth. Under severe conditions, a small leaf-like outgrowth on the lower side of the infected leaves (enations) may also be visible. This disease is transmitted by insect vector whitefly (*Bemisia tabaci*). Currently, this disease is prevalent only in north India including Punjab, Haryana and Rajasthan (Rajagopalan *et al.*, 2012).

Washing test

The washing test is a qualitative test approved by ISTA that is used for testing seed health and for presence of externally seed-borne pathogens; the inoculum or propagules may be loosely present on the surface of seed (Mathur and Kongsdal, 2003).

Protocol: 2 gm of seed are mixed with 10 ml of water in a test tube, shaken for up to 10 mins with a mechanical shaker. Then, the suspension sample is examined or spores are concentrated by centrifugation at 3000 rpm for 15–20 minutes. The suspension is thrown and again spores are suspended in 2 ml lacto phenol dye in ratio 1:1:1:2, containing a mixture of lactic acid, phenol, water and glycerol. This suspension is examined under the microscope for the presence of any spores, conidia, fruiting structures. This test is effective for presence and for examining spores/conidia of *Alternaria*, *Macrophomina*, *Fusarium*, *Corynespora* etc., fungal pathogens from cotton seed lot.

Integrated Disease Management (IDM) Strategies

1. Field sanitation
2. Exclusion of the pathogen(s)
3. Use of tolerant and resistant cultivars
4. Cultural practices: (i) crop rotation, (ii) elimination of alternate hosts and weeds from the fields, (iii) weed and water management, (iv) application of balance nutrition and avoidance of indiscriminate use of nitrogenous and phosphatic fertilizers, (v) restriction of rank/excess vegetative growth of cotton crop by good management practices, (vi) facilitation of proper drainage in the field, (vii) soil incorporation of FYM/composts for enhancing the soil health and disease management and (viii) optimum spacing to manage crop canopy.
5. Insect vector management
Infestation of piercing/sucking bugs/insects should be monitored during squaring, flowering and boll-development stages and their timely management with recommended practices is key, e.g. management of whitefly for CLCuD, thrips for TSV and stink bug and red cotton bug for inner boll rot.
6. Biological control
Utilization and application of biological control agents (BCAs) like PGPR, mycoparasites etc. for disease management, e.g. *Bacillus*, *Pseudomonas*, *Trichoderma*, *Streptomyces* etc.

7. Chemical control
a. For the management of angular leaf spot disease, seed treatment with carboxin 75% WP @ 1.5 g/kg seed or carboxin 37.5% + Thirum 37.5% DS @ 3 g/kg seed or Fluxapyroxad 333 g/l FS @ 1.5 ml/kg seed for seedling disease or Tetraconazole 11.6% w/w (12.5% w/v) SL @2 ml/kg seed against root rot (*R. solani*) disease management is recommended.
b. As a precautionary measure for inner boll rot, prophylactic sprays of copper oxychloride 50 WP @ 25 g+ [Streptomycin sulphate IP 90% w/w + Tetracycline hydrochloride IP 10% w/w] @ 1 g in 10 l water is suggested during early boll-developmental stages (60–90 DAS) at 15-day intervals for the management of inner boll rot and bacterial blight.
c. To prevent infection of leaf spot/fungal boll rot, spray Carbendazim 50% WP @ 20 g or Kresoxim-methyl 44.3% SC @ 10 ml or Pyraclostrobin 20% WP @ 20 g or Propineb 70% WP @ 25–30 g (Pyraclostrobin 5% + Metiram 55% WG) @ 20 g or Propiconazole 25% EC @ 10 ml or (Fluxapyroxad 167 g/l + Pyraclostrobin 333 g/l SC) @ 6 g or (Azoxystrobin 18.2% w/w + Difenoconazole 11.4% w/w SC) @ 10 ml in 10 l water is recommended.
d. Use recommended dose of insecticides for the management of insect vector-transmitting viral

diseases like cotton leaf curl virus transmitted by whitefly (*Bemicia tabaci*) and tobacco streak virus transmitted by thrips (*Thrips tabaci*) (Nagrare *et al.*, 2019b).

Cotton in India – The Future

Historically, cotton is an important commercial crop in India. Currently it has the highest acreage and production in the world but is marked by the lowest productivity among the major cotton-growing countries. Traditionally, cotton cultivation in India is beset with a number of handicaps; the most important one is being rain-dependent cultivation and uneven distribution of rainfall. Genetically modified cotton (Bt cotton) adapted in the country two decades ago is dominant, covering an area of about 95% under its cultivation. Bt technology has been adapted in the form of hybrids, the seeds of which farmers need to buy every year unlike varieties. The Bt technology has provided benefits of protection of cotton crop against bollworms, increased seed cotton yield and reduced insecticide use during early years of Bt adoption. However, susceptibility of most of the Bt hybrids to sucking pests and subsequent breakdown of resistance/susceptibility of Bt hybrids to pink bollworm has resulted in an increased use of insecticides in cotton.

The cottonseed sector is currently dominated by Bollgard-II hybrids developed by the private sector. The public sector institutes, including State Agricultural Universities in India, are continuing research on improvement of non-GM cotton. However, non-GM cotton is being grown in limited areas, preferred mostly by organic growers and as a choice by some farmers for desi/diploids as well as upland cotton. Recently, ICAR-CICR released and notified seven Bt varieties containing cry1Ac (MON 531) for different zones. Varieties with Bt would provide protection against American bollworm and possess better tolerance/resistance to sucking pests. Compact Bt varieties may also fit well in the high-density planting system as to realize better yield in the rain-dependent cotton ecosystem. Seed multiplication of varieties would be easier and cheaper than for hybrid seeds, and farmers may retain seeds produced by them for planting in the subsequent crop season, reducing dependency on private seed companies. Promotion of GM in the form of varieties is the priority for ICAR-CICR.

Indian farmers are open to new technologies. After the recent breakdown of resistance in Bt cotton to pink bollworm, farmers are looking for newer GM varieties/hybrids. Failure of BGII cotton to control pink bollworm is a warning signal for researchers, policymakers and users alike. For the success of any technology, however good it may be, adequate measures need to be taken and implemented in letter and spirit. Also, imparting training and creating awareness among farmers is needed so they can acquire the necessary skills for monitoring changes in technology, insect pests, and disease reaction and development of resistance in pests.

References

Armes, N.J., Wightman, J.A., Jadhav, D.R. and Ranga Rao, G.V. (1997) Status of insecticide resistance in *Spodoptera litura* in Andhra Pradesh, India. *Pesticide Science* 50, 240–248.

Atkinson, C.F. (1892) Some diseases of cotton: 3. *Frenching. Bulletin – Alabama Agricultural Experiment Station* 41, 19–29.

Bagla, P. (2010) India. Hardy cotton-munching pests are latest blow to GM crops. *Science* 327, 1439.

Balyan, R.S., Bhan, V.M. and Singh, S.P. (1983) Chemical and cultural weed control studies in cotton. *Tropical Pest Management* 29, 56–59.

Bautista, B.N., Guillermo, F., Lankenau, D., Guitron, S.P., Jennings, B.D. *et al.* (2017) Design of an integrated cotton picking system for small-scale Indian agriculture. Volume 4: *22nd Design for Manufacturing and the Life Cycle Conference*; 11th International Conference on Micro- and Nanosystems.

Bhattiprolu, S.L. (2012) Estimation of crop losses due to grey mildew (*Ramularia areola* Atk.) disease in Bt cotton hybrid. *Journal of Cotton Research and Development* 26, 109–112.

Blaise, D. (2011) Tillage and green manure effects on Bt transgenic cotton (*Gossypium hirsutum* L.) hybrid grown on rainfed Vertisols of central India. *Soil and Tillage Research* 114, 86–96.
Blaise, D. and Kranthi, K.R. (2019) Cotton production in India. In: Jabran, K. and Chauhan, B.S. (eds) *Cotton Production*. Available at: https://doi.org/10.1002/9781119385523.ch10 (accessed 19 July 2021).
Blaise D. and Prasad, R. (2005) Integrated plant nutrient supply: an approach to sustained cotton production. *Indian Journal of Fertilisers* 1, 37–46.
Blaise, D., Venugopalan, M.V. and Raju, A.R. (2014) Introduction of Bt cotton hybrids in India: did it change the agronomy? *Indian Journal of Agronomy* 59, 1–20.
Blaise, D., Bonde, A.N. and Reddy, D.D. (2016) Nutrient management options for rainfed cotton grown on Vertisols. *Indian Journal of Fertilisers* 12, 46–52.
Brar, J.S., Sidhu, B.S., Sekhon, K.S. and Buttar, G.S. (2008) Response of Bt cotton to plant geometry and nutrient combinations in sandy loam soil. *Journal of Cotton Research and Development* 22, 59–61.
Chattannavar, S.N., Kulkarni, S. and Khadi, B.M. (2006) Chemical control of *Alternaria* blight of cotton. *Journal of Cotton Research and Development* 20, 125–126.
Chaudhari, V.K., Desai, H.R. and Patel, N.M. (2015) Assessment of the insecticide resistance build up on cotton leafhopper *Amrasca bigutulla bigutulla* (Ishida). *International Journal of Advanced Multidisciplinary Research* 2, 4–8.
Chidambaram, P. and Kannan, A. (1989) Grey mildew of cotton. Technical Bulletin of Central Institute for Cotton Research, Regional Station, Coimbatore, India.
Chohan, S., Perveen, R., Abid, M., Tahir, M.N. and Sajid, M. (2020) Cotton diseases and their management. In: *Ahmad, S.* and Hasanuzzaman, M. (eds) *Cotton Production and Uses*. Springer, Singapore.
Choudhary, R., Singh, P., Sidhu, H.S., Nandal, D.P., Jat, H.S., Singh, Y.S. and Jat, M.L. (2016) Evaluation of tillage and crop establishment methods integrated with relay seeding of wheat and mungbean for sustainable intensification of cotton-wheat system in South Asia. *Field Crops Research.* Available at: https://doi.org/10.1016/j.fcr.2016.08.011 (accessed 19 July 2021).
CICR (2011) Annual Report 2010–11. Central Institute for Cotton Research. Available at: http://www.cicr.org.in/cicr_annual_reports-0203.htm (accessed 17 December 2020).
CICR (2018) Annual Report 2017–18. Central Institute for Cotton Research. Available at: http://www.cicr.org.in/cicr_annual_reports-12-13.htm (accessed 17 December 2020).
Datta, A., Ullah, H., Ferdous, Z., Santiago-Arenas, R. and Attia, A. (2019) Water management in cotton. In: Jabran, K. and Chauhan, B.S. (eds) *Cotton Production*. Wiley Blackwell, Chichester, UK.
Dhara Jothi, B., Surulivelu, T., Gopalakrishnan, N. and Manjula, T.R. (2009) Occurrence of papaya mealybug, *Paracoccus marginatus* Williams and Granara de Willink (Hemiptera: Pseudococcidae), on cotton. *Journal of Biological Control* 23, 321–323.
Dhara Jothi, B., Prakash, A.H., Venkatesan, R. and Gnana Prasuna, J. (2018) Tea mosquito bug *Helopeltis theivora* Waterhouse (Hemiptera: Miridae): a new pest on cotton. *Cotton Research Journal* 9, 61–63.
Dhara Jothi, B., Nagarajan, T. and Karthikeyan, A. (2011) Cotton stem weevil and its management. *Madras Agricultural Journal* 98, 308–313.
Dhurua, S. and Gujar, G.T. (2011) Field-evolved resistance to Bt toxin Cry1Ac in the pink bollworm, *Pectinophora gossypiella* (Saunders) (Lepidoptera: Gelechiidae), from India. *Pest Management Science* 67, 893–903.
Fabrick, J.A., Ponnuraj, J., Singh, A., Tanwar, R.K., Unnithan, G.C. *et al.,* (2014) Alternative splicing and highly variable cadherin transcripts associated with field-evolved resistance of pink bollworm to Bt cotton in India. *PLOS One* 9, e97900.
Fand, B.B., Kumar, M., Kamble, A.L., Fand, B.B., Nagrare, V.S. *et al.* (2019) Widespread infestation of pink bollworm, *Pectinophora gossypiella* (Saunders) (Lepidoptera: Gelechidae) on Bt cotton in central India: a new threat and concerns for cotton production. *Phytoparasitica* 47, 313–325.
Fand, B.B., Nagrare, V.S., Bal, S.K., Naik, V.C.B., Naikwadi, B.V., Mahule, D., Gokte-Narkhedkar, N. and Waghmare, V.N. (2020) Degree day-based model predicts pink bollworm phenology across geographical locations. *Scientific Reports. Available at*: https://doi.org/10.1038/s41598-020-80184-6 (accessed 19 July 2021).
Gawande, S.P., Raghavendra, K.P., Monga, D., Nagrale, D.T. and Kranthi, S. (2019) Rapid detection of Tobacco Streak Virus (TSV) in cotton (*Gossypium hirsutum*) based on Reverse Transcription Loop Mediated Isothermal Amplification (RT-LAMP). *Journal of Virological Methods* 270, 21–25.
Ghadekar, I.R. and Patil, V.R. (1990) Climatological water requirements of some rabi and summer crops under Nagpur agro-climatological condition. *PKV Research Journal* 14, 1–41.

Halappa, B. and Patil, R.K. (2016) Detoxifying enzyme studies on cotton leafhopper, *Amrasca biguttula biguttula* (Ishida), resistance to neonicotinoid insecticides in field populations in Karnataka, *India. Journal of Proteome Research* 56.

Hegde, M., Oliveira, J.N., da Costa, J.G., Bleicher, E., Santana, A.E. *et al.* (2011) Identification of semiochemicals released by cotton, *Gossypium hirsutum*, upon infestation by the cotton aphid, *Aphis gossypii*. *Journal of Chemical Ecology* 37, 741–750.

Hemant, P., Fand, B.B., Sawai, H.R. and Lavhe, N.V. (2020) Estimation and validation of developmental thresholds and thermal requirements for cotton pink bollworm *Pectinophora gossypiella*. *Crop Protection* 127, 104984. DOI: 10.1016/j.cropro.2019.104984

Hillocks, R.J. (1992) *Cotton Diseases*. Redwood Press, Melksham, UK.

Hudson, J. (2000) *Seed Rot Hits South Carolina Cotton*. Southeast Farm Press. Available at: http://southeast farmpress.com/mag/farming_seed_rot_hits/ (accessed June 2018).

Jalota, S.K., Buttar, G.S., Sood, A., Chahal, G.B.S., Ray, S.S. and Panigrahy, S. (2008) Effects of sowing date, tillage and residue management on productivity of cotton (*Gossypium hirsutum* L.) – wheat (*Triticum aestivum* L.) system in North west India. *Soil and Tillage Research* 99, 76–83.

Jayaraj, S., Rangarajan, A.V., Murugesan, S., Santharamj, G., Jayaraghavan, S.V. and Thangaraj, D. (1986) Studies on the outbreak of whitefly, *Bemisia tabaci* (Gennadius) on cotton in Tamil Nadu. In: Jayaraj, S. (ed.) *Resurgence of Sucking Pests*. Proceedings of National Symposium. Tamil Nadu Agricultural University, Coimbatore, India, pp. 225–240.

Kairon, M.S. and Venugopalan, M.V. (1999) Nutrient management research in cotton – achievements under all India coordinated cotton improvement project. *Fertilizer News* 44, 137–144.

Kranthi, K.R. (2015a) Pink bollworm strikes Bt-cotton. *Cotton Statistics and News*. Cotton Association of India.

Kranthi K. R. (2015b) Whitefly – the black story. *Cotton Statistics and News*. No. 23, 8 September. Cotton Association of India.

Kranthi K.R. (2020) Cotton in India: long-term trends and way forward. *ICAC Recorder* 38, 3–40.

Kranthi, K.R., Jadhav, D.R., Kranthi, S., Wanjari, R.R., Ali, S.S. and Russell, D.A. (2002) Insecticide resistance in five major insect pests of cotton in India. *Crop Protection* 21, 449–460.

Kranthi, S., Ghodke, A.B., Raghavendra, K.P., Mandle, M., Nandanwar, R. *et al.* (2017) Mitochondria COI-based genetic diversity of the cotton leafhopper *Amrasca biguttula biguttula* (Ishida) populations from India. *Mitochondrial DNA Part A*, DOI:10.1080/24701394.2016.1275595

Krishna Ayyar, P.N. and Margabandhu, V. (1941) Biology of the cotton stem-weevil, *Pempherulus affinis*, Fst., under controlled physical conditions. *Bulletin of Entomological Research* 32, 61–82.

Kumar, R., Monga, D., Naik, V.C.B., Singh, P. and Waghmare, V.N. (2020) Incipient infestations and threat of pink bollworm *Pectinophora gossypiella* (Saunders) on Bollgard-II cotton in the northern cotton-growing zone of India. *Current Science* 118, 1454–1456.

Majumdar, S. and Prakash, N.B. (2018) Prospects of customized fertilizers in Indian agriculture. *Current Science* 115, 242–248.

Malavath, R., Naik, R., Pradeep, T. and Chauhan, S. (2014) Performance of Bt cotton hybrids to plant population and soil types under rainfed condition. *Agrotechnology* 3, 120.

Mathur, S.B. and Kongsdal, O. (2003) *Common Laboratory Seed Health Testing Methods for Detecting Fungi*. International Seed Testing Association, Bassendorf, Switzerland.

Menon, M. and Uzramma (2017) *A Frayed History: The Journey of Cotton in India*. Oxford University Press.

Mohan, K.S., Ravi, K.C., Suresh, P.J., Sumerfordc, D. and Head, G.P. (2015) Field resistance to the *Bacillus thuringiensis* protein Cry1Ac expressed in Bollgard® hybrid cotton in pink bollworm, *Pectinophora gossypiella* (Saunders), populations in India. *Pest Management Science*. DOI:10.1002/ps.4047

Monga, D., Kranthi, K.R., Gopalakrishnan, N. and Mayee, C.D. (2011) Changing scenario of cotton diseases in India: the challenge ahead. Lead Paper presented at WCRC-5 held at Mumbai, 7–11 November, pp. 272–280. Available at: https://www.yumpu.com/en/document/read/6733036/changing-scenario-of-cotton-diseases-in-india-the-icac (accessed 19 July 2021).

Mukherjee, A.K., Chahande, P.R., Meshram, M.K. and Kranthi, K.R. (2012) First report of *Polerovirus* of the family *Luteoviridae* infecting cotton in India. *New Disease Reports* 25, 22.

Nagrale, D.T., Gawande, S.P., Gokte-Narkhedkar, N. and Waghmare, V.N. (2020) Association of phytopathogenic *Pantoea dispersa* inner boll rot of cotton (*Gossypium hirsutum* L.) in Maharashtra state, India. *The European Journal of Plant Pathology* 158, 251–260.

Nagrare, V.S., Kranthi, S., Biradar, V.K., Zade, N.N., Sangode, V. *et al.* (2009) Widespread infestation of the exotic mealybug species, *Phenacoccus solenopsis* (Tinsley) (Hemiptera: Pseudococcidae), on cotton in India. *Bulletin of Entomological Research* 99, 537–541.

Nagrare, V.S., Kranthi, S., Kranthi, K., Chinna Babu Naik, V., Kumar R., Dhara Jothi, B. *et al.* (2013) *Handbook of Cotton Plant Health*. Central Institute for Cotton Research, Nagpur, India.

Nagrare, V.S., Kumar, R. and Dhara Jothi, B. (2014) A record of five mealybug species as minor pests of cotton in India. *Journal of Entomology and Zoology Studies* 2, 110–114.

Nagrare, V.S., Fand, B.B., Naikwadi, B. and Deshmukh, V. (2019a) Potential risk of establishment and survival of cotton aphid *Aphis gossypii* in India based on simulation of temperature-dependent phenology model. *International Journal of Pest Management* 67, 187–202.

Nagrare, V.S., Naik, V.C.B., Fand, B.B., Gawande, S.P., Nagrale, D.T., Gokte-Narkhedkar, N. and Waghmare, V.N. (2019b) Cotton: integrated pest, disease and nematode management. ICAR-CICR Technical Bulletin No. 1/2019, pp. 1–40. Available at: https://www.cicr.org.in/pdf/WA/30_2020/english.pdf (accessed 20 July 2021).

Naik, V.C.B., Kumbhare., Kranthi, S., Satijia, U. and Kranthi, K.R. (2018) Field-evolved resistance of pink bollworm, *Pectinophora gossypiella* (Saunders) (Lepidoptera: *Gelechiidae*), to transgenic *Bacillus thuringiensis* (Bt) cotton expressing Cry1Ac and Cry2Ab in India. *Pest Management Science* 74, 2544–2554.

Naik, V.C.B., Pusadkar, P.P., Waghmare, S.T., Raghavendra, K.P., Kranthi, S. *et al.* (2020) Evidence for population expansion of cotton pink bollworm *Pectinophora gossypiella* (Saunders) (Lepidoptera: Gelechiidae) in India. *Scientific Reports* 10, 4740.

Ojha, A., Sree, K.S., Sachdev, B., Rashmi, M.A., Ravi, K.C. *et al.* (2014) Analysis of resistance to Cry1Ac in field-collected pink bollworm, *Pectinophora gossypiella* (Lepidoptera:*Gelechiidae*), populations. *GM Crops Food* 5, 280–286.

Parameswaran, S. and Chelliah, S. (1984) Damage potential and control of cotton stem weevil, *Pempherulus affinis*. *Tropical Pest Management* 30, 121–124.

Patil, B.V., Bheemanna, M., Patil, S.B., Udikeri, S.S. and Hosmani, A. (2006) Record of mirid bug, *Creontiades biseratense* (Distant.) on cotton from Karnataka, India. *Insect Environment* 11, 176–177.

Patil, S.K. (2020) Modernization and automation in ginning and pressing industries for performance enhancement. *International Journal of Engineering Research & Technology* 9, 227–232.

Prabhakar, M., Prasad, Y.G., Thirupathi, M., Sreedevi, G., Dhara Jothi, B. and Venkateswarlu, B. (2011) Use of ground based hyperspectral remote sensing for detection of stress in cotton caused by leafhopper (Hemiptera: Cicadellidae). *Computers and Electronics in Agriculture* 79, 189–198.

Puri, S.N., Murthy, K.S. and Sharma, O.P. (1999) Integrated pest management for sustainable cotton production. In: Basu, A.K., Narayanan, S.S., Krishna Iyer, K.R. and Rajendran, T.P. (eds) *Handbook of Cotton in India*. Indian Society for Cotton Improvement, Mumbai, pp. 223–255.

Rageshwari, S., Renukadevi, P., Malathi, V.G., Amalabalu, P. and Nakkeeran, S. (2017) DAC-ELISA and RT-PCR based confirmation of systemic and latent infection by *Tobacco Streak Virus* in cotton and parthenium. *The Journal of Plant Pathology* 99, 469–475.

Rajagopalan, P.A., Naik, A., Katturi, P., Kurulekar, M., Kankanallu, R.S. and Anandalakshmi, R. (2012) Dominance of resistance-breaking cotton leaf curl Burewala virus (CLCuBuV) in northwestern India. *Archives of Virology* 157, 855–868.

Ramakrishnan, N., Saxena, V.S. and Dhingra, S. (1984) Insecticide resistance in the population of *Spodoptera litura* (Fab.) in Andhra Pradesh. *Pesticides* 18, 23–27.

Rane, M.S. and Patel, M.K. (1956) Diseases of cotton in Bombay 1. Alternaria leaf spot. *Indian Phytopathology* 9, 106–113.

Rao, A.N., Wani, S.P. and Ladha, J.K. (2014) Weed management research in India – an analysis of the past and outlook for future. In: *Souvenir* (1989–2014). DWR Publication No. 18. Directorate of Weed Research, Jabalpur, India, pp. 1–26.

Reddy, A.R. (2020) MSP of cotton in India: will it distort international prices? *Cotton Statistics & News* 28, 1–4.

Rohinim, R.S., Mallapur, C.P. and Udikeri, S.S. (2009) Incidence of mirid bug, *Creontiades biseratense* (Distant) on Bt cotton in Karnataka. *Karnataka Journal of Agricultural Sciences* 22(3-Spl. Issue), 680–681.

Rothrock, C.S. (1996) Cotton diseases incited by *Rhizoctonia solani*. In: Sneh, B., Jabaji-Hare, S., Neate, S. and Dijst, G. (eds) *Rhizoctonia Species: Taxonomy, Molecular Biology, Ecology, Pathology, and Disease Control*. Kluwer Academic Publishers, Boston, Massachusetts, pp. 269–277.

Sagar, D., Balikai, R. and Khadi, B. (2013) Insecticide resistance in leafhopper, *Amrasca biguttula biguttula* (Ishida) of major cotton growing districts of Karnataka, India. *Biochemical and Cellular Archives* 13, 261–265.

Salunkhe, V.N., Gawande, S.P., Nagrale, D.T., Hiremani, N.S., Gokte-Narkhedkar, N. and Waghmare, V.N. (2019) First report of Corynespora leaf spot of cotton caused by *Corynespora cassiicola* in central India. *Plant Disease* 103(7), 1785. Available at: https://doi.org/10.1094/PDIS-05-18-0823-PDN (accessed 20 July 2021).

Sandhu, R.K. and Kang, B.K. (2015) Status of insecticide resistance in leafhopper, *Amrasca biguttula biguttula* (Ishida) on cotton. *Bioscan* 10, 1441–1444.

Sawhney, K. and Sikka, S.M. (1960) Agronomy. In: Dastur, R.H., Asana, R.D., Sawhney, K., Sikka, S.M., Vasudeva, R.S. *et al.* (eds) *Cotton in India*. Vol. II, ICCC, Bombay, pp. 106–163.

Sethi B.L. (1960) History of cotton. In: Sethi, B.L., Sikka, S.M., Dastur, R.H. *et al.* (eds) *Cotton in India: A Monograph I*. Indian Central Cotton Committee, Bombay, pp. 1–39.

Sharma, O.P. and Bambawale, O.M. (2008) Integrated management of key diseases of cotton and rice. In: Ciancio, A. and Mukerji, K.G., (eds) *Integrated Management of Diseases Caused by Fungi, Phytoplasma and Bacteria*. Springer, Dordrecht, The Netherlands, pp. 271–302.

Sharma, R. and Sharan, L. (2016) Evaluation of various synthetic insecticides against Thrips (*Thrips tabaci*) in Bt cotton. *The International Journal of Agriculture and Environmental Research* 2, 35–37.

Siddagangamma, K.R. and Channabasavanna, A.S. (2018) A review on weed management in Bt-cotton. *International Journal of Current Microbiology and Applied Sciences* (2018) Special Issue-7, 900–913.

Singh, M., Sharma, K. and Suryavanshi, V.R. (2014) Field evaluation of portable handheld type cotton picking machines for different cotton varieties. *Journal of Cotton Research and Development* 28, 82–87.

Singh, R. and Jaglan, R.S. (2005) Development and management of insecticide resistance in cotton whitefly and leafhopper – a review. *Agricultural Reviews* 26, 229–234.

Sivanappan, R.K. (2004) Irrigation and rain water management for improving water use efficiency and production in cotton crop. In: Proceedings of the International Symposium on Sustainable Cotton Production, UAS, Dharwad, India.

Solaippan, U., Mani, L.S. and Sharif, N.M. (1992) Weed management in cotton. *Indian Journal of Agronomy* 37, 878–888.

Sridhar, J., Naik, V.C.B., Ghodke, A., Kranthi, S., Kranthi, K.R., Singh, B.P., Choudhary, J.S. and Krishna, M.S.R. (2017) Population genetic structure of cotton pink bollworm, *Pectinophora gossypiella* (Saunders) (Lepidoptera: *Gelechiidae*) using mitochondrial cytochrome oxidase I (COI) gene sequences from India. Mitochondrial DNA Part A: *DNA Mapping, Sequencing, and Analysis* 28, 941–948. Available at: https://doi.org/10.1080/24701394.2016.1214727 (accessed 20 July 2021).

Tanwar, R.K., Jeyakumar, P. and Vennila, S. (2010) Papaya mealybug and its management strategies. *Technical Bulletin* 22. National Centre for Integrated Pest Management, New Delhi.

Textile Exchange (2020) *Organic Cotton Market Report 2020*. Available at: https://textileexchange.org/wp-content/uploads/2020/08/Textile-Exchange_Organic-Cotton-Market-Report_2020-20200810.pdf (accessed 20 July 2021).

Tiwari, S.P., Nema, S. and Khare, M.N. (2013) Whitefly – a strong transmitter of plant viruses. *The Journal of Plant Pathology* 2, 102–120.

Udikeri, S.S., Kranthi, K.R., Patil, S.B. and Khadi, B.M. (2011) Emerging pests of Bt cotton and dynamics of insect pests in different events of Bt cotton. Paper presented in 5th Asian Cotton Research and Development Network Meeting, Lahore, Pakistan, 23–25 February.

Venugopalan, M.V. and Pundarikakshudu, R. (1999) Long term effects of nutrient management and cropping system on cotton yield and soil fertility in rainfed vertisols. *Nutrient Cycling in Agroecosystems* 55, 159–164.

Venugopalan, M.V., Sankarnarayanan, K., Blaise, D. *et al.* (2009) Bt cotton (*Gossypium* sp.) in India and its agronomic requirements – a review. *Indian Journal of Agronomy* 54, 343–360.

Venugopalan, M.V., Blaise, D. and Kranthi, S. (2013) Cotton. In: Prasad, R. (ed.) *Text Book of Field Crops Production, vol.* 2 Commercial Crops. ICAR Pub, New Delhi, 305–344.

Venugopalan, M.V., Blaise, D., Lakde, S. *et al.* (2016) Evaluation of agro-techniques to improve boll weight of cotton under high density planting system. *Cotton Research Journal* 7, 12–16.

Venugopalan, M.V., Reddy, A.R., Kranthi, K.R., Yadav, M.S., Satish, V. and Pable, D. (2017) A decade of Bt cotton in India: land use changes and other socio-economic consequences. In: Reddy, G.P., Patil, N.G. and Chaturvedi, A. (eds) *Sustainable Management of Land Resources: An Indian Perspective.* CRC Press, Boca Raton, Florida.

Vinodkumar, S., Nakkeeran, S., Malathi, V.G., Karthikeyan, G., Amala Balu, P., Mohankumar, S. and Renukadevi, P. (2017) Tobacco streak virus: an emerging threat to cotton cultivation in India. *Phytoparasitica* 45, 729–743.

Vithal, B.M. (2018) Mechanical picking in India. *Cotton Statistics & News* 50, 1–4.
Ward, A. and Mishra, A. (2019) Addressing sustainability issues with voluntary standards and codes: a closer look at cotton production in India. In: Arora, B., Budhwar, P. and Jyoti, D. (eds) *Business Responsibility and Sustainability in India.* Palgrave Studies in Indian Management. Palgrave Macmillan, Cham, Switzerland.

4 Cotton Growing in Pakistan, Bangladesh and Myanmar

Abid Ali*, Zeeshan Ahmed and Zheng Guo

As indicated in Chapter 1, the first use of cotton fibres from wild cotton to make cloth was detected in the Indus Valley (currently Sindh province of Pakistan) (Baffes and Ruh, 2005) and over the centuries, cloth from *Gossypium arboretum*, also referred to as desi cotton was produced and traded within Asia and spread to Europe and Africa. Following the voyages of Columbus and others, cottonseeds were introduced from the New World and in the 16th century, the Portuguese introduced cottonseeds into Goa (a state on the south-western coast of India) by importing from Mexico and Brazil. Subsequently the British East India Company started the introduction of American cottonseeds around the beginning of the 19th century, and early in the 20th century American cotton was first planted in Sargodha district (now in Punjab province of Pakistan) (Afzal, 1946).

The first cotton research station was established at Lyallpur (now Faisalabad) for the improved breeding types by making selections from the available cotton varieties/genotypes. In Punjab, the first cotton line of *arboretum*, '15-Mollisone', was approved in 1930 followed by '39-Mollisone', and by 119-Sanguineum in 1941 and 231-R in 1959 (Afzal, 1947). For a long period, the *hirsutum* varieties could not compete with the indigenous desi cotton.

As in southern Africa, there was concern about the impact of jassids on cotton plants grown from seeds, imported from the USA, so selection from upland cotton, *Gossypium hirsutum*, was studied and plants with hairy leaves showed great success against jassids, *Empoasca spp.* (Ghani *et al.*, 1945). Overall, much of the research on root system of cotton, plant growth, cotton leaf, buds, flowers, bolls and jassids resistance was conducted on both Mollisoni (*arboreum*) and 4F, 289F, 289F/43 (*hirsutum*) varieties (Afzal, 1947) as upland cotton (*G. hirsutum*) and desi cotton (*G. arboretum*) competed with respect to their staple length and yield potential. Further comprehensive studies showed that insect populations in *hirsutum* (4F) were higher than in *arboreum* (Mollisoni) (Afzal *et al.*, 1943). Cotton was planted during April–June in different parts of the country and picking continued from mid-September until mid-January depending on the growing regions and cropping pattern, according to Afzal (1969). It was not until 1969 that plant protection was considered a main priority for cotton.

Establishing Pakistan as an independent state enabled the government to set up the Pakistan Central Cotton Committee (PCCC) in 1948. This was modified in 1951 but there has not been any detailed function related to cotton-pest

*Corresponding author: abid_ento74@yahoo.com

DOI: 10.1079/9781800620216.0004

management, despite the traditional desi cotton (*G. arboreum*) being heavily infested by spotted bollworm (SBW), pink bollworm (PBW) and whitefly, while the early growing of American upland cotton (*G. hirsutum*) was mainly infested by jassids, in addition to PBW, SBW and whitefly (Afzal and Ali, 1983). The major pests have undoubtedly affected cotton production and have led to greater requirements for integrated pest management, and in the 1960s, farmers realized that yield cannot be enhanced without (i) proper pest management strategy; (ii) development of new varieties; (iii) improved agronomic practices; (iv) availability of sufficient amount of fertilizers; and (v) canal irrigation systems, which were supplemented by tube wells. The expansion of irrigation and application of fertilizer coincided with the availability of synthetic insecticides, such as DDT, that brought a major change to protecting cotton plants.

GM Cotton

Bt technology entered Pakistan's cottonseed market in the mid-2000s without the requisite biosafety approvals (Rana, 2014; Spielman *et al.*, 2015). It is doubtfully reported that local varieties containing the MON531 transgenic event (commercially known as Bollgard I, developed by Monsanto) proliferated widely as a means of controlling bollworm infestations (Ishtiaq and Saleem, 2011). Official approval for a set of Bt cotton varieties that were already under cultivation came from the National Biosafety Committee in 2010, with a second batch of approvals following in 2014. Since then, Pakistan's seed industry has continued to rely on the MON531 event. Newer and more effective events (such as Bollgard II) are not yet commercially available in Pakistan (ISAAA, 2017).

By November 2020, cotton was the only transgenic crop that had been commercialized in Pakistan. For genetic gain, since the establishment of the Center of Excellence in Microbiology (CEMB) and the National Institute of Biotechnology and Genetic Engineering (NIBGE) in 1987, the government of Pakistan provided liberal support to biotechnology research in the last two decades. So far, the infrastructure and expertise in biotechnology is 53 public sector institutes/departments working in Pakistan. Moreover, research projects in agricultural biotechnology including cotton is being sponsored by the top funding agencies of Pakistan: (i) Agricultural Linkage Program (ALP) by Pakistan Agricultural Research Council (PARC) (n = 15); (ii) National Research Program for Universities (NRPU) by Higher Education Commission of Pakistan (HEC) (n = 46), (3) Natural Sciences Linkages Program (NSLP) by Pakistan Science Foundation (PSF) (n = 14); and (iv) Punjab Agricultural Research Board (PARB) by provincial government (n = 14). For cotton, particularly, research is not only being conducted on traits related to insect resistance, cotton leaf curl virus resistance and biopesticides to control cotton bollworm but also on biotic and abiotic stress resistance, drought-stress tolerance, reduction of toxic gossypol production and glyphosate resistance.

The private sector established very robust research and development sections in Four Brothers Group Pakistan, Everest Pakistan Pvt Ltd., Auriga Group of Companies, Bayer Pakistan and Corteva Agriscience. Overall, 224 cases of GMOs are under review through the National Biosafety Center (NBC) that has been working since 2017 under Environmental Protection Agency – Pakistan. Globally, to combat Covid-19, the government is pushing to speed up the process of genetically engineered technology-based vaccine development and approval, but in many other countries around the world, political interest to permit genetically engineered crops, or not, is vital.

For Pakistan, there is an urgent need to upgrade the biosafety guidelines to strengthen the biotechnology potential in Pakistan (Ali *et al.*, 2019). As a result of the federal government's revolutionary steps for the betterment of agriculture and the seed industry, in order to protect the rights of plant breeders, the Plant Breeders Rights Registry (PBRR) has been operational since 15 February 2021, where cotton is of prime importance. SANIFA – a joint venture of Saphire, Nishat and Fatima Group in Pakistan to uplift the agriculture base in the country, particularly with reference to the cotton crop – has been launched with the aim of cotton sustainability. SANIFA consultants will establish a laboratory in Pakistan to train scientists in managing the cotton leaf curl virus (CLCUV) disease using molecular tools by revalidating the 1990s developed tools. Peanut, Mung bean and Sesame intercropping with cotton in Pakistan will be introduced by Chinese cotton scientists to enhance the yield and maintain the crop rotation policy, in order to improve the fibre characteristics and resistance against chewing insects and herbicides, by genetic manipulation of *G. arboreum* genome. For this, on leaf-feed-based insect bioassay, almost 99% mortality was observed for *Helicoverpa armigera* on the transgenic cotton plant (L3) (Tahir *et al.*, 2021).

Before Bt cotton was commercially approved, local market, seed companies and home-retained Bt cottonseeds were among the main sources of Bt cotton (Arshad *et al.*, 2009a). Due to higher pesticide use, higher bollworm infestation, low or stagnant yield and higher labour costs, farmers in Pakistan adopted Bt cotton (Arshad *et al.*, 2007; Arshad *et al.*, 2009a). The major pests have been pink bollworm *Pectinophora gossypiella*, American bollworm *Heliothis armigera*, Spiny bollworm *Earias insulana* and *E. fabia*, jassids *Amrasca devastans*, cotton whitefly *Bemisia tabaci* and aphids *Aphis gossypii*. The red cotton bug, or cotton stainer *Dysdercus koenigii*, has also caused crop damage (Afzal and Ali, 1983; Arshad *et al.*, 2009a). Other insects like cotton mealybug, *Phenacoccus solenopsis*, cotton dusky bug, *Oxycarenus* species and red spider mites also have been recorded.

When insecticides were first used on cotton, the chlorinated hydrocarbon insecticides were applied initially in some areas by aircraft (during the locust plague in 1950). Later, knapsack sprayers were most commonly used and these sprays had a noticeable impact on beneficial insects, which led to certain pests being more serious. These were, notably, the bollworm *Helicoverpa armigera*, as well as heavy infestations of pink bollworm noted for the first time. The increasing abundance of spiny bollworm was also noted (Afzal and Ali, 1983). The Cotton Research Section, Multan, was renamed Cotton Research Institute in 1970, and with support from UNDP experts in 1976, entomologists were trained for the development of a pest-management strategy in cotton. In 1977, the first insecticide spray was applied about mid-August to control the sucking pests and was followed by 2–3 sprays to control bollworm.

Jassids

Amrasca devastans (Hemiptera: Cicdellidae) was first reported as a major pest when it devastated an upland cotton variety in Punjab province of subcontinent in 1913. Toughness of cuticle of leaf veins with hairiness was determined as a major breakthrough for the determination of cotton varieties resistant to jassids, in particular the growth, length and density of hairs (Afzal and Ghani, 1953). Late sowing and heavy irrigation induced higher jassid infestations, these being more serious on upland cotton. Late reporting for these findings is, likewise, the studies in South Africa including cotton varieties from India, studied in 1929 but not published until 1949 (Parnell *et al.*, 1949). Saeed *et al.*, (2016a) observed that imidacloprid applied at the recommended dose of 5 g/kg^{-1} seed is effective against *A. devastans* and appears to be safer than thiamethoxam for natural enemies, and also enhances plant growth directly, but without reporting the possible sublethal negative effects on individual beneficial arthropods.

As sucking pests have not been affected by the Bt toxins in new varieties of cotton, multiple studies on jassids on cotton have been reported, e.g.:

- patterns of genetic differentiation among populations of *A. biguttula biguttula* (Shiraki) (Cicadellidae: Hemiptera) (Akmal *et al.*, 2018);
- evaluation of action thresholds for *A. devastans* management on transgenic and conventional cotton across multiple planting dates (Saeed *et al.*, 2018);
- toxicity and resistance of the cotton leaf hopper, *A. devastans*, to neonicotinoid insecticides in Punjab, Pakistan (Saeed *et al.*, 2017);
- evaluation of insecticide spray regimes to manage cotton leafhopper, *A. devastans*: their impact on natural enemies, yield and fibre characteristics of transgenic Bt cotton (Saeed *et al.*, 2016b);
- the importance of alternative host plants as reservoirs of the cotton leaf hopper, *A. devastans*, and its natural enemies (Saeed *et al.*, 2015);
- evaluation of neonicotinoids and conventional insecticides against cotton jassid on cotton (Razaq *et al.*, 2005; Karar *et al.*, 2013);
- detection of resistance to pyrethroids in field populations of cotton jassid (Ahmad *et al.*, 1999a);

- estimated losses in cotton yield in Pakistan attributable to the jassid (Ahmad *et al.*, 1985);
- varietal resistance against jassid in cotton and role of abiotic factors in population fluctuation (Aheer *et al.*, 2006);
- comparative effectiveness of some latest spray schedules against cotton jassid on cotton variety FH-672 (Razaq *et al.*, 1998);
- comparative efficacy of some insecticides against cotton jassid and their effect on non-target insects in cotton (Yazdani *et al.*, 2000); and
- importance of alternative host plants as reservoirs of the cotton leaf hopper and its natural enemies (Saeed *et al.*, 2015).

Mealybug

Cotton mealybug, *Phenacoccus solenopsis* Tinsely, 1898 (Hemiptera: Pseudococcidae) has been described as a serious and invasive pest of cotton from 2005 after significant crop losses were reported, particularly in areas of Punjab and Sindh provinces (Hodgson *et al.*, 2008; Arif *et al.*, 2009; Abbas *et al.*, 2010; Ashfaq *et al.*, 2010). Functional role of the venom of *Aenasius bambawalei* Hayat (Hymenoptera: Encyrtidae) provides comprehensive analysis of the role of the venom of the parasitoid on the regulation processes in cotton mealybug (Abbas *et al.*, 2014). Biological characteristics and fitness role of *A. bambawalei* is explored by Zain-ul-Abdin *et al.* (2012, 2013). Synthetic insecticides were extensively used and considered an effective tool for the control of this pest insect (Saeed *et al.*, 2007). Ejaz *et al.* (2019) reported characterization of inheritance and preliminary biochemical mechanisms of spirotetramat resistance in *P. solenopsis*.

Whitefly

A heavy infestation of whitefly *B. tabaci* had been observed for the first time in the current Punjab province of Pakistan during 1915. Early studies showed the complete life cycle was 14–27 days in April–September (Hussain, 1930) with possibly ten generations during the year. During 1931, damage caused by whitefly included honeydew on leaves. Whitefly eggs were predominantly laid on the top (51.5%) and middle (46.7%) of plants through May–September when temperatures can range between 33 and 41°C. Desi varieties were more infested till the end of August and then upland cotton later in the season. Trehan (1944) studied the role of natural enemies and observed chalcids (33% pupae in September and third instar nymphs), chrysopids, coccinellids-brumus spp. However latter studies conducted by Haq (1968) reported no significant impact of natural enemies in cotton whitefly control. However, foliar application of silicon was reported as an eco-friendly approach to control *B. tabaci* if integrated with ongoing pest management programmes (Abbasi *et al.*, 2020).

Whitefly was significantly controlled when triazophos was applied as compared to deltamethrin and cypermethrin, by applying through knapsack sprayer in contrast to spinning-disc ULV sprayer (Attique and Shakeel, 1983). Whitefly damage is not only caused directly by sucking plant sap but also by virus transmission. Whitefly was first noted as a vector of cotton leaf curl disease in 1967 and then in 1985, which led to great epidemic losses in the early 1990s and in 1999. Whitefly damaged cotton crop on a large scale, although some resistance varieties were introduced in addition to chemical control.

Recently, the Punjab government established biological control laboratories in major cotton-growing areas for the mass-rearing of chrysopids to release in the cotton fields, in the absence of a single chemical product available in Pakistan to control the nymphs of whitefly. When whitefly is attacking later in the season it is on the undersides of leaves and spraying downwards over the crop is ineffective as there is a translaminar impact. However, some new insecticides are systemic and will control whitefly. Some progressive farmers say that if we can only control the whitefly, we can save the cotton in Pakistan.

Whitefly control was mainly focused among the farming community due to its vector nature during the early 1990s because whitefly developed resistance against conventional insecticides (Ahmad *et al.*, 2001). New varieties were released, but needed the introduction of 'new chemistry' insecticides applied as seed treatment to avoid the early establishment of whitefly populations up to 70–90 days after sowing. In addition, eradication of weeds, better health of the plant through providing balanced doses of fertilizers, biological agents etc. were among the immediate measures taken to control this disease via whitefly eradication (Basit *et al.*, 2013b).

Despite the fact that old ETLs are being used in Pakistani cotton pest management, their ETL (5–8 and/or 4–5 whitefly per leaf) is recommended for chemical control irrespective of the fact that a single viruliferous whitefly can transmit virus from one plant to the other (Rahman *et al.*, 2017). But whitefly also developed resistance against new chemistry insecticides (Basit *et al.*, 2011; Basit *et al.*, 2012; Basit *et al.*, 2013a, b). The application of biopesticides is another control measure to control the whitefly population (Sarwar and Sattar, 2016); however, its impact is yet to be realized. Clean cultivation is important for controlling the whitefly population. Similarly, avoiding the cultivation of other crops off-season in cotton-growing areas may help in breaking the life cycle of whitefly (Rafiq *et al.*, 2008).

Six species of the *B. tabaci* complex are present in Pakistan including Asia II 1, Asia II 5, Asia II 7, Asia 1, Middle East-Asia Minor 1 (MEAM 1) and 'Pakistan', while four of these species (Asia II 1, Asia II 5, Asia 1, MEAM 1) were found on cotton in Pakistan. Asia II 7 was only collected in a Malaise trap in Pakistan but has been recorded on cotton in India (Simon *et al.*, 2003; Ahmed *et al.*, 2011; Ashfaq *et al.*, 2014). All these species members are associated with the transmission of cotton leaf curl disease (CLCuD). MEAM 1 was restricted to Sindh and Asia II 1 to the Punjab, whereas Asia 1 was found in both regions (Ahmed *et al.*, 2011). Forecasting and monitoring of whitefly incidence and its parasitism in relationship with abiotic factors among different cotton-growing areas of Sindh province was also explored by Ahmad *et al.* (2018). Recently, efficacy of biorational insecticides against *Bemisia tabaci* (Genn.) and their selectivity for its parasitoid *Encarsia formosa* Gahan on Bt cotton has been reported by Gogi *et al.* (2021).

Cotton Leaf Curl Disease

In addition to insect pests, the cotton crop is greatly damaged by cotton leaf curl disease (CLCuD) that is transmitted by whitefly (*Bemisia tabaci*). It is one of the single-stranded DNA viruses belonging to the genus *Begomovirus* (family, *Geminiviridae*) in association with satellite molecules. The taxonomy of the *Geminiviridae* divides the viruses into four genera. The division is on the basis of insect vector (either whitefly (Begomovirus), leafhopper (Mastrevirus and Curtovirus) or treehopper (Topocuvirus) and genome arrangement. Most of the economically important geminiviruses are in the genus *Begomovirus*, which consists of more than 70 species that are all transmitted by the whitefly *B. tabaci* and infect only dicotyledonous plants (Briddon and Markham, 2000). CLCuD affects okra, hibiscus, hemp, tomato, ageratum, sunflower, tobacco and many weeds in addition to cotton (Nour and Nour, 1964). Cotton leaf curl disease is characterized by small and large veins thickening and upward or downward curling of the leaves. Severe disease infection results in the development of leaf enation and finally retards growth of the cotton plant, which substantially reduces cotton yield by 15–70% (Idris, 1990; Brown, 2001).

The CLCuD epidemic in the early 1990s and in 2001 was first noted in 1967 and restricted to Multan area until 1986 (Hussain and Mahmood, 1988; Hussain *et al.*, 1991; Khan and Ahmad, 2005) but the first time it drew the attention of experts was in 1988 when approximately 60 ha (cotton variety S12, newly released, covered 46% area in Punjab as plant breeders were keen on increasing the ginning outturn by 2%) were affected at Moza Khokhran near Multan (a major cotton-growing area).

The severity increased from affecting 60 ha in 1988 to 810 ha in 1990, 14,000 ha in 1991, 121,000 ha in 1992 and 202,000 ha in 1993 (Khan and Khan, 1995; Briddon and Markham, 2000) (Fig. 4.2). CLCuD continued to spread until it occurred throughout the cotton-growing area of the Punjab. During the early phase of the epidemic, the disease spread rapidly and predominantly to the north-east, assisted by the strong prevailing winds ('Daccan'). The spread westwards into Baluchistan and north-west into the North West Frontier province and southwards into the Sindh province was slower, due in part to the environmental conditions and the prevailing winds, although significant infection was recorded in the Sindh in 1996–1997. The losses to the Pakistan economy are calculated to

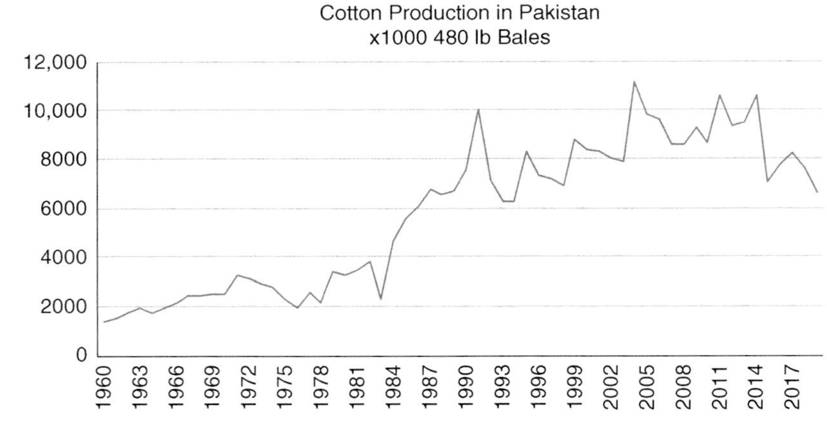

Fig. 4.1. Increasing cotton production from 1984.

have been US$5 billion in the five years between 1992 and 1997 and around 30–35% yield reduction (Hussain and Ali, 1975; Briddon and Markham, 2000; Rahman *et al.*, 2017; Biswas *et al.*, 2020). In Pakistan, resistant cotton varieties were developed by crossing resistant sources with local cultivated susceptible varieties but the resistance in the newly released cotton varieties remained intact until the emergence of a Burewala viral strain of the disease that appeared in Vehari district of Punjab in 2001 (Mansoor *et al.*, 2003a, b).

All resistant cotton varieties got infected with the newly evolved resistant-breaking strain. This strain was named cotton leaf curl Burewala virus (CLCuBuV). These viruses are of two types: monopartite and bipartite begomovirus. Bipartite begomoviruses infect different cotton species in Pakistan (CLCrV; Idris and Brown, 2004). The development of leaf enation as a consequence of severe infection of this disease retards the cotton plant growth, which significantly reduces the yield of cotton by 15–70% (Idris, 1990; Brown, 2001). In Pakistan, a great diversity of CLCuV was identified by molecular characterization (Amrao *et al.*, 2010), i.e. cotton leaf curl Kokharan virus (CLCuKoV); cotton leaf curl Shahdadpur virus (CLCuShV); cotton leaf curl Gezira virus (CLCuGeV); cotton leaf curl Burewala virus (CLCuBuV); cotton leaf curl Multan virus (CLCuMuV) and CLCuRaV (Nawaz-ul-Rehman *et al.*, 2012; Muhire *et al.*, 2014; Saleem *et al.*, 2016; Zubair *et al.*, 2017). A recent study revealed that non-cultivated cotton species (*G. raimondii*, *G. thurberi* and *G. mustelinum*), grown at the Central Cotton Research Institute Multan, act as reservoirs for CLCuD-associated begomoviruses. Where an extreme resistance to cotton leaf curl disease was reported in two wild *G. hirsutum* L. cultivars (Ullah *et al.*, 2017), the *G. arboreum* desi cotton showed multiple resistance mechanisms against CLCuD (Naqvi *et al.*, 2017). These results provide novel insights into understanding the spread of begomoviruses and associated satellites in new-world cotton species introduced into the old world (Briddon, 2003; Briddon *et al.*, 2014; Sattar *et al.*, 2017; Shakir *et al.*, 2019).

Although cotton varieties being used in Pakistan have played a major role in the epidemic, currently no cotton variety resistant to CLCuD was available. So, the virus became a potential threat to numerous cultivars of the cotton (Hussain and Ali, 1975; Idris, 1990). Efforts are underway to develop transgenic cotton plants exhibiting high resistance to the disease (Rahman *et al.*, 2017). Thus, sustainability of cotton production is at possible risk in numerous cotton-growing countries due to CLCuD, highlighting the need for control measures to manage this problem in different countries.

Globally, cotton leaf curl disease (CLCuD), after its first epidemic in 1912 in Nigeria, has spread to different cotton-growing countries including the USA, Pakistan, India and China. The disease is of viral origin, transmitted by the whitefly *Bemisia tabaci*, which is difficult to control because of the prevalence of multiple virulent viral strains or related species. The problem is further complicated as the CLCuD-causing virus complex has a higher recombination rate. The availability of alternate host crops like tomato, okra etc. and practising mixed-type farming systems have further exaggerated the situation by adding synergy to the evolution of new viral strains and vectors. Efforts to control this disease using host-plant resistance remained successful using two-gene-based resistance that was broken by the evolution of a new resistance-breaking strain called Burewala virus. Development of transgenic cotton using both pathogen and non-pathogen-derived approaches are in progress. In future, screening for new forms of host resistance, use of DNA markers for the rapid incorporation of resistance into adapted cultivars overlaid with transgenics, and using genome editing/engineering technologies by the CRISPR/Cas9 system will be instrumental in adding multiple layers of defence to control the disease. Thus cotton-fibre production will be sustained. Desi cotton, *Gossypium arboreum*, having inherent ability to withstand remarkable resistance to sucking pests and cotton leaf curl virus (Venugopalan *et al.*, 2021) could be further used for qualitative improvement of cotton fibre through plant-breeding techniques (Tahir *et al.*, 2021).

Pink Bollworm

Platyedra gossypiella (now *Pectinophora gossypiella*) was first reported in the Punjab province of subcontinent in 1894, with severe attacks occurring in 1917 and 1922, so research was started in 1923 at Faisalabad (currently in Punjab province of Pakistan) that was intensified in 1926 (Afzal and Ghani, 1953). High humidity (RH 60–80%) was more favourable for the development of pupae and oviposition (Khan, 1938), so areas with the highest boll damage (15%) were observed where rainfall exceeded 15 inches (Khan, 1938; Haq, 1968). Despite this, pink bollworm was only considered a major pest of cotton in 1961 when the PCCC meeting discussed its potential threat after an increase in its population was linked with the irrigation water supply and use of fertilizers delaying crop maturity and reducing yields. Cheema *et al.* (1980a, 1980b) gave details of the biology of parasitoids and predators of pink bollworm in Pakistan. Pearson and Darling (1958) pointed out that pink bollworm can attack all cottons, whether cultivated or wild, and many species of *Abelmoschu*, particularly cultivated okra, kenaf and also jute. However, through DNA barcoding, a single species of PBW is reported in Pakistan (Webinar, 2020). Smaller populations of PBW were observed in Bt cotton as compared to non-Bt cotton fields (Arshad *et al.*, 2015a).

Cultural control of PBW by goats

Grazing small animals in fields after cotton picking proved to be an effective management tool

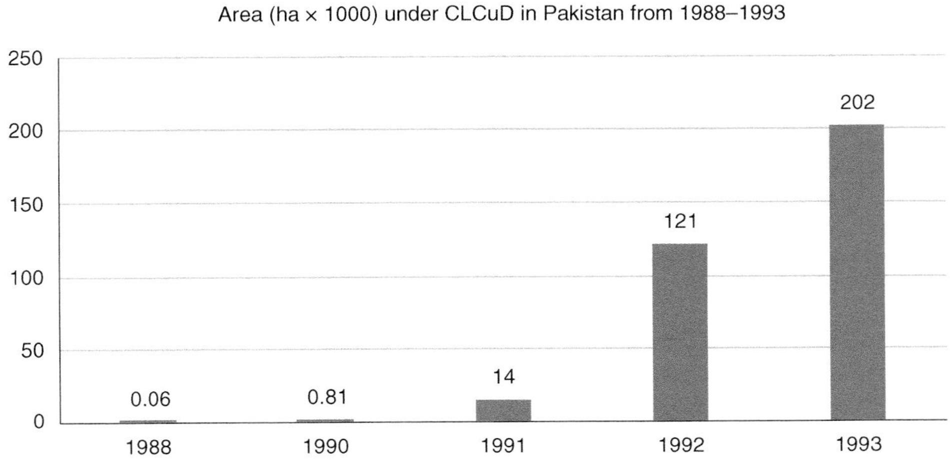

Fig. 4.2. CLCuD outbreak area (ha) during 1988–1993 in Pakistan.

for pink bollworm. A study at the Central Cotton Research Institute Multan (Pakistan) revealed that over 70% of infested bolls were reduced in fields after grazing of goats or sheep. Punjab Agriculture Department encouraged growers to let the animals graze in cotton fields after picking and before cutting cotton sticks, which were stored as firewood in the rural areas. If the sticks stored were not grazed, they were a prime source of pink bollworm, which remained inside infested cotton bolls, and resulted in a severe early attack on a new cotton crop (Munshi and Mecci, 1976; Ahmed, 1980; Mallah *et al.*, 2000). Overall, this recommendation would have controlled PBW by allowing goats to eat the remaining foliage on cotton plants after harvest to reduce survival between seasons. However, during 1974/75, a study revealed that the percentage of double seed (with PBW larvae) found in the ginning factories was too small to be a source of PBW in the following season's crop (Ahmad, 1976). All of this emphasizes the importance of a closed season, banning ratooning of cotton.

Pheromone trapping

In Pakistan, initially, pheromones for the control of cotton bollworm, including PBW, were imported from Japan. Currently, the only major threat among cotton bollworm is PBW and, therefore, PB ropes and pheromone traps were recommended and distributed with subsidy to cotton-growing farmers during the 2020/21 season. In fact, it should have been implemented long before 1990 when hollow-fibre, microencapsulated and twist-tie formulations of PBW pheromone-based field trials were conducted from 1986 to 1988 (PBW pheromone formulations) and later during 1993 (PB ropes). As a result, an effective control of PBW was obtained at flowering and fruiting stage, as it reduced the need for two insecticide sprays for the bollworm complex (ABW, SBW and PBW) by mating disruption, as compared to conventional insecticide spray programme in terms of infested bolls and yield (Attique, 1985; Campion and Murlis, 1985; Critchely *et al.*, 1987; Qureshi and Ahmed, 1989; Critchely *et al.*, 1991; Chamberlain *et al.*, 1992, 1993; Attique *et al.*, 2000). Late distribution of PB ropes and pheromone traps was made after the recommendations of the International Entomological Congress, an exclusive day on the current scenario and management of PBW, held during December 2016 in the University of Agriculture, Faisalabad, where Thomas A. Miller (UC Davis, USA) suggested to widely adopt the above said control methods. The delayed reason could be possibly due to presence of higher population density of SBW and ABW in cotton fields in previous years.

Chemical control

Lambda cyhalothrin, triazophos and gamma cyhalothrin are recommended special products available in Pakistan for PBW control; however, triazophos is preferred by farmers even though it is costly compared to others, and in peak seasons it is scarce on the market. Laboratory bioassays and field experiments showed that triazophos produced higher mortality of PBW in Bt cotton (Rajput *et al.*, 2017; Akhtar *et al.*, 2018). Although, through webinars, few experts from CCRI, NIGBE and UAF in Pakistan claimed to have a laboratory population of PBW, it is difficult to rear and to conduct resistance studies. During an international webinar, NIBGE claimed PBW resistance against CryAc1 and susceptibility to Cry1Ac+Cry2Ab. Similar findings by Four Brothers Pvt Ltd have been presented in cotton stalkholder meetings where hexagene cotton is under field trials; however, no online published work is so far available. In addition to cultural practices and collaborative area-wide control, regardless of size of individual cotton plots, Bollgard-II and multi-gene cotton transgenic varieties are recommended to develop in Pakistan (and upcoming lines by NIBGE in field trials for breeding) to combat this pest in order to increase the cotton yield and area under cultivation (personal communication, cotton stakeholder meeting, 2019; Webinar, 2020). Currently, in Pakistan, cotton research, particularly on insect resistance, is going on in 15 research institutes or universities (Webinar, 2020). Funnel traps were recommended after their cotton field trials, which are still being utilized in Pakistan for moth capture, and also this study provided proof in Pakistan that large-scale areas provide better PBW moth control compared to small landholders' fields (Attique, 1985; Qureshi and Ahmed, 1989; Chamberlain *et al.*, 1993; Attique *et al.*, 2000).

More recently, in the major cotton-growing areas (Multan, DG Khan, Bahawalpur, Sahiwal, Sargodha and Faisalabad divisions in Punjab), good-quality cottonseeds have been freely distributed among farmers with 1–25 acres of cotton fields. Because the adults live longer, the use of pheromone formulations resulted in significant yield increase and decrease in the number of sprays used as control measures (usually six per season for bollworm complex).

American Bollworm (*Helicoverpa armigera*)

In Pakistani cotton, major outbreaks of American bollworm (ABW), *Helicoverpa armigera* (Hubner) (Lepidoptera: Noctuidae), were recorded in 1977, 1983, 1990, 1994 and 1997. Earlier, organophosphates, carbamates and pyrethroids were applied, and now new chemistries, such as spinosyns, avermectins, diamides and insect growth regulators are commonly used in Pakistan (Ahmad *et al.*, 2019). The Pakistani populations of *H. armigera* have developed resistance to broad-spectrum insecticides, including organophosphates, carbamates and pyrethroids (Ahmad *et al.*, 1995; Ahmad *et al.*, 1997; Ahmad *et al.*, 1998), including those marketed more recently, which include spinosyns, avermectins and diamides. Insect growth regulators for chewing pests are commonly used in Pakistan (Ahmad *et al.*, 2003; Qayyum *et al.*, 2015; Ahmad *et al.*, 2019). For the last ten years, ABW has become a non-pest, either due to (i) Bt cotton (Arshad *et al.*, 2009b; Ahmad *et al.*, 2019); (ii) the general population has been low (Government of the Punjab, 2020); and/or (iii) (perhaps complex) interaction due to the fact that it has multi-hosts so cotton is not preferred if other crops provide food (personal communication, Graham Matthews, 2020).

Spiny Bollworm (also called Spotted Bollworm)

Before independence, spiny bollworm was first reported as a serious pest in 1905 in central and south-western Punjab (Haq, 1968). Research started in 1918 and more extensive work during 1934–40 identified *Earias insulana* and *E. fabia* as cotton pests in Punjab, with most activity from June to December, with *E. insulana* as the dominant species (Khan, 1944). *Earias insulana* preferably feeds on flower buds and pods of other host plants – kanghi, hollyhock, sankukra, sonchal, kuchiri and bhindi – while *E. fabia* feeds on bhindi as a first preference in the absence of cotton.

As a cultural control, since 1968, cutting cotton sticks 2–3 inches below the soil surface to eliminate regrowth allowing SBW to feed was recommended (Haq, 1968). Three larval parasites, including ectoparasites *Elasmus johnstoni* and *Bracon lefroyi* and endoparasite *Rogas tstaceus*, and two pupal parasites *Brachymeria techardiae* and *Goryphus nursei* were recorded (Khan, 1938). Biology of spiny bollworm *Earias vitella* was reported by Rehman and Ali (1981). Less population was observed in Bt cotton fields as compared to non-Bt cotton (Arshad *et al.*, 2015a). Additive interactions of some reduced-risk biocides and two entomopathogenic nematodes suggest implications for integrated control of *Spodoptera litura* (Lepidoptera: Noctuidae), as reported by Khan *et al.* (2021).

Armyworm

In 2003, an outbreak of *Spodoptera litura* (F.) occurred in Pakistan throughout the cotton belt and it devastated the mid- to final stage of the crop. Most of the insecticides, especially pyrethroids and carbamates, failed to provide adequate control. Insecticide resistance was implicated as the major cause of its control failures (Ahmad *et al.*, 2007a). In the first study after the outbreak, which caused enormous losses in many economically important crops including cotton, the genetics and mechanisms of the resistance of *S. litura* to deltamethrin were investigated (Ahmad *et al.*, 2007a; Ahmad *et al.*, 2007b). Field populations of *S. litura* from Pakistan were evaluated for their resistance to conventional insecticides, i.e. organochlorine (endosulfan), organophosphates (chlorpyrifos, phoxim, quinalphos, profenofos), carbamates (methomyl, thiodicarb) and pyrethroids (bifenthrin, cyfluthrin), during 1997–2005 using a leaf-dip bioassay method (Ahmad *et al.*, 2007b). Integrated pest management tactics aimed at reducing pesticidal applications, rotating chemicals of diverse modes of action and conserving natural enemies are recommended (Ahmad *et al.*, 2007b). Transgenic Cry1Ac Bt cotton showed no significant effect under field and laboratory conditions against *S. exigua* (Arshad and Suhail, 2011).

Integrated Pest Management

Despite a closed season, insecticides became a major way of controlling insect pests, but greater efforts are now aimed at reducing their use by delaying the first spray to build the natural enemies population or using selective insecticides, improved quality of seeds and other cultural practices or mating-disruption techniques. Bt cotton was released to mitigate the effect of lepidopterans (PBW, SBW and ABW) excluding armyworm, but due to increases in the population density of sucking insect pests and their natural enemies (Arshad and Suhail, 2010; Arshad *et al.*, 2014; Arshad *et al.*, 2015b; Razaq *et al.*, 2019), their updated economic threshold levels are suggested for future integrated pest management strategies in Bt cotton ecosystems (Arshad *et al.*, 2018; Anees and Shad, 2020). Although spiders as generalist predators have a significant role in pest reduction (Li *et al.*, 2011; Wu *et al.*, 2015; Ali *et al.*, 2018), especially in cotton fields (Wu and Guo, 2005; Lu *et al.*, 2012; Ali *et al.*, 2016), further studies are needed to characterize their role in IPM. Cotton sowing may continue till the end of June depending upon the recommended seed variety and planting area. The picking of early planted varieties begins in August and continues until harvesting of cotton sticks during October and November of every season (Government of the Punjab, 2020; PACRA, 2020). According to the latest advisory for cotton, red cotton bug (RCB) and dusky cotton bug (DCB), among sucking insect pests, and pink bollworm and armyworm, among bollworms, are major insect pests, but no ETL is recommended for RCB and DCB in the cotton-production technology plans (Government of the Punjab, 2020).

Pesticide use

In the early days of DDT being used to control vectors of disease, scientists were mainly concerned about possible effects of the new insecticides on humans. DDT was remarkably safe for humans to use and it was only later that adverse impact on birds was recognized. Nevertheless, post-Second World War, many toxic chemicals became available. The new insecticides provided quick insect pest control, so being relatively inexpensive, spraying insecticides became the plant-protection practice adopted globally although little was then known about their impact on non-target organisms (e.g. natural enemies and human beings, i.e. farmers) (Desneux *et al.*, 2007).

Spray equipment being utilized in Pakistan

Pesticides were first applied with aircraft in the 1950s (Mazari, 2005; Ali, 2018), but due to the presence of trees within fields, the aircraft usually flew too high above the cotton. Farmers then mostly used manually operated knapsack sprayers. Farmers are most vulnerable to potential health risks (Jabbar and Mallick, 1992; Iqbal *et al.*, 1997) if pesticides are used without information regarding their toxicity and proper usage (Fig. 4.12), including the spray machines and personal protective equipment (PPE) (Saeed *et al.*, 2017). During the hot weather in cotton-growing seasons in Pakistan, the spray volume per ha with a lever-operated knapsack sprayer was 200–250 litres/ha and 350–500 litres/ha depending upon the crop height (Attique and Shakeel, 1983). These sprayers were fitted with a lance and hollow cone nozzles (Figs 4.3, 4.4, 4.5, 4.6). After the outbreak of cotton mealybug during 2005, a curved mealybug lance was introduced by Agricultural Machinery Research Institute, Multan (AMRI) to provide better coverage as compared to traditional lances. During the early 1980s there were experimental trials to promote the use of spinning disc ULV sprayers applying 2.5 litres/ha by ULV in Pakistan. These sprayers covered a larger area per day and were cheaper than knapsack sprayers during field trials in 1980–1981 (Attique and Sahkeel, 1983). A mixture of abamectin + bifenthrin applied as a ULV spray gave better control of *H. armigera* than the conventional insecticides (Ahmed *et al.*, 2003).

Figs. 4.3 and 4.4. Spraying cotton using a manually operated knapsack sprayer. (Photos: Graham Matthews)

A tractor-mounted Jetoo mistblower sprayer applying 300–400 litres/per ha and a boom sprayer using 400–500 litres/ha were used on larger farms (Fig. 4.11). These sprayers have been used by farmers before and after the introduction of Bt cotton (Figs 4.7, 4.8 and 4.9).

Pesticides legislation in Pakistan

The introduction of chemical pesticides in Pakistan was linked with the locust control efforts during 1950, commenced in 1952 by aerial spraying of major crops including cotton. However, in 1954, the first-ever import of formulated pesticides (254 metric tonnes) was reported, evidence of the beginning of the pesticides business in Pakistan (Nasira, 1996). Pesticide usage in Pakistan started with aerial spraying on major crops (Ali, 2018). Due to farmers trust of the new solution to control insect pests, mainly on cotton, pesticide imports increased with every passing year (Fig. 4.10).

Before 1980, importation of pesticides was under the jurisdiction of the government of

Figs. 4.5 and 4.6. Spray men using lever-operated knapsack sprayer machine in Okara district of Punjab, Pakistan. (Photos: Noman Ahmad, 2020)

Pakistan, which distributed them among the farming community either via the agriculture ministry or through the private sector at subsidized rates. After 1980, the import and sale were permitted by the private sector, which actually set the foundation of the local pesticide business in Pakistan (Farid-u-ddin, 1985). Pesticide use then increased during the 1990s as farmers

Fig. 4.7. Ciba Geigy training course spraying downwind, near Multan. (Photo: Graham Matthews)

were using them to combat either bollworm or CLCuD. Farmers depended on the recommendations of pesticide dealers, as there was no effective extension service to advise farmers. Now, under new regulations, the provincial governments have more responsibility to implement the rules, especially about agriculture, as the Punjab Agriculture Policy 2018 and the Punjab Agricultural Pesticides Rules 2018 (Government of the Punjab, 2020) have come into effect.

In Pakistan, registered insecticides increased from 108 (Zia *et al.*, 2009; Anwar *et al.*, 2011) to more recently 196 (Ali, 2018). Punjab province used pesticides (88.3%) followed by the provinces Sindh (8.2%), Khyber Pakhtunkhwa (KPK) (2.8%) and Baluchistan (0.76%) (Khan *et al.*, 2010). Overall pesticides applied on cotton crop range between 70–85% (Shahid *et al.*, 2016) and 60–65% (PACRA, 2020). Chemical insecticides hold ~60% of the pesticides market in Pakistan (PACRA, 2020) while 90% (Anon, 1993) or 75% (Ali, 2018) of insecticides are used in cotton alone. Malathion replaced the use of DDT and BHC in Pakistan during the mid-1970s. The government of Pakistan distributed subsidized pesticides among farmers until 1981 when the private sector started to market and promote their sale and use (Beg, 2004). It is reported that pesticides were applied on 5–10% of cotton-growing areas in the Punjab before 1983 and 100% by 1997 (MinFA, 1995; Anon, 1995–6; Tariq, 2005). Approximately 90% of pesticides are imported in raw form and then further formulated, while only a maximum of 10% are imported as final product to focus on the quality in addition to developing more strict policies for pesticide dealers and companies.

Recently (Webinar, 2020), the Pakistani government is working with the Chinese government under the China–Pakistan Economic Corridor (CPEC) to launch pesticides manufacturing units in special economic zones in order to decrease chemical pesticide use that was actively promoted through the public extension service, farmer subsidy schemes and other interventions. Not surprisingly, farmers quickly found themselves on a pesticide 'treadmill' that forced them to combat emerging resistance with a succession of increasingly toxic pesticides such as chlorinated hydrocarbons, organophosphates and pyrethroids (Kouser *et al.*, 2019). For this, legislations has been passed to punish farmers with a

Figs 4.8 and 4.9. Spray men using lever-operated knapsack spray machine equipped with non-recommended eight-holes shower disc in Bahawalnagar district of Punjab, Pakistan. (Photos: Muhammad Asif Farooq, 2020)

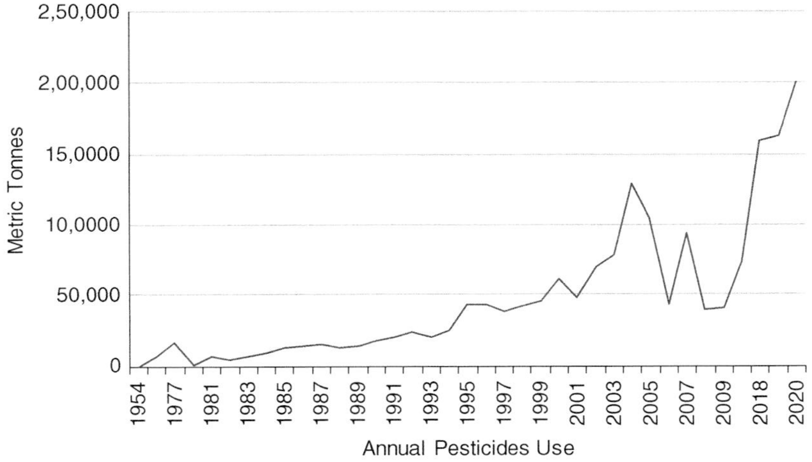

Fig. 4.10. Annual pesticides use trend in Pakistan. (Data collected from Khooharo *et al*., 2008; GOP, 2011; PCPA, 2020)

Fig. 4.11. Tractor-mounted boom sprayer in Ghotki district of Sindh province, Pakistan. (Photo: Noman Ali, 2020)

few months to many years in prison along with heavy fines. Farmers usually select a pesticide on the basis of its availability on credit with the local dealer or which is suggested by the dealer or they follow other farmers, or whichever one is provided by the landlord in the case of tenant farmers. Farm workers are highly exposed to pesticide residues in the fields (Ali *et al.*, 2014; Bakhsh *et al.*, 2016). Overall, mostly males take responsibility for spraying while females act as cotton pickers (Beg, 2004). Pesticide poisoning among cotton pickers is reported in Pakistan (Baloch, 1985; Khwaja, 2001) particularly in Punjab (Masud and Baig, 1992; Ahmad *et al.*, 2004) and Sindh (Rizwan *et al.*, 2005). For application of pesticides, farmers usually follow the dose recommended by the pesticide dealer and they neglect the advisories of respective departments. Mostly, they apply higher amounts of pesticides than are recommended without taking into account the lethal consequences. Governments should take strict action and devise policies in this respect. To minimize the negative impacts of pesticides, farmers should be properly educated and fully trained regarding pesticide use and application (Khan *et al.*, 2010; Ahmed *et al.*, 2011). Farmers are recommended to use power or hand sprayers (boom sprayers are not recommended) for sucking insect pests (e.g. mealybugs) and seed treatment is recommended to delay the whitefly population for 30 days.

Communication to Farmers

In the past, farmers' guidelines were regularly published by the Pakistan Central Cotton Committee to educate the farmers (Chaudhry, 1987). The other choices for farmers were radio programmes and then television. Based on previous years' experiences and their adaptive research field trials, the Directorate of Agriculture (Extension and Adaptive Research) publish the updated next season cotton production technology and plan in detail extensive advertising through their local workers and companies to guide farmers (Government of the Punjab, 2020). By November 2020, due to declines in cotton production (Fig. 4.1) and areas under cultivation, the government, through the involvement of public sector institutes and organizations, has conducted several webinars and taken some important steps to regain farmers' trust with cotton. Pakistan has the biggest, and a strong chain of cotton markets as compared to other crops (e.g. rice, maize, sugarcane etc.) and currently, cotton has much higher prices (Webinar, 2020). Selected farms with good road connectivity demonstrated higher yield outcomes after doing blind sprays without following the recommended pesticide application procedures and not realizing the importance of using personal protective equipment. It is very common practice among farmers that they believe their personal experiences rather than a private company's representative or government extension service providers. Moreover, farmers are getting agricultural advice from different social media platforms like WhatsApp and Facebook agricultural groups based on their own experiences. This kind of

Fig. 4.12. Spray man mixing insecticides without PPE in the Bahawalnagar district of Punjab, Pakistan. (Photo: Muhammad Asif Farooq, 2020)

practice must be discouraged at all levels, as the Cotton Advisory Committee (government level) conducts a fortnightly meeting and makes decisions after discussion with cotton stakeholders and issues advice for the next 15 days. This advice is then distributed to farmers either through electronic or print media. Compared to the last decade, extension services and pest-warning departments in Pakistan are now becoming very efficient in mobilizing advisory recommendations through conducting farmers' meetings and providing demonstrations, brochures and books (Government of the Punjab, 2020). Economic threshold levels are the same as they were before the introduction of Bt cotton in Pakistan (Younas, 1973; Chaudhry, 1987; Government of the Punjab, 2020).

Delinted seed

Farmers are recommended to purchase delinted cottonseed and this is usually done by seed sellers with sulfuric acid at the rate of 1kg/10kg seed. Hydrochloric acid is also used in rare cases. However, in the 2020/21 season, due to heavy rain in Sindh and southern Punjab of Pakistan, the collection of good-quality seeds for the next season is primarily a task for the public research organizations. A report on the internet (Webinar, 2020) estimates 40–43% less production than last year, but that was already less than the target. In the major cotton-growing areas, good-quality cottonseeds have been freely distributed among farmers with 1–25 acres of cotton fields.

Closed season

In Punjab province, cotton sowing is prohibited before 1 April and after 31 May every season (generally in April and May), but in Sindh province, sowing is done from March to April.

Cotton Growing in Bangladesh

Traditionally, tree cotton (*Gossypium arboreum*) has been grown and local spinners have produced fine fabrics, notably muslin, but now 95% of the crop is American upland (*Gossypium hirsutum*); but with a thriving textile industry needing some 9 million bales, only about 3% of the cotton is grown locally, so 97% has to be imported from other cotton-growing countries. A Cotton Development Board was established in 1972 to boost cotton production and in 1976/77, commercial growing of upland cotton began with medium-staple-length upland cotton varieties. The Cotton Development Board (CBD) is now conducting cotton research, extension, training, seed production and distribution, marketing and ginning, and providing small-scale credit facilities to cotton farmers.

The aim has been to increase cotton production by expanding the area and yield of the crop, by getting farmers to adopt profitable cotton, replacing tobacco cultivation, and extending into other areas, growing high-yielding modern varieties, hybrids, transgenic cotton, along with improved management practices sowing higher-quality seeds (Fig. 4.13).

The introduction of Bt hybrid cotton is expected to double production from the same area of land and save production costs with varieties' resistance to bollworm and other chewing insects, which will enable farmers to increase their net income, contributing to the economy of the rural people. To develop own hybrid varieties, the plan has been to strengthen the research facilities within the CDB in collaboration with other research institutions inside and outside of the country.

Among the insect pests, jassids (*Amrasca devastans, Amrasca biguttula* (Homoptera: Ciccadellidae)) have been regarded in the Indian subcontinent as the most common and most devastating major insect pest of cotton (*Gossypium hirsutum* L.) as they suck sap from plant leaves and also inject toxic saliva, which can cause stunted plant growth, with leaves curling downwards and becoming yellow and then brown and dry; and, in severe cases, causing the shedding of fruiting bodies. Cotton yield losses have ranged from 37% to 67%. Farmers have relied on chemical insecticides to manage *A. devastans*, even though it had been recognized that frequent spraying could adversely affect the natural enemy fauna (Saeed *et al.*, 2015). However, the bollworm, *Helicoverpa armigera* and *Earias* spp., is considered a major pest. Other notable pests are pink bollworm and cotton aphid.

Among the diseases of cotton, fusarium wilt, bacterial blight, angular leaf spot, damping

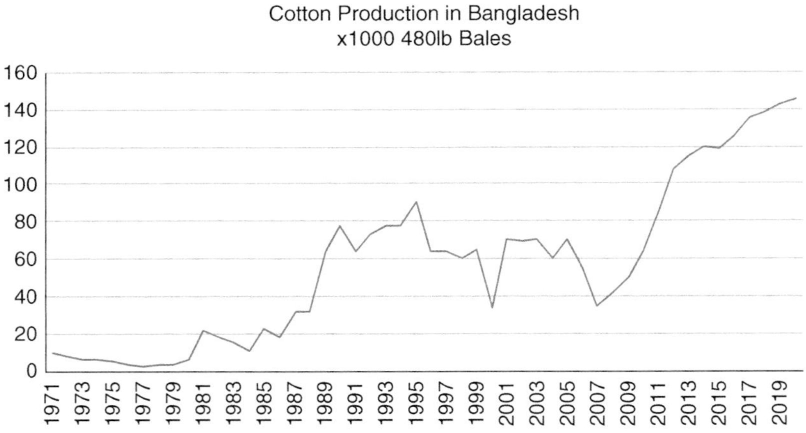

Fig. 4.13. Cotton production in Bangladesh showing increase from 2010.

off, leaf spot, and anthracnose are notable, but compared to other cotton-growing countries of the world, cotton diseases are less prevalent in Bangladesh.

Over the last ten years the textile industry has expanded by supplying a range of cotton garments to meet orders for fashionable goods within the European Union and the USA. However, in 2020, the global spread of the Covid-19 virus resulted in closure of textile factories as orders for these goods were cancelled due the lockdown in numerous areas, which resulted in a sharp decrease in production of cotton goods.

According to a BBC news item, a precious fabric known as Dhaka muslin was imported into Europe from what is now Bangladesh, formerly Bengal, in the late 18th century. The muslin thread counts were in the range of 800–1200. It was made by an elaborate 16-step process with a rare cotton that only grew along the banks of the holy Meghna river. Recently, it was possible to sequence its DNA on samples of preserved leaves from the 19th century at the Royal Botanic Gardens, Kew. Back in Bangladesh, a search was made for wild plants that resembled old drawings in an area close to the Meghna, determined by examining historical maps and comparing them to modern satellite images to see how the course of the river has changed over the last 200 years and to find the best spots for potential candidates. They searched and found wild plants that resembled old drawings. In a project to resurrect the fabric, they planted seeds in 2015 to grow rows of the wild cotton *pluti karpas* (*Gossypium arboreum var. neglecta*) by the river and have since made a hybrid thread combining ordinary and *phuti karpas* cotton and made several saris from their hybrid muslin.

Cotton Production in Myanmar

Agriculture in Myanmar (also known as Burma) is the main industry in the country, accounting for 60% of GDP and employing some 65% of the labour force. Burma was once Asia's largest exporter of rice, and rice remains the country's most crucial agricultural commodity. Cotton is a traditional crop grown in Myanmar and occupies about 350,000 ha, primarily in the central zone of the country, which receives 600–1000 mm of rainfall.

Cotton research was initiated in central farms in Mandalay, Myanmar, along with the establishment of the Department of Agriculture by the government in 1906. Based on the research findings, systematic line sowing in cotton cultivation was introduced in 1927. Until then, the broadcasting sowing method had been practised (Chit, 2005). Traditionally, cotton farmers grew indigenously developed varieties of *Gossypium arboreum* around the 1920s until the large-scale commercial adoption of upland cotton varieties of *G. hirsutum* in the 1960s (Win, 2008).

The government agency, the Agricultural and Rural Development Corporation (ARDC), was established in parallel with the Department of Agriculture under the same ministry in 1953/54. The corporation introduced upland cotton, *G. hirsutum*, from the USA and former USSR in 1957. After conducting adaptive field trials, the corporation distributed some *hirsutum* varieties commercially in 1960. Furthermore, the corporation set up two central farms, Hlaingdet in 1957 and Lungyaw in 1963, to conduct field research and seed multiplication of *hirsutum* cotton. ARDC carried out research on varietal development and agronomic management for *hirsutum* cotton in those farms. Due to the introduction of upland cotton, the provision of basic farm inputs and appropriate technology packages, swift progress in seed-cotton production was made for both quality and quantity (Fig. 4.14).

Yield loss ranging from 30% to 70% in cotton production in Myanmar is attributed to the damage inflicted by bollworm such as *Heliothis armigera* and *Pectinophora gossypiella*. So, seeking improved plant protection measures in harmony with an IPM system is always the priority in cotton research even though the implementation of IPM at farm level is still poor. One of the efficient ways to resist bollworm attack is supposed to be the adoption of genetically engineered (GE) cotton, i.e. Bt cotton cultivation. Although the issue needs further examination, a field trial relating to feasibility of Bt cotton cultivation was conducted in Lungyaw cotton research and seed farm, where the bollworm pressure is higher than at other research sites. Tested Bt cotton cultivars were so glabrous that some chemical sprays were still needed to check the sucking pest population growth. Biological control of a serious sucking pest, aphids, would be helpful to minimize the sprays. Moreover, in conventional *hirsutum* varieties cultivation, a minimum number, or avoidance, of chemical sprays on early-season sucking pests would spare the beneficial insects that contribute to the suppression of bollworm in the later stages. To avoid the sprays on the early-season serious sucking pest *Aphis gossypii*, the ladybird beetle, *Menochilus sexmaculatus*, has been reported as beneficial for biological control of cotton aphids. Methods of mass-rearing of *M. sexmaculatus* have been developed. To maintain the sustainable biological control of cotton aphids, the predators should be provided with a certain number of aphids to retain them in the cotton ecosystem. The release of the predators is in progress in cotton research and seed farms.

In co-operation with the Myanmar Agriculture Service, mass-rearing of another potent predator – the bugs of the genus *Eocanthecona furcellata* (Wolff) – has been carried out at the biological control lab of Myanmar Agriculture Service. Test-release of the bugs has been initiated in two research sites named Lungyaw and Shwedaung.

For plots close to the pump stations and the main canal (i.e. lands of Pyat Ywar village), irrigation water is sufficient for double paddy cropping with monsoon paddy grown between July and December (with rainfall and supplementary irrigation) and summer paddy between February and June (with irrigation). However, for plots further from the pump stations and the main canal, irrigation water is insufficient to grow summer paddy. Upland cash crops including oil seeds (groundnuts, sesame), pulses (green gram, pigeon pea) and cotton are grown in the summer after the monsoon paddy cycle. For plots with poor water access (above canals and further away), paddy cultivation is impossible year-round and farmers often grow two cycles of upland cash crops.

Regarding pest control, a study on biology and identification of cotton stem weevil, *Pempherulus affinis*, which was very sporadic, was conducted as the infestation become common and damage was obvious in some cotton-growing areas, especially in research and seed farms where cotton is grown continuously. Pouring the diluted insecticide solution to the crown portion of the young plant and soil was found to be effective to control the stem weevil but it is impractical for large areas. Thus, finding control measures for this pest is another aspect of future research.

In 2019, a UK government-funded programme invested over two years to support the silk and organic cotton production and textiles in Myanmar. Sericulture, or silk farming, is the cultivation of silkworms to produce silk. Although there are several commercial species of silkworm, *Bombyx mori* (the caterpillar of the domestic silkmoth) is the most widely used and intensively studied silkworm.

By introducing new, resistant hybrid cotton varieties, training farmers in organic agriculture methods, the aim is to have third-party organic certification and ensure premium prices for the final product. By the end of 2020, the aim was to contract at least 500 farmers to grow organic cotton.

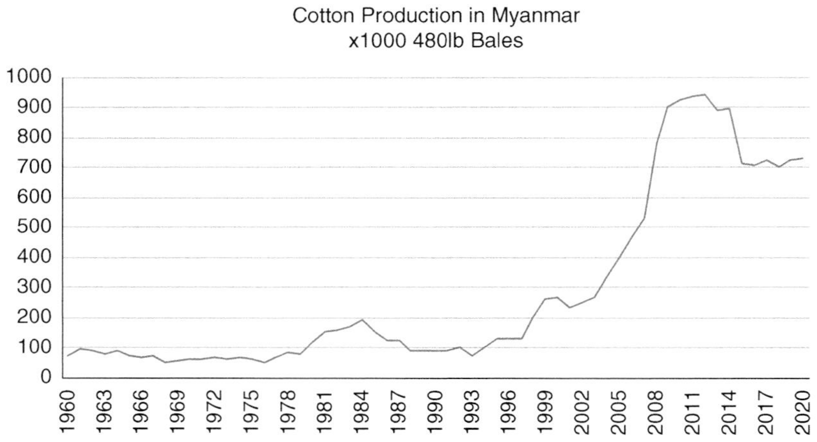

Fig. 4.14. Cotton Production in Myanmar.

References

Abbas, G., Arif, M.J., Ashfaq, M., Aslam, M. and Saeed, S. (2010) Host plants, distribution and overwintering of cotton mealybug, *Phenacoccus solenopsis* (Hemiptera: Pseudococcidae). *International Journal of Agriculture and Biology* 12, 421–425.

Abbas, S.K., Abdin, Z.U., Arif, M.J. and Jamil, A. (2014) Functional analysis of the venom of mealybug parasitoid *Aenasius bambawalei* (Hymenoptera: Encyrtidae). *Biologia* 69, 1046–1050.

Abassi, A., Sufyan, M., Arif, M.J. and Sahi, S.T. (2020) Effect of silicon on oviposition preference and biology of *Bemisia tabaci* (Gennadius) (Homoptera: Aleyrodidae) feeding on *Gossypium hirsutum* (Linnaeus). *International Journal of Pest* Management 66.

Afzal, M. (1946) American cottons in India, their introduction and development. *Indian Farming* 7, 457–462.

Afzal, M. (1947) Problems of cotton improvement in the Punjab. *Indian Cotton Growers Review* 1, 50–59.

Afzal, M. (1969) *The Cotton Plant in Pakistan*. Scientific Monograph. I. Pakistan Central Cotton Committee, Karachi-I.

Afzal, M. and Ali, M. (1983) *Cotton Plant in Pakistan*, 2nd edn. Ismail Aiwan-i-Science, Shahrah-i-Roomi, Lahore.

Afzal, M. and Ghani, M.A. (1953) Cotton jassid in the Punjab. The Mall, Lahore: The Pakistan Association for the Advancement of Science, University Institute of Chemistry.

Afzal, M., Rajaraman, S. and Abbas, M. (1943) Studies on the cotton jassid in the Punjab. III. Effect of jassid infestation on the development and fiber properties of the cotton plant. *Indian Journal of Agriculture Sciences* 13, 192–203.

Aheer, G.M., Ali, A. and Hussain, S. (2006) Varietal resistance against jassid, *Amrasca devastans* Dist. in cotton and role of abiotic factors in population fluctuation. *Journal of Agricultural Research* 44, 299–305.

Ahmad, F., Sanjrani, M.W., Khuhro, S.N., Sajjad, A., Ali, A. *et al.* (2018) Spatial field survey of cotton whitefly and its pupal parasitism in relation to temperature and humidity in southern Pakistan. *Pakistan Journal of Zoology* 50, 1231–1236.

Ahmad, M., Arif, M.I. and Ahmad, Z. (1995) Monitoring insecticide resistance of *Helicoverpa armigera* (Lepidoptera: Noctuidae) in Pakistan. *Journal of Economic Entomology* 88, 771–776.

Ahmad, M., Arif, M.I. and Attique, M.R. (1997) Pyrethroid resistance of *Helicoverpa armigera* (Lepidoptera: Noctuidae) in Pakistan. *Bulletin of Entomological Research* 87, 343–347.

Ahmad, M., Arif, M.I., Ahmad, Z. and Attique, M.R. (1998) *Helicoverpa armigera* resistance to insecticides in Pakistan. Proceedings of Beltwide Cotton Conference, National Cotton Council, Memphis, Tennessee, pp. 1138–1140.

Ahmad, M., Arif, M.I. and Ahmad, Z., (1999a) Detection of resistance to pyrethroids in field populations of cotton jassid (Homoptera: Cicadellidae) from Pakistan. *Journal of Economic Entomology* 92, 1246–1250.

Ahmad, M, Arif, M.I. and Ahmad, Z. (1999b) Patterns of resistance to organophosphate insecticides in field populations of *Helicoverpa armigera* in Pakistan. *Journal of Pesticide Science* 55, 626–632.

Ahmad, M., Arif, M.I. and Ahmad, Z. (2001) Reversion of susceptibility to methamidophos in the Pakistani populations of cotton whitefly, *Bemisia tabaci*. In: *Proceedings of the Beltwide Cotton Conferences*, vol. 2. 9–13 January 2001, pp. 874–876. National Cotton Council of America, Memphis, Tennessee.

Ahmad, M., Arif, M.I. and Ahmad, Z. (2003) Susceptibility of *Helicoverpa armigera* (Lepidoptera: Noctuidae) to new chemistries in Pakistan. *Crop Protection* 22, 539–544.

Ahmad, M., Arif, M.I. and Ahmad, M. (2007a) Occurrence of insecticide resistance in field populations of *Spodoptera litura* (Lepidoptera: Noctuidae) in Pakistan. *Crop Protection* 26, 809–817.

Ahmad, M., Sayyed, A.H., Crickmore, N. and Saleem, S.A. (2007b) Genetics and mechanism of resistance to deltamethrin in a field population of *Spodoptera litura* (Lepidoptera: Noctuidae). *Pest Management Science* 63, 1002–1010.

Ahmad, M., Rasool, B., Ahmad, M. and Russell, D.A. (2019) Resistance and synergism of novel insecticides in field populations of cotton bollworm *Helicoverpa armigera* (Lepidoptera: Noctuidae) in Pakistan. *Journal of Economic Entomology* 112, 859–871.

Ahmad, R., Baloach, M.K., Ahmad, A., Rauf, R., Siddiqui, H. and Khokar, M.Y. (2004) Evaluation of toxicity due to commercial pesticides in female workers. *Pakistan Journal of Medical Sciences* 20, 392–396.

Ahmad, Z. (1976) Source of infestation of pink bollworm. *Proceedings of Cotton Production Seminar, Sukkur, 1976*. ESSO. Pakistan Fertilizer Company, Karachi.

Ahmad, Z., Attique, M.R. and Rashid, A. (1985) An estimate of the loss in cotton yield in Pakistan attributable to the jassid *Amrasca devastans* Dist. *Crop Protection* 5, 105–108.

Ahmed, M.Z., De Barro, P.J., Greeff, J.M., Ren, S.X., Naveed, M. *et al.* (2011) Genetic identity of the *Bemisia tabaci* species complex and association with high cotton leaf curl disease (CLCuD) incidence in Pakistan. *Pest Management Science* 67, 307–317.

Ahmed, S., Saleem, M.A., Rauf, I. and Rahi, S. (2003) Efficacy of high volume (HV) vs ultra low volume (ULV) spraying of talstar 10EC (bifenthrin), mustang 380 EC (zetacypermethrin + ethion) and vovastar 56EC (abamectin + bifenthrin) against different larval stages of *Helicoverpa armigera* (Hub). *International Journal of Agriculture & Biology* 4, 621–624.

Ahmed, Z. (1980) Incidence of major cotton pests and diseases in Pakistan with special reference to pest management. In: *Proceedings, 1st International Consultation on Cotton Production Research with Focus on the Asian Region*. Manila, pp. 156–179.

Akhtar, Z.R., Saeed, Z., Anjum, A.A., Hira, H., Ihsan, A., Noreen, A. and Salman, M.A. (2018) Pink boll worm resistance evaluation against organophosphate in Cry1Ac expressing transgenic cotton. *Journal of Entomology and Zoology Studies* 6, 821–824.

Akmal, M., Freed, S., Dietrich, C.H., Mehmood, M. and Razaq, M. (2018) Patterns of genetic differentiation among populations of *Amrasca biguttula biguttula* (Shiraki) (Cicadellidae: Hemiptera). *Mitochondrial DNA* Part A 29, 897–904.

Ali, A., Desneux, N., Lu, Y.H., Liu, B. and Wu, K.M. (2016) Characterization of the natural enemy community attacking cotton aphid in the Bt cotton ecosystem in Northern China. *Scientific Reports* 6, 24273.

Ali, A., Desneux, N., Lu, Y.H. and Wu, K.M. (2018) Key aphid natural enemies showing positive effects on wheat yield through biocontrol services in northern China. *Agriculture, Ecosystems & Environment* 266, 1–9.

Ali, M.A. (2018) *A handbook for agriculture extension agents on the pesticides registered with recommendations for safe handling and use in Pakistan*. Pakistan Agricultural Research Council, Ministry of National Food Security and Research, Islamabad.

Ali, M.A., Farooq, J., Batool, A., Zahoor, A., Azeem, F., Mahmood, A. and Jabran, K. (2019) Cotton production in Pakistan. In: Jabran, K. and Chauhan, B.S. (eds) *Cotton Production*. John Wiley & Sons Ltd.

Ali, U., Syed, J.H., Malik, R.N., Katsoyiannis, A., Li, J., Zhang, G. and Jones, K.C. (2014) Organochlorine pesticides (OCPs) in south Asian region: a review. *Science of the Total Environment* 476, 705–717.

Amrao, L., Akhter, S., Tahir, M.N., Amin, I., Briddon, R.W. and Mansoor, S. (2010) Cotton leaf curl disease in Sindh province of Pakistan is associated with recombinant begomovirus components. *Virus Research* 153, 161–165.

Anees, M. and Shad, S.A. (2020) Insect pests of cotton and their management. In: Ahmad, S. and Hasanuzzaman, M. (eds) *Cotton Production and Uses – Agronomy, Crop Protection, and Postharvest Technologies*. Springer Nature, Singapore Pte Ltd. Available at: https://doi.org/10.1007/978-981-15-1472-2 (accessed 25 July 2021).

Anon (1993) *Report of Prime Minister's Task Force.* Ministry of Finance, Revenue and Economic Affairs, Government of Pakistan, Islamabad.

Anon (1995–1996) *Agricultural Statistics of Pakistan*. Ministry of Food, Agriculture and Livestock, Government of Pakistan, Islamabad.

Anwar, T., Ahmad, I. and Tahir, S. (2011) Determination of pesticide residues in fruits of Nawabshah district, Sindh, Pakistan. *Pakistan Journal of Botany* 43, 1133–1139.

Arif, M.I., Rafiq, M. and Ghaffar, A. (2009) Host plants of cotton mealybug (*Phenacoccus solenopsis*): a new menace to cotton agro-ecosystem of Punjab, Pakistan. *International Journal of Agriculture and Biology* 11, 163–167.

Arshad, M. and Suhail, A. (2010) Studying the sucking insect pests community in transgenic Bt cotton. *International Journal of Agriculture and Biology* 12, 764–768.

Arshad, M. and Suhail, A. (2011) Field and laboratory performance of transgenic Bt cotton containing Cry1Ac against beet armyworm larvae (Lepidoptera: Noctuidae). *Pakistan Journal of Zoology* 43, 529–535.

Arshad, M., Suhail, A., Asghar, M., Tayyab, M. and Hafeez, F. (2007) Factors influencing the adoption of Bt cotton in the Punjab, Pakistan. *Journal of Agricultural and Social Sciences* 3, 121–124.

Arshad, M., Suhail, A., Gogi, M.D., Yaseen, M., Asghar, M., Tayyib, M., Karar, H., Hafeez, F. and Ullah, U.N. (2009a) Farmers' perceptions of insect pests and pest management practices in Bt cotton in the Punjab, Pakistan. *International Journal of Pest Management* 55, 1–10.

Arshad, M., Suhail, A., Arif, M.J. and Khan, M.A. (2009b) Transgenic Bt and non-transgenic cotton effects on survival and growth of *Helicoverpa armigera. International Journal of Agriculture and Biology* 11, 473–476.

Arshad, M., Arif, M.J., Gogi, M.D., Abdu-ur-Rehman, M., Zain-ul-Abdin, Wakil, W. and Saeed, N.A. (2014) Seasonal abundance of non-target natural enemies in transgenic Bt and conventional cotton. *Pakistan Entomologist* 36, 37–40.

Arshad, M., Zain-ul-Abdin, Gogi, M.D., Arif, M.J. and Khan, R.R. (2015a) Seasonal pattern of infestation by spotted bollworm, *Earias insulana* (Boisd.) and pink bollworm, *Pectinophora gossypiella* (Saund.) in field plots of transgenic Bt and non-Bt cottons. *Pakistan Journal of Zoology* 47, 177–186.

Arshad, M., Khan, H.A.A., Abdul-ur-Rehman, M. and Saeed, N.A. (2015b) Incidence of insect predators and parasitoids on transgenic Bt cotton in comparison to non-Bt cotton varieties. *Pakistan Journal of Zoology* 47, 823–829.

Arshad, M., Rashad, R.R., Aslam, A. and Akbar, W. (2018) Transgenic Bt cotton: effects on target and non-target insect diversity. In: Mehboob-Ur-Rahman and Zafar, Y. (eds) *Past, Present and Future Trends in Cotton Breeding.* Intech Open.

Ashfaq, M., Shah, G.S., Noor, A.R., Ansari, S.P. and Mansoor, S. (2010) Report of a parasitic wasp (Hymenoptera: Encyrtidae) parasitizing cotton mealybug (Hemiptera: Pseudococcidae) in Pakistan and use of PCR for estimating parasitism levels. *Biocontrol Science and Technology* 20, 625–630.

Ashfaq, M., Hebert, P.D.N., Mirza, M.S., Khan, A.M., Mansoor, S. *et al.* (2014) DNA barcoding of *Bemisia tabaci* complex (Hemiptera: Aleyrodidae) reveals southerly expansion of the dominant whitefly species on cotton in Pakistan. *PLOS One* 9, e104485.

Attique, M.R. (1985) Pheromones for the control of cotton pests in Pakistan. *Pakistan Cottons* 29, 133–136.

Attique, M.R. and Shakeel, M.A. (1983) Comparison of ULV with conventional spraying on cotton in Pakistan. *Crop Protection* 2, 231–234.

Attique, M.R., Ahmad, M.M. and Ahmad, Z. (2000) Efficacy of different sex pheromone traps for monitoring and control of pink bollworm. *Pakistan Journal of Biological Sciences* 3, 309–312.

Baffes, J. and Ruh Paul, A. (2005) Part 1: The Cotton Trade: History and Background. In: Townsend, T. (ed.) *Cotton Trading Manual*. Woodhead Publishing, Boca Raton, Florida.

Bakhsh, K., Ahmad, N., Kamran, M.A., Hassan, S., Abbas, Q., Saeed, R. and Hashmi, M.S. (2016) Occupational hazards and health cost of women cotton pickers in Pakistani Punjab. *BMC Public Health* 16, 961.

Baloch, U.K. (1985) Problems associated with the use of chemicals by agricultural workers. *Basic Life Science* 34, 63–78.

Basit, M., Sayyed, A.H., Saleem, M.A. and Saeed, S. (2011) Cross resistance, inheritance and stability of resistance to acetamiprid in cotton whitefly, *Bemisia tabaci* (Genn) (Hemiptera: Aleyrodidae). *Crop Protection* 30, 705Ð712.

Basit, M., Saleem, M.A., Saeed, S. and Sayyed, A.H. (2012) Cross resistance, genetic analysis and stability of resistance to buprofezin in cotton whitefly, *Bemisia tabaci* (Homoptera: Aleyrodidae). *Crop Protection* 40, 16–21.

Basit, M., Shafqat, S., Saleem, M.A. and Sayyed, A.H. (2013a) Can resistance in *Bemisia tabaci* (Homoptera: aleyrodidae) be overcome with mixtures of neonicotinoids and insect growth regulators? *Crop Protection* 44, 135–141.

Basit, M., Saeed, S., Saleem, M.A., Denholm, I. and Shah, M.(2013b) Detection of resistance, cross-resistance, and stability of resistance to new chemistry insecticides in *Bemisia tabaci* (Homoptera: Aleyrodidae). *Journal of Economic Entomology* 106, 1414–1422.

Beg, M.A.A. (2004) Report on Pesticide Usage in Pakistan. *Ummat*, a National Language Daily. DOI:10.13140/RG.2.1.4536.5284

Biswas, K.K., Bhattacharyya, U.K., Palchoudhury, S., Balram, N., Kumar, A. *et al.* (2020) Dominance of recombinant cotton leaf curl Multan-Rajasthan virus associated with cotton leaf curl disease outbreak in northwest India. *PLOS ONE.*

Briddon, R.W. (2003) Cotton leaf curl disease, a multicomponent begomovirus complex. *Molecular Plant Pathology* 4, 427–434.

Briddon, R.W. and Markham, P.G. (2000) Cotton leaf curl virus disease. *Virus Research* 71, 151–159.

Briddon, R.W., Akbar, F., Iqbal, Z., Amrao, L., Amin, I., Saeed, M. and Mansoor, S. (2014) Effects of genetic changes to the begomovirus/betasatellite complex causing cotton leaf curl disease in South Asia post-resistance breaking. *Virus Research* 186, 114–119.

Brown, J.K. (2001) Viral and phytoplasma disease. In: Kirkpatrick, T.L. and Rothrock, C.K. (eds) *Cotton Leaf Curl Disease*. 2nd edn. American Psychopathological Society, St Paul, Minnesota, 52–54.

Campion, D.G. and Murlis, J. (1985) Sex pheromones for the control of insect pests in developing countries. *Mededelingen van de Faculteit Landbouwwetenschappen, Rijksuniversiteit Gent* 50, 203–209.

Chamberlain, D.J., Critchley, B.R., Campion, D.G., Attique, M.R., Rafique, M. and Arif, M.I. (1992) Use of a multi-component pheromone formulation for control of cotton bollworms (Lepidoptera: Gelechiidae and Noctuidae) in Pakistan. *Bulletin of Entomological Research* 82, 449–458.

Chamberlain, D.J., Ahmad, Z., Attique, M.R. and Chaudhry, M.A. (1993) The influence of slow release PVC resin pheromone formulations on the mating behaviour and control of the cotton bollworm complex (Lepidoptera: Gelechiidae and Noctuidae) in Pakistan. *Bulletin of Entomological Research* 83, 335–343.

Chaudhry, G.Q. (1987) *Cultural control of cotton pests*. In: Proceedings of Seminar on Insect Pests of Cotton. Cotton Research Institute, Multan. Organized by Pakistan Central Cotton Committee, Karachi, pp. 82–86.

Cheema, M.A., Muzaffar, N. and Ghani, M.A. (1980a) Biology, host range and incidence of parasites of *Pectinophora gossypiella* (Saunders) in Pakistan. *The Pakistan Cottons* 24, 37–73.

Cheema, M.A., Muzaffar, N. and Ghani, M.A. (1980b) Investigation on phenology, distribution, host range and evaluation of predators of *Pectinophora gossypiella* (Saunders) in Pakistan. *The Pakistan Cottons* 24, 139–176.

Chit, D. (2005) Scientific approaches to cotton cultivation and production (in Myanmar language). Myanmar Academy of Agricultural, Forestry, Livestock and Fishery Science, Yangon, Myanmar.

Critchley, B.R., Campion, D.G., Cavanagh, G.G., Chamberlain, D.J. and Attique, M.R. (1987) Control of three major bollworm pests of cotton in Pakistan by a single application of their combined sex pheromone. *Tropical Pest Management* 34, 374.

Critchley, B.R., Chamberlain, D.J., Campion, D.G., Attique, M.R., Ali, M. and Ghaffar, A. (1991) Integrated use of pink bollworm pheromone formulations and selected conventional insecticides for the control of the cotton pest complex in Pakistan. *Bulletin of Entomological Research* 81, 371–378.

Desneux, N., Decourtye, A. and Delpuech, J.M. (2007) The sublethal effects of pesticides on beneficial arthropods. *Annual Review of Entomology* 52, 81–106.

Ejaz, M., Ullah, S., Shad, S.A., Abbas, N. and Binyameen, M. (2019) Characterization of inheritance and preliminary biochemical mechanisms of spirotetramat resistance in *Phenacoccus solenopsis* Tinsley: an economic pest from Pakistan. *Pesticide Biochemistry and Physiology* 156, 29–35.

Farid-u-ddin, A. (1985) Review of agro-pesticide consumption and its impact on crop protection in Pakistan. *Pakistan Agriculture*, p. 28.
Ghani, M.A., Afzal, M. and Nanda, D.N. (1945) Studies on cotton jassid in the Punjab. VII: Age of leaf and jassid susceptibility. *Proceedings of Indian Academy of Science* 22, 219–224.
Gogi, M.D., Syed, A.H., Atta, B., Sufyan, M., Jalal, M. *et al.* (2021) Efficacy of biorational insecticides against *Bemisia tabaci* (Genn.) and their selectivity for its parasitoid *Encarsia formosa* Gahan on Bt cotton. *Scientific Reports* 11, 2101.
GOP (Government of Pakistan) (2011) *Agriculture Statistics of Pakistan 2010–11*. Statistics Division, Pakistan Bureau of Statistics. Available at: http://www.pbs.gov.pk/content/agriculture-statistics-pakistan-2010-11 (accessed 14 September 2020).
Government of the Punjab, Agriculture Department (2018) The Punjab Agricultural Pesticides Rules 2018. *The Punjab Gazette*, Registered no. L-7532. Available at: http://www.agripunjab.gov.pk/system/files/The%20Punjab%20Agricultural%20Pesticieds%20Rules%202018.pdf (accessed 14 September 2020).
Government of the Punjab, Agriculture Department (2020) *Cotton Production Technology 2019–20*. (in Urdu language) (accessed September 14, 2020).
Haq, K.A. (1968) A focus of insect pest control problem in cotton, with special reference to central region of west Pakistan. Paper read at the Cotton Seminar held at Multan, 18–19 November.
Hodgson, C., Abbas, G., Arif, M.J. and Saeed S. (2008) *Phenacoccus solenopsis* (Tinsley) (Sternorrhyncha: Coccoidea: Pseudococcidae), an invasive mealybug damaging cotton in Pakistan and India, with a discussion on seasonal morphological variation. *Zootaxa* 1913, 1–35.
Hussain, M.A. (1930) A preliminary note on white fly of cotton in the Punjab. *Agricultural Journal of India* 25, 508–525.
Hussain, T. and Ali, M. (1975) A review of cotton diseases of Pakistan. *Pakistan Cotton* 19, 71–86.
Hussain, T. and Mahmood, T. (1988) A note on leaf curl disease of cotton. *Pakistan Cotton* 32, 248–251.
Hussain, T., Tahir, M. and Mahmood, T. (1991) Cotton leaf curl virus. *Pakistan Journal of Plant Pathology* 3, 57–61.
Idris, A.M. (1990) Cotton leaf curl virus disease in the Sudan. *Mededelingen van de Faculteit Landbouwwetenschappen, Rijksuniversiteit Gent* 55, 263–267.
Idris, A.M. and Brown, J.K. (2004) Cotton leaf crumple virus is a distinct Western Hemisphere begomovirus species with complex evolutionary relationships indicative of recombination and reassortment. *Phytopathology* 94, 1068–1074.
Iqbal, Z., Zia, K. and Ahrnad, A. (1997) Pesticide abuse in Pakistan and associated human health and environmental risks. *Pakistan Journal of Agricultural Sciences* 34, 1–4.
ISAAA (2017) *Global Status of Commercialized Biotech/GM Crops*. ISAAA Briefs No. 53. International Service for the Acquisition of Agri-Biotech Applications, Ithaca, New York.
Ishtiaq, M. and Saleem, M.A. (2011) Generating susceptible strain and resistance status of field populations of *Spodoptera exigua* (Lepidoptera: Noctuidae) against some conventional and new chemistry insecticides in Pakistan. *Journal of Economic Entomology* 104, 1343–1348.
Jabbar, A. and Mallick, S. (1992) Pesticides and environmental situation in Pakistan. SDPI Working Paper Series. SDPI, Islamabad.
Karar, H., Babar, T.K., Shahazad, M.F., Saleem, M., Ali, A. and Akram, M. (2013) Performance of novel vs traditional insecticides for the control of *Amrasca biguttula biguttula* (Homoptera, Cicadellidae) on cotton. *Pakistan Journal of Agriculture Sciences* 50, 223–228.
Khan, J.A. and Ahmad, J. (2005) Diagnosis, monitoring and transmission characters of cotton leaf curl virus. *Current Science* 88, 1803–1809.
Khan, M.H. (1938) Studies on *Platyedra gossypiella* Saunders (the pink bollworm of cotton) in the Punjab. Part IV: The incidence of *Platyedra gossypiella* in relation to climate (1926–31). *Indian Journal of Agricultural Sciences* 8, 191–214.
Khan, M.H. (1944) Studies on *Earias* species (the spotted bollworm of cotton) in the Punjab. Part I: *Earias fabia and Earias insulana* Boisd in various parts in relation to environmental conditions. *Indian Journal of Entomology* 6, 15–27.
Khan, M.J., Zia, M.S. and Qasim, M. (2010) Use of pesticides and their role in environmental pollution. *World Academy of Science, Engineering and Technology* 4, 122–128.
Khan, R.R., Arshad, M., Aslam, A. and Arshad, M. (2021) Additive interactions of some reduced-risk biocides and two entomopathogenic nematodes suggest implications for integrated control of *Spodoptera litura* (Lepidoptera: Noctuidae). *Scientific Reports* 11, 1268.

Khan, W.S. and Khan, A.G. (1995) Cotton situation in the Punjab: an overview. *Proceedings of National Seminar 'Strategies for Increasing Cotton Production'*, 26–27 April. Government of Punjab, pp. 1–29.
Khooharo, A.A., Memon, R.A. and Mallah, M.U. (2008) An empirical analysis of pesticide marketing in Pakistan. *Pakistan Economic and Social Review*, 57–74.
Khwaja, M.A. (2001) Impact of pesticides on environment and health. *SDPI Research and News Bulletin* 8(2).
Kouser, S., Spielman, D.J. and Qaim, M. (2019) Transgenic cotton and farmers' health in Pakistan. *PLOS One* 14, e0222617.
Li, X.J., Guo, Z., Wang, S.X., Xing, X., Li, Y., Yu, G.W. and You, G.L. (2011) The population dynamics and control effect of important natural enemies of the soybean aphid *Aphis glycines*. *Chinese Journal of Applied Entomology* 48, 1613–1624.
Lu, Y., Wu, K., Jiang, Y., Guo, Y. and Desneux, N. (2012) Widespread adoption of Bt cotton and insecticide decrease promotes biocontrol services. *Nature* 487, 362–365.
Malik, S.J. (2015) Agriculture policy in Pakistan – what it is and what it should be. Innovative Development Strategies (Pvt.) Ltd. Pakistan Institute of Development Economics, Islamabad. Available at: https://www.pide.org.pk/pdf/Seminar/AgriculturePolicyPakistan.pdf (accessed on 14 September 2020).
Mallah, G.H., Soomro, A.R., Soomro, A.W., Kourejo, A.K. and Kalhoro, A.D. (2000) Studies on the leftover standing cotton as carry-over sources of pink bollworm in Sindh. *Pakistan Journal of Biological Science* 3, 147–149.
Mansoor, S., Amin, I., Iram, S., Hussain, M., Zafar, Y., Malik, K.A. and Briddon, R.W. (2003a) Breakdown of resistance in cotton to cotton leaf curl disease in Pakistan. *Plant Pathology* 52, 784.
Mansoor, S., Briddon, R.W., Zafar, Y. and Stanley, J. (2003b) Geminivirus disease complexes: an emerging threat. *Trends in Plant Science*, 128–134.
Masud, S.Z. and Baig, M.H. (1992) *Annual Report of Tropical Agricultural Research Institute*. Pakistan Agricultural Research Council, Karachi.
Mazari, R.B. (2005) Pakistan (report). In: Proceedings of the Asia Regional Workshop: Implementation, Monitoring and Observance: International Code of Conduct on the Distribution and Use of Pesticides. RAP Publication 2005/29. FAO, Bangkok.
MinFA (1995) Agricultural Statistics of Pakistan. *Islamabad: Ministry of Food and Agriculture*. Government of Pakistan.
Muhire, B.M., Varsani, A. and Martin, D.P. (2014) A virus classification tool based on pairwise sequence alignment and identity calculation. *PLOS ONE* 9, e108277.
Munshi, G.H. and Mecci, A.K. (1976) Emergence and carryover of the pink bollworm *Pectinophora gossypiella (*Saunders) through cotton lint and double seeds at Tandojam. *Agriculture Pakistan* 27, 107–111.
Naqvi, R.Z., Zaidi, S.S., Akhtar, K.P., Strickler, S., Woldemariam, M. *et al.* (2017) Transcriptomics reveals multiple resistance mechanisms against cotton leaf curl disease in a naturally immune cotton species, *Gossypium arboreum. Scientific Reports* 7, 15880.
Nasira, H., (1996) *Invisible Farmers in Pakistan: A Study on the Role of Women in Agriculture.* KHOJ Research and Publication Centre, Lahore, Pakistan.
Nawaz-ul-Rehman, M.S., Briddon, R.W. and Fauquet, C.M. (2012) A melting pot of old world begomoviruses and their satellites infecting a collection of *Gossypium* species in Pakistan. *PLOS ONE* 7, e40050.
Nour, M.A. and Nour, J.J. (1964) Identification, transmission and host range of leaf curl viruses infecting cotton in the Sudan. *Empire Cotton Growing Review* 41, 27–37.
PACRA (2020) *Pakistan Credit Rating Agency – Pesticides Sector Analysis*. 1–10. Available at: https://www.pacra.com.pk/uploads/doc_report/Pesticides%20Sector%20Feb20%20Upload%20Version_1582895472.pdf (accessed 11 September 2020).
Parnell, F.R., King, H.E. and Ruston, D.F. (1949) Jassid resistance and hairiness of the cotton plant. *Bulletin of Entomological Research* 39, 539.
PCPA (Pakistan Crop Protection Association) (2020) Personal communication.
Pearson, E.O. and Darling, R.C.M. (1958) *The Insect Pests of Cotton in Tropical Africa*. Empire Cotton Growing Corporation and Commonwealth Institute of Entomology, London.
Qayyum, M.A., Wakil, W., Arif, M.J., Sahi, S.T., Saeed, N.A. and Russell, D.A. (2015) Multiple resistances against formulated organophosphates, pyrethroids, and newer-chemistry insecticides in populations of *Helicoverpa armigera* (Lepidoptera: Noctuidae) from Pakistan. *Journal of Economic Entomology* 108, 286–293.

Qureshi, Z.A. and Ahmed, N. (1989) Efficacy of combined sex pheromones for the control of three major bollworms of cotton. *Journal of Applied Entomology* 108, 386–389.
Rafiq, M., Ghaffar, A. and Arshad, M. (2008) Population dynamics of whitefly (*Bemisia tabaci*) on cultivated crop hosts and their role in regulating its carry-over to cotton. *International Journal of Agriculture* 10, 577–580.
Rahman, M., Khan, A.Q., Rahmat, Z., Iqbal, M.A. and Zafar, Y. (2017) Genetics and genomics of cotton leaf curl disease, its viral causal agents and whitefly vector: a way forward to sustain cotton fiber security. *Frontiers in Plant Science* 8, 1157.
Rajput, I.A., Syed, T.S., Gilal, A.A., Ahmed, A.M., Khoso, F.N., Abro, G.H. and Rustamani, M.A. (2017) Effect of different synthetic pesticides against pink bollworm *Pectinophora gossypiella* (Saund.) on Bt. and non-Bt. cotton crop. *Journal of Basic & Applied Sciences* 13, 454–458.
Rana, M.A. (2014) *The Seed Industry in Pakistan: Regulation, Politics and Entrepreneurship*. Pakistan Strategy Support Program Working Paper 19. International Food Policy Research Institute, Washington, DC.
Razaq, M., Tufail, M., Afzal, M. and Ali, S. (1998) The comparative effectiveness of some latest spray schedules against cotton jassid, *Amrasca devastans* (Dist.) and aphid *Aphis gossypii* G. on cotton variety FH-672. *Pakistan Entomologist* 20, 59–61.
Razaq, M., Suhail, A., Aslam, M., Arif, M.J., Saleem, M.A. and Khan, H.A. (2005) Evaluation of neonicotinoides and conventional insecticides against cotton jassid, *Amrasca devastans* (Dist.) and cotton whitefly, *Bemisia tabaci* (Genn.) on cotton. *Pakistan Entomologist* 27, 75–78.
Razaq, M., Mensah, R. and Athar, H. (2019) Insect pest management in cotton. In: Jabran, K. and Chauhan, B.S. (eds) *Cotton Production*. John Wiley & Sons Ltd. Published by John Wiley & Sons Ltd.
Rehman, M.H. and Ali, H. (1981) Biology of spotted bollworm of cotton, *Earias vittella* (F.). *Pakistan Journal of Zoology* 13, 105–110.
Rizwan, S., Ahmad, I., Ashraf, M., Aziz, S., Yasmine, T. and Sattar, A. (2005) Advance effect of pesticides on reproduction hormones of women cotton pickers. *Pakistan Journal of Biological Sciences* 8, 1588–1591.
Saeed, M., Zafar, Y., Randles, J.W. and Rezaian, M.A. (2007) A monopartite begomovirus associated DNA β satellite substitutes for the DNA B of a bipartite begomovirus to permit systemic infection. *Journal of Genetic Virology* 88, 2881–2889.
Saeed, M.F., Shaheen, M., Ahmad, I., Zakir, A., Nadeem, M., Chishti, A.A., Shahid, M., Bakhsh, K. and Damalas, C.A. (2017) Pesticide exposure in the local community of Vehari district in Pakistan: an assessment of knowledge and residues in human blood. *Science of the Total Environment* 587–588, 137–144.
Saeed, R., Razaq, M. and Hardy, I.C.W. (2015) The importance of alternative host plants as reservoirs of the cotton leaf hopper, *Amrasca devastans*, and its natural enemies. *Journal of Pest Science* 88, 517–531.
Saeed, R., Razaq, M. and Hardy, I.C.W. (2016a) Impact of neonicotinoid seed treatment of cotton on the cotton leaf hopper, *Amrasca devastans* (Hemiptera: Cicadelliae), and its natural enemies. *Pest Management Science* 72, 1260–1267.
Saeed, R., Razaq, M., Rafiq, M. and Naveed, M. (2016b) Evaluating insecticide spray regimes to manage cotton leafhopper, *Amrasca devastans* (Distant): their impact on natural enemies, yield and fiber characteristics of transgenic Bt cotton. *Pakistan Journal of Zoology* 48, 703–711.
Saeed, R., Razaq, M., Abbas, N., Jan, M.T. and Naveed, M. (2017) Toxicity and resistance of the cotton leaf hopper, *Amrasca devastans* (Distant) to neonicotinoid insecticides in Punjab, Pakistan. *Crop Protection* 93, 143–147.
Saeed, R., Razaq, M., Rehman, H.M.U., Waheed, A. and Farooq, M. (2018) Evaluating action thresholds for *Amrasca devastans* (Hemiptera: Cicdellidae) management on transgenic and conventional cotton across multiple planting dates. *Journal of Economic Entomology* 111, 2182–2191.
Saleem, H., Nahid, N., Shakir, S., Ijaz, S., Murtaza, G., Ali, A.K. *et al.*, (2016) Diversity mutation and recombination analysis of cotton leaf curl geminiviruses. *PLOS ONE* 11, e0151161.
Sarwar, M. and Sattar, M. (2016) An analysis of comparative efficacies of various insecticides on the densities of important insect pests and the natural enemies of cotton, *Gossypium hirsutum* L. *Pakistan Journal of Zoology* 48, 131–136.
Sattar, M.N., Iqbal, Z., Tahir, M.N. and Ullah, S. (2017) The prediction of a new CLCuD epidemic in the Old World. *Frontiers in Microbiology* 8, 631.
Shahid, M., Ahmad, A., Khalid, S., Siddique, H.F., Saeed, M.F., Ashraf, M.R., Sabir, M., Niazi, N.K., Bilal, M. and Naqvi, S.T.A. (2016) Pesticides pollution in agricultural soils of Pakistan. In: Hakeem, K.H.,

Akhtar, J. and Sabir, M. (eds) *Soil Science: Agricultural and Environmental Prospectives*. Springer, pp. 199–229.

Shakir, S., Zaidi, S.S., Atiq-ur-Rehman, Farooq, M., Amin, I., Scheffler, J., Scheffler, B., Nawaz-ul-Rehman, M.S. and Mansoor, S. (2019) Non-cultivated cotton species (*Gossypium* spp.) act as a reservoir for cotton leaf curl begomoviruses and associated satellites. *Plants* 8, 127.

Simon, B., Cenis, J.L., Beitia, F., Khalid, S., Moreno, I.M. *et al.*, (2003) Genetic structure of field populations of begomoviruses and of their vector *Bemisia tabaci* in Pakistan. *Phytopathology* 93, 1422–1429.

Spielman, D.J., Nazli, H., Ma, X., Zambrano, P. and Zaidi, F. (2015) Technological opportunity, regulatory uncertainty, and Bt cotton in Pakistan. *AgBioForum* 18, 98–112.

Tahir, M.S., Latif, A., Bashir, S., Shad, M., Khan, M.A.U. *et al.* (2021) Transformation and evaluation of broad-spectrum insect and weedicide resistant genes in *Gossypium arboreum* (desi cotton). *GM Crops and Food* 12, 292–302.

Tariq, M.I. (2005) Leaching and degradation of cotton pesticides on different soil series of cotton growing areas of Punjab, Pakistan in Lysimeters. PhD thesis, University of the Punjab, Lahore, Pakistan.

Trehan, N.K. (1944) Further notes on the bionomics of *Bemisia gossypiperda* M. & L., the white fly of cotton in Punjab. *Indian Journal of Agricultural Sciences* 14, 53–63.

Ullah, R., Akhtar, K.P., Moffett, P., Akbar, F., Hassan, I., Mansoor, S., Briddon, R.W., Hassan, H.M., and Saeed, M. (2017) Examination by grafting of the extreme resistance to cotton leaf curl disease in two wild *Gossypium hirsutum* L. cultivars. *The Journal of Plant Pathology* 99.

Venugopalan, M.V., Baig, K.S. and Chinchane, V.N. (2021) The era of long linted *G. arboreum* cotton has begun. *Cotton Innovations* 1, 4.

Webinar (2020) Fourth national seminar on pink bollworm management. 16 November, organized by Central Cotton Research Institute, Multan; and Global impact of biotech crops: economic and environmental effects, 1996–2018: focus on Pakistan, 17 November, organized by Pakistani Academy of Sciences, Pakistan.

Win, T. (2008) *Cotton Research in Myanmar – An Overview*. 4th Meeting of the Asian Cotton Research and Development Network. International Cotton Advisory Committee (ICAC), Myanmar.

Wu, K.M. and Guo, Y.Y. (2005) The evolution of cotton pest management practices in China. *Annual Review of Entomology* 50, 31–52.

Wu, Y., Guo, Z., Li, X.J., Yu, G.W. and Yan, L. (2015) Predatory function of *Philodromus cespitum* on soybean aphids. *Soybean Science* 34, 470–473.

Yazdani, M.S., Suhail, A., Razaq, M. and Khan, H.A. (2000) Comparative efficacy of some insecticides against cotton jassid, *Amrasca devastans* (Dist.) and their effect on non-target insects in cotton. *International Journal of Agriculture and Biology* 2, 18–20.

Younas, M. (1973) Threshold of economic injury of insects and mite pests of cotton. In: Proceedings of Seminar on Insect Pests of Cotton, 28–29 April. Cotton Research Institute, Multan. Organized by Pakistan Central Cotton Committee, Karachi, pp. 61–66.

Zain-ul-Abdin, Arif, M.J., Gogi, M.D., Arshad, M., Hussain, F. *et al.* (2012) Biological characteristics and host stage preference of mealybug parasitoid *Aenasius bambawalei* Hayat (Hymenoptera: Encyrtidae). *Pakistan Entomologist* 34, 47–50.

Zain-ul–Abdin Hussain, F., Khan, M.A., Abbas, S. K., A. Manzoor, A. *et al.* (2013) Reproductive fitness of mealybug parasitoid, *Aenasius bambawalei* Hayat (Hymenoptera: Encyrtidae). *Pakistan Journal of Agricultural Sciences* 26, 1198–1203.

Zia, M.S., Khan, M.J., Qasim, M. and Rehman, A. (2009) Pesticide residue in the food chain and human body inside Pakistan. *Journal of the Chemical Society of Pakistan* 31, 284–291.

Zubair, M., Zaidi, S.S., Shakir S., Farooq, M., Amin, I., Scheffler, J.A., Scheffler, B.E. and Mansoor, S. (2017) Multiple begomoviruses found associated with cotton leaf curl disease in Pakistan in early 1990 are back in cultivated cotton. *Scientific Reports* 7, 680.

5 Growing Cotton in China

Lu Zhaozhi*, Li Xueyue, Zhang Wangfeng, Zheng Juyun, Liang Fei, Yang Desong, Tian Jingshan, Gao Guizhen, Wang Juneduo and Abid Ali

China is the second largest producer and consumer of cotton in the world and an important cotton-trading country. Cotton is one of the main cash crops, and also an important raw material for its textile industry, with comparative advantages and development potential. Cotton production is characterized by multi-labour input and high input–output, contributing to rural employment and alleviating rural poverty. Cotton was grown in the Yellow River and Yangtze River valleys, but the planting area was reduced gradually because of environmental factors, such as climate and topography. In contrast, the area of cotton grown in Xinjiang (northwestern China) has increased and now accounts for more than 80% of the country's total production (Fig. 5.1).

The area of cotton increased from 4.8 to 5.3 million ha (2008), but as inputs of pesticides and fertilizer cost more, so farmers grew less cotton, especially in the Yangtze valley, as they switched to other crops, such as maize, to increase their income. In 2011, China began to implement the fixed price for domestic cotton open purchase and storage system, which stabilized the area of cotton and which gradually increased with an overall improvement in planting efficiency.

Cotton output increased from 2167 to 7232 million tons between 1978 and 2008. However, after the implementation of the subsidy and price-acquisition policy for the cotton-cultivation industry, cotton output gradually stabilized. The yield per unit area of cotton has been increasing steadily since 1978 from 445 to 1819 kg/ha, quadrupling the overall yield alongside improved cultivated land (Fig. 5.2). After China entered the WTO, the import and export trade of cotton was seriously impacted by the import of a large quantity of low-priced cotton, which led to the great fluctuation of the domestic cotton price and affected the normal development of the cotton industry and cotton farmers.

In recent years, the import value of China's cotton trade has increased rapidly, while the export has been shrinking, and the deficit of cotton trade has been expanding. Since 2000, China's cotton import has been about 50,000 tons, and then it kept surging, reaching a peak value of 5.13 million tons in 2012. Imports then fell to 1.57 million tons by 2018. Cotton exports fell from 292,475 million tons in 2000 to 47,349 tons in 2018.

Introduction and Genetic Improvement of Bt Transgenic Cotton Varieties

The outbreak of cotton bollworm in the Yellow River Basin in 1992–1993 was extremely serious,

*Corresponding author: zhaozhi_lv@sina.com

(eds G.A. Matthews and T. Miller)

DOI: 10.1079/9781800620216.0005

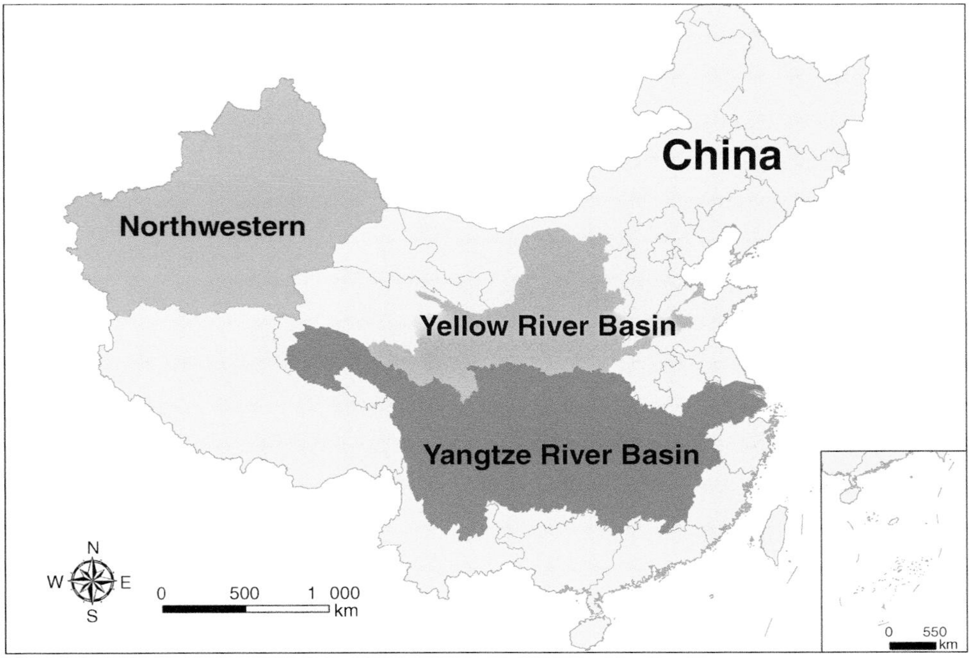

Fig. 5.1. The three cotton-growing areas in China. (Map courtesy of Lu Zhaozhi)

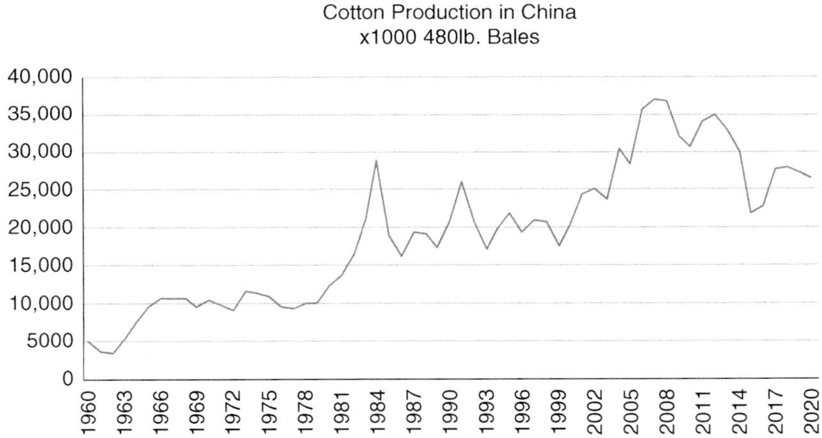

Fig. 5.2. Production of cotton in China. (Graph courtesy of Lu Zhaozhi)

and the economic loss caused by cotton bollworm every year amounted to US$ 2 billion (Wu *et al.*, 2008). Cotton insect resistance has become one of the main directions of breeding new varieties.

Since 1998, China has approved the first batch of domestic transgenic cotton varieties with monovalent insect-resistant gene, such as GK12, Jinmian 26 and GKZ1 (National Anti-hybrid No. 1). In 1999, China approved the first national insect-resistant cotton variety, Cotton Institute 38, which has set out the steps of cultivating and promoting transgenic cotton varieties in China. By 2013, China had

Table 5.1. Evolution of cotton varieties in different regions of China. (Table courtesy of Zhaozhi Lu)

Year	Cotton region	Variety	Comments
1978–1980	Yellow River and Yangtze River basins	CCRI NO.10	The growth period of cotton was shortened from 140 days to 110 days, which solved the problem of long growth period at that time.
	Xinjiang	Xinluzao NO.1	Planted in the early-maturing cotton area of northern Xinjiang for nearly 20 years
1980–2000	Yellow River and Yangtze River basins	CCRI NO.16, Lumian NO.10, Yumian NO.7, Yumian NO.9, Wanmian NO.6, Emian NO.13	Preliminary realization of early maturity, high-yield, high-quality, disease resistance and many other characteristics of the aggregation
	Xinjiang	Xinluzao NO.5, Xinluzao NO.6, Xinluzao NO.7, Xinluzao NO.8	Improving cotton yield and quality
2000–2005	Yellow River and Yangtze River basins	Lumianyan NO.19, CCRI NO.50, CCRI NO.58, Lumianyan NO.35, CCRI NO.73, CCRI NO.74	Early-maturing, high-yield, transgenic insect-resistant cotton
	Xinjiang	Xinluzao NO.25 to Xinluzao NO.49	Early maturity, high-yield and close planting tolerance
2008–2018	Yellow River and Yangtze River basins	CCRI NO.68, CCRI NO.73, CCRI NO.74, CCRI NO.84, Zhongxiaza NO.06, CCRI NO.88, CCRI NO.92, CCRI NO.94A361, CCRI NO.94A213, CCRI NO.94A915, CCRI NO.103, CCRI NO.94A1822, CCRI NO.61, Yuzaomian NO.9110	Transgenic insect-resistant, high-quality, early-maturing
	Xinjiang	Xinluzao NO.52, Xinluzao NO.55, Xinluzao NO.59, Xinluzao NO.64, Xinluzao NO.65, Xinluzao NO.66, Xinluzao NO.67, Xinluzao NO.68, Xinluzao NO.69, etc.	The growth period is about 120 days, and the fibre quality is above Shuang 30.
2019	Yellow River and Yangtze River basins	CCRI NO.425	The growth period is 98 days, and the fibre strength is above 31cnN/ tex

134 nationally approved transgenic insect-resistant cotton varieties, 96 approved in the Yellow River Basin, 36 approved in the Yangtze River basin, and two approved in Xinjiang. Among them, medium-mature varieties with the largest production demand were the seven early-maturing varieties with a growth period of less than 110 days, and one coloured cotton variety.

Since the popularization and application of transgenic insect-resistant cotton, Shandong Cotton Research Centre has bred 60 varieties, including transgenic insect-resistant hybrid cotton, and conventional insect-resistant cotton varieties Lumianyan NO.15, 16, 17, 21, 22 and 23 (Wang *et al.*, 1999). Lumianyan NO.28, a conventional insect-resistant cotton variety, has realized the collaborative improvement of insect resistance, high yield and stable yield and heterosis management. Approved by the National Crop Variety Approval Committee in 2006, this variety has been designated as the leading variety in the Yellow River Basin of China for ten consecutive years.

Cotton Insect Pests in China

China has long history in cotton production in various geographical areas with distinctive climatic conditions. More than 23 species of pests have been reported (Table 5.3), but some are invasive species, such as *Pectinophora gossypiella* that originated from multiple sources (Liu *et al.*, 2009). Whitefly (*Bemisia tabaci*) is a typical invasive pest which occurred in all of cotton-production areas (Zhang *et al.*, 2019). The increasing area of crops protected in greenhouses boosted its spread, particularly in northern China where the protection areas provide a refuge for overwinter of whitefly. Several other insects are occasionally pests in cotton, namely the beet armyworm (*Spodoptera exigua*), tobacco cutworm (*Spodoptera litura*), and spotted cotton bollworm (*Earias cupreoviridis*, *Earias insulana* and *Earias fabia*), particularly in southern China. The cotton semi-looper (*Anomis flava*) is occasionally a local pest in the Yangtze and Yellow River valleys.

A national programme for monitoring, supported by the Ministry of Agriculture since the 1950s, provided a productive understanding of the status of pest population dynamics with data on crop phenology and management, weather and natural enemies. The automatic monitoring equipment provided an effective cotton information system to local government, industries and farmers through the internet and various media and helped in the development of computer modelling.

Table 5.2. Transgenic Bt cotton varieties and their characteristics. (Table courtesy of Zhaozhi Lu)

Period	Variety	Characteristics
1998	GK12, Jinmian NO.26, GKZ1	The first batch of transgenic cotton varieties with univalent insect resistance gene
1999	CCRI NO.38	The first national insect-resistant cotton variety
2000	Lumianyan NO.15, CCRI NO.29, Xinmian NO.33b, etc.	467,000 ha, accounting for 90% of the cotton area in Shandong province
2006–2008	Yuza NO.35	For three consecutive years, it has planted more than 1 million mu in Henan province.

Figs 5.3 and 5.4. Farmer preparing to spray and then using a knapsack sprayer to apply an insecticide. (Photos: Graham Matthews)

Figs 5.3 and 5.4. Continued.

Table 5.3. The main pests and their status in cotton-production areas. (Table courtesy of Zhaozhi Lu)

Insect	Distribution	Damage	Comments
Helicoverpa armigera	All areas, except extremely cold	Buds and bolls damaged where 3–6 generations occur	Low incidence pre-1980, then increased and more resistant to insecticides, followed by decline with Bt cotton after 1997
Pectinophora gossypiella	Mostly in Yangtze River valley, fewer in Yellow River valley	3–4 generations in most areas in Yangtze River	Serious in the mid-1980s, and declined after the commercial release of Bt-cotton
Aphis gossypii	All areas	20+ generations in one season. Two peaks – spring and late summer	Serious, so chemical control is required
Bemisia tabaci	All areas	10+ generations	Usually present July–August – survive only in protected crops in winter
Thrips tabaci	All serious in Xinjiang	6–10 generations	Serious 1970–80, but controlled dressing seed
Lygus pratensis	Mostly in north-west	3–4 generations	Serious in 1980s
Adelphocoris fasciaticollis	Yangtze and Yellow River valleys	3 generations in Yellow River valley	After Bt cotton release, the pest is more serious in Yellow River valley

Chemical Control and Pest Resistance to Pesticides

A range of insecticides was recommended to control cotton pests. These were generally applied using locally made knapsack sprayers (Figs 5.3 and 5.4). Recently, there has been an increase in the use of drones (Figs 5.5 and 5.6) to avoid the farmer having a heavy sprayer on his back. Regardless of these application techniques, repeated use of insecticides with the same mode of action has led to the pests becoming resistant to them, as illustrated in Table 5.4.

Table 5.4. The resistance of selected main pests to different chemicals in cotton-production areas of China. (Z, Zhang *et al.*, 2016; A, An *et al.*, 2018; D, Dang *et al.*, 2019; P, Patima *et al.*, 2020; H, Hu *et al.*, 2016)

Cotton areas	Insect	Chemicals	Resistance ratio
Yellow River	*Aphis gossypii*	Imidacloprid	9.6–32.1 (D) 9.60–32.13 (A) 18.3–2207.6 (Z)
		Clothianidin	4.3–6.7 (D) 13.14–74.29 (A) 25.0–30.3 (D)
		Omethoate	20.31–30.32 (A) 149.4–2206.7 (Z)
		Abamectin	7.5–108.7 (D)
		Thiamethoxam	4.34–7.14 (A)
	Helicoverpa armigera	Cypermethrin	11485.1–19633.7 (Z) 24.2–65.7 (H)
		Phoxim	9.4–35.1 (Z)
		Emamectin benzoate	2.7–6.9 (Z) 10.6–92.8 (H)
		Cyhalothrin	11.4–205.1 (Z) 0.3–4.4 (Z)
Yangtze River	*Aphis gossypii*	Avermectin	19.6–220 (P)
		Emamectin benzoate	18.4–130 (P)
		Deltamethrin	728–4153 (P)
		Cypermethrin	353–4932 (P)
		Omethoate	2137–9501 (P)
	Helicoverpa armigera	Emamectin benzoate	0.07 (Z)
		Cyhalothrin	12.0 (Z)
		Phoxim	2.3 (Z)

Matrine-based biopesticides, which have demonstrated acaricidal and insecticidal activity in both field and laboratory studies have been used for the control of *Aphis gossypii* and cotton mirids, as well as the neonicotinoids nitenpyram and thiamethoxam. To control *Helicoverpa armigera* and *Spodoptera exigua*, preference was for the *H. armigera* multiple nuclearpolyhedrosis virus and *Mamestra brassicae* multiple nuclearpolyhedrosis virus, as well as growing Bt cotton. Other insecticides used included emamectin benzoate and cyantraniliprole. Avermectin and other biological insecticides are used to control cotton spider mites.

The Effect on Insect Pests of Growing Bt Cotton

Widespread adoption of Bt-cotton with a single Cry1Ac gene in the Huanghe River valley (HRV) and Yangtze River valley (YRV) resulted in a significant reduction in damage due to *H. armigera*, as sampling indicated the number of eggs was three times lower in 2006 than before Bt cotton was sown and fewer larvae were detected on various other crops in the same area. Pink bollworm was reduced in six provinces in the YRV between 2000 and 2010 (Wan *et al.*, 2012). From 1997 to 2012, reduced use of pesticides saved US$ 8.46 billion and boosted the biodiversity of natural enemies providing bio control of cotton aphid (*Aphis gossypii*) (Lu *et al.*, 2012). However, farmers were spraying insecticides to control secondary pests as a consequence of a pest complex shift phenomenon (Wang and Fok, 2018). The reduced spray frequency in Bt cotton for bollworm control was associated with a significant and widespread increase in mirid pests in mainland China (Lu *et al.*, 2010), as in other cotton-growing countries.

Resistance of *H. armigera* to Bt cotton increased significantly as the frequency of dominant inherited resistance to the Bt protein increased 100-fold from 2006 to 2016 in most cotton areas (Jin *et al.*, 2018), whereas the susceptibility to Bt in *H. armigera* remained high in the north-western region, although no mandatory refuge policy had been promulgated in this cotton-growing area.

Where farmers have returned to non-Bt cotton, the abundance and damage due to *H. armigera* has rebounded. However, the cotton industry has been shifting to north-western China (Xinjiang) (Fig. 5.1) due to an advantageous production environment (land, weather and low costs), although Bt cotton planting is not officially sanctioned in this area.

The matrine, nitenpyram and thiamethoxam can be selected for the control of *Aphis gossypii* and cotton mirids. For the control of *Helicoverpa armigera* and *Spodoptera exigua*, preference is given to *Helicoverpa armigera* multiple nuclear polyhedrosis virus, *Mamestra brassicae* multiple nuclear polyhedrosis virus, Bt, etc. Chemicals can be emamectin benzoate, cyantraniliprole etc. Avermectin and other biological insecticides are used to control cotton spider mites.

Integrated Pest Management in Cotton-production Areas

A pilot study examined the impact of reducing fertilizers and pesticides in a cotton area in

Figs 5.5 and 5.6. Using a drone to spray cotton. (Photos: Yang Desong)

Figs 5.5 and 5.6. Continued.

China. Novel technologies were developed to reduce the reliance on pesticides and protect the environment. The management tactics and version are changed from field to landscape. More studies have been carried out for understanding the relationship between targeted pests and associated natural enemies in landscape in cotton areas in China recently. Trap cropping was used to attract the natural enemies, and agents of baited food were employed to control *H. armigera*. However, the distribution of pest insects and time of presence are complex in cotton fields, so the implementation of IPM needs to be considered on an area-wide basis and not individual small farms. The target of pest management in different cotton areas is also distinctive, due to weather and other factors. In view of the main diseases and insect pests in different growth stages of cotton, the measures of 'prevention first and comprehensive control' were adopted to prevent the diseases and insect pests at the seedling stage, to control the pests during the growth period, and to protect the boll and yield at the boll stage.

The use of resistant (tolerant) varieties, agronomic and biological control technologies should be the priority in achieving the sustainable and safe control in cotton fields. Additionally, the high-quality and high-yield cotton varieties were selected in cotton area, particularly the Yellow River and Yangtze River valleys. Chemical insecticides were not used to control aphids during the cotton seedling to bud stage in the Yangtze River Basin. Low-toxicity and environmentally friendly chemicals should be selected rationally and correctly (Zhang and Chen, 1991).

In north-west China, the application of ecological control technology in the cotton area has been adopted with alfalfa and other crops planted in surrounding fields and under the forest belt to encourage survival of natural enemies to control cotton aphids, bollworms and cotton spider mites (Lin *et al.*, 2010). Planting corn and *Abutilon theophrasti* in strips is to lure bollworm and kill them in the area where bollworm is frequently occurring, before they enter cotton fields. Areas with irrigation are kept moist in autumn and in the winter so as to reduce the overwintering of certain diseases and insect pests. IPM needs to be improved to reduce the severity of sucking pests in areas where Bt cotton is grown to limit the damage caused early in the season.

Cotton Diseases and Their Management (Type of Disease Factors according to Regions, Washing Tests)

Main diseases of cotton in China

Verticillium wilt and fusarium wilt are the main diseases on cotton and are widespread, causing severe damage since the 1930s, especially in certain years (Tables 5.5 and 5.6). With the development of Bt-cotton, those diseases are more serious due to lack of varieties resistant to pathogens. Verticillium wilt has been more than four times as severe as fusarium wilt in Xinjiang in 2015–2020.

More than 25 cotton diseases were recorded in cotton area, but except for verticillium wilt and fusarium wilt, their occurrence and damage declined with improved management and treating cottonseeds. The key diseases are in Table 5.6.

Table 5.5. The disease index of cotton main disease in Henan province (1978–2007) (Tang *et al.*, 2011).

Year	Fusarium wilt	Verticillium wilt
1978–1984	0.1	35.1
1985–1988	4.9	10.5
1989–1993	2.4	2.1
1994–1998	2.6	18.2
1999–2003	2.7	19.3
2004–2005	7.5	25.2
2006–2007	6.7	28.2
Average	4.5	17.2

Management of cotton diseases in China

Germplasm with resistance to the two main diseases is rare, so further research and national and international collaboration is needed to improve variety resistance. Agronomic rotations are appropriate approaches for suppressing the occurrence of fusarium wilt and verticillium wilt, examples being wheat–cotton and rice–cotton. The rice–cotton rotation pattern is widely used in northern and southern China and has been highly effective for suppressing these diseases. Furthermore, with more than 70% of cotton in Xinjiang with drip irrigation, this approach can help to suppress the occurrence of both diseases.

The technology of the dressing of cottonseed has been extended since the 1980s, and diseases and pests (such as aphid and thrips) in the cotton seedling period have declined.

Weeds and Weed Management

As cotton is grown in different parts of the country, the distribution and community structure of weeds in cotton fields varies with the ecological environment. Weeds are a factor affecting yields 2–3 times during a season (Wang *et al.*, 2015) Initially, there is a peak period of emergence, mainly in seedling stage and then when there are flower buds, and lastly weeds need to be controlled prior to harvest.

Table 5.6. The main diseases in China cotton production areas. (Table courtesy of Yang Desong)

Common name	Latin name of pathogen	Distribution and physiological strain of pathogens	The state of disease
Verticillium wilt	*Verticillium dahliae*	All cotton-production areas in China with three physiological types: I: strongest II: weak III: moderate	Widespread in all cotton areas. More serious with outbreaks in 1997, 2003, 2005 and 2009 in the Huanghe River valley and Yangtze River valley
Fusarium wilt	*Fusarium oxysporum vasinfectum*	All cotton-production areas. There are three physiological strains: 3 = light, in north-west 7 = in all areas 8 = very light, in south	Serious In 1950–1970, with tremendous loss in yield and economy. In 1980–2000, effective control was obtained with resistant varieties. After Bt cotton release in 1997, more serious but effective control was obtained with varieties more resistant to fusarium wilt.
Boll blight	*Phytophthora boehmeriae*	in Yangtze River valley and Yellow River valley, occasionally in Xinjiang	More serious in Yangtze River valley and Yellow River valley. In the boll stage in those areas, the percentage of rotted bolls is more than 20% due to heavy rainfall.
Seedling damping-off disease	*Pythium aphanidermatum*	All cotton-production areas	Reported in all of cotton areas, and seriously occurs in wet and rainy areas and seasons in cotton seedling stage.
Anthracnose	*Glomerella gossypii. Coll-etotrichum indicum*	All cotton-production areas	Yield loss 10–30% before 1990s. Now the damage of this disease has declined because of the technology of seed coating.
Rhizoctoniosis	*Rhizoctonia solani*	All cotton-production areas	The percentage of this disease has declined because of cottonseed dressing.

Table 5.7. The occurrence regularity and distribution of weeds in cotton fields in the main cotton areas (Ma *et al*., 2010, 2012).

Cotton areas	Main weeds	Occurrence regularity of weeds
Yangtze River valley	*Malachium aquaticum* (L.) *Leptochloa chinensis* (L.) Nees and *Alternanthera philoxeroides* (Mart.) Griseb; *Digitaria sanguinalis* (L.) Scop., *Echinochloa crusgalli* (L.), *Cynodon dactylon* (L.) Pers., *Portulaca oleracea* (L.)	There are three peaks of weed occurrence, the first peak is in mid-May, the second peak in cotton fields occurs in late June, and the third peak occurs in late July, with a small number of weeds.
Yellow River valley in northern China	The main weeds are drought-tolerant, including *Digitaria sanguinalis* (L.), *Malus japonicus, Alternanthera philoxeroides, Amaranthus retroflexus* and *Amaranthus fovea*	There are two peaks of weeds in cotton fields. The first peak appears in mid–late May. The second peak occurs in July.
North-western China	Weed community composed of drought-tolerant and salt-tolerant weeds such as *Solanum nigrum, Chenopodium glaucum* (L.), *Amaranthus retroflexus* (L.)	There are two peaks of weeds in this cotton area, the first occurs in late May, and the second peak occurs from early July to early August.

The aim is to apply integrated management of weeds in cotton fields, but chemical control is more practical, and manual weeding is used in smallholder farming.

Agricultural weeding control

Adopting scientific crop rotation and a stubble-turning system can change the ecological environment, which is conducive to controlling weeds, avoiding the spread of weeds associated with various crops and, obviously, reducing weeds. In the process of crop stubble turning, herbicides were naturally changed, thus avoiding single applications of similar herbicides within a long time. For example, the rotation of wheat and cotton can effectively reduce the occurrence of *Solanum nigrum*. The occurrence of weeds can be reduced by 50% by implementing rice–cotton alternation. Therefore, crop rotation is an effective way to eliminate weeds in cotton fields. Soil management, such as burying and ploughing, also can be used to eliminate the sprouts and plants of weeds in different degrees, cut off the vegetative propagation organs of perennial weeds and reduce the seed bank.

Physical control

Physical control of weeds refers to the method of weed control with physical measures or forces, such as mechanical or manual, which results in the injury or death of individual weeds. Weeds around cotton fields, such as weed seeds in canals, are one of the main sources of weeds in the field. These seeds are brought into the field by wind, running water, human and animal activities, or spread to the field through underground roots. Weeds around cotton fields must be carefully removed to prevent their spread, especially before the seeds of weeds are mature. The cover with mulch in cotton fields is an economical approach for suppressing the weeds, especially in cotton seedlings (Liu and Liu, 1997). In fact, this cover with mulch is to increase temperature and moisture in northern China. Deep ploughing of soil is an effective measure for weeding. Soil tillage in cotton fields can be roughly divided into deep ploughing before winter, harrowing before sowing and inter-tillage after sowing. Deep ploughing of cotton fields before winter can not only turn the topsoil grass to the lower layer, reducing the germination opportunity of grass, but also cut off and expose the roots of perennial weeds and reduce the number of overwintering weeds. Harvesting before sowing can not only preserve soil moisture before sowing, but also kill weeds that have germinated and emerged. Inter-row tillage weeding effectively kills weeds in the seedling stage and the bud and boll stage. The younger the weed's age, the more suitable the time for inter-tillage, and generally 2–3 times are best for suppressing weeds.

Chemical control of weeds in cotton fields

Spraying is an economic, efficient, timely and labour-saving approach in cotton fields. In recent years, with a greater variety of herbicides, chemical technology has made great progress. Spraying the soil before sowing is predominantly to control the early emergence of weeds in cotton fields, as a plastic film is used to warm the soil and enable earlier sowing. After soil preparation and before sowing, the herbicides applied are 48% trifluralin at 1800–3000 mL/ha, 33% pendimethalin at 2250–3000 mL or 90% acetochlor, mixed in water at 600 litres/ha. Within four hours after spraying, deep raking for 8–10 cm will be conducted. Later spray treatments are not always used in fields.

Chemical defoliation

In order to reduce the foreign matter content of machine-harvested cotton and improve the fibre quality, a chemical defoliant is applied in mechanical harvestings. This has a great impact on improving the quality of machine-harvested cotton by reducing the foreign matter content. Timing defoliation is important, so a spray is recommended after about 30%–40% of bolls have opened to promote good leaf abscission of cotton seven days after application. To obtain more than 55% leaf abscission, the maximum temperature and growing degree days should be more than 27.2°C and 4.6°C per day, respectively.

In general, the defoliant application time is from 6–17 September. One application is sufficient in small cotton fields, but in fields with a high plant population and vigorous growth, a single treatment with a 50% dose may not be adequate as the defoliant is effective only on leaves where defoliant is deposited, so a second spray is applied ten days after the first with 70% of the normal dose, aimed at the middle and lower leaves.

Harvesting and Fibre Quality

Harvesting technology in different regions in China

In the Yangtze River Basin and the Yellow River Basin, cotton is mainly picked by hand, with the advantages of high quality and the disadvantages of high cost and low benefit. The cotton harvesting period is relatively short in the Xinjiang region, where use of machine harvesters has increased in recent years to reduce cotton-production costs. Cotton harvesting is promoted late in Hubei province (Yellow River Basin area), where the machine-harvesting area is only 5–18% of the cotton-planting area (Fig.5.7). The type of harvester has an important influence on fibre quality. Fibre quality is better in spindle-harvested cotton than in stripped cotton. According to the technical standards of cotton pickers, the leaf abscission rate must be >90%, boll opening content >95%, and the foreign matter content <12%. The harvest rate of machine-harvested cotton must be 90–93%, moisture content <12%, and the loss content (falling, hanging, leftover etc.) should be <5–6%. Machine-harvested seed cotton is free of oil pollution and chlorophyll pollution.

In the Xinjiang region, the harvest rate of machine-harvested cotton can be up to 99%, but at high harvest rates, foreign-matter content and fibre damage increases with fibre strength decreasing by up to 11%, and the short-fibre index of the cotton picker increased 51% compared with hand-harvested cotton. The problem is how to co-ordinate the harvest rate and impurity rate to reduce fibre damage during cotton cleaning and machine-harvesting in Xinjiang.

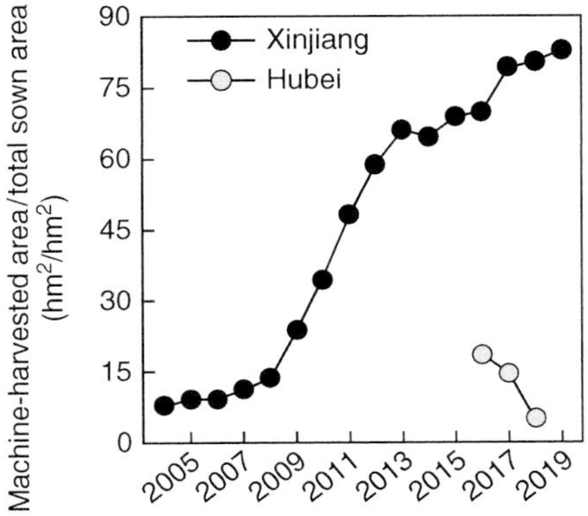

Fig. 5.7. Changes of machine-harvested area on the Xinjiang (north-west inland) and Hubei (Yellow River Basin).

Basic Situation of Germplasm Resources

In the 21st century, China's cotton industry has been confronted with problems facing farmers and the requirements of the textile industry. There has been a major genetic improvement with innovation technology leading to the selection of more than 200 types of cotton varieties, with early maturity, high fibre quality and high yields. Good progress has been made in the demonstration and popularization of these new varieties and new technologies to achieve high yields.

Distribution, Production Potential and Quality of Varieties and Resources

In 1980, the Cotton Research Institute of the Chinese Academy of Agricultural Sciences successfully selected the cotton variety CCRI 10, which shortened the plant growth period from 40 days to 110 days. Other early-maturing varieties were bred by crossing CCRI 10 with other varieties, and CCRI 14, CCRI 16, Yumian 5 and Lumian 10 have been widely used in production. In the 1990s, the approved early-maturing cotton varieties had the same average growth period as CCRI 10, and their lint percentage, fibre quality and disease resistance were higher than those of CCRI 10.

Fig. 5.8. Harvested cotton. (Photo: Zheng Juyun.)

Fig. 5.9. Machine-harvesting. (Photo: Li Xueyuan.)

Figs 5.10 and 5.11. Manual harvesting. (Photos: Wang Juneduo and Zheng Juyun).

Since 2000, early maturing, high yield and insect resistance have been the main objectives of new varieties grown in the Yellow and Yangtze River basins. Most of the varieties have been insect-resistant transgenic cotton, such as Lumianyan 19, CCRI 50 and CCRI 58, with direct seeding after wheat promoted in some areas. Good results were achieved and in 2008–18, a total of 13 new transgenic varieties, including CCRI 68, CCRI 73 and CCRI 74, were bred. Breeding units have successively cultivated high-quality varieties with growth period less than 100 days, such as Jinke 707, which was approved in 2016 with a growth period of 98 days and improved fibre quality.

In the Xinjiang cotton region, as early as 1978, the first early-maturing cotton variety, Xinluzao No. 1, was approved, and increased the yield by more than 10% compared with the control varieties at that time. It has been widely planted in northern Xinjiang for nearly 20 years. In the 1990s, the selected cultivars of Xinluzao No. 5, No. 6, No. 7 and No. 8 played a decisive role in improving the yield per unit area and quality of cotton and replaced many cotton varieties in northern Xinjiang. Since 2005, with the great increase of cotton-planting area in Xinjiang, the breeding unit has selected and bred early-maturing upland cotton varieties from Xinluzao 25 to Xinluzao 49; and Xinluzao 33, aiming for close planting tolerance suitable machine picking is one of the main cotton varieties promoted in northern Xinjiang. Xinluzao 42 has high yield and anti-blight characteristics and is suitable for planting early cotton in northern and southern Xinjiang.

High-yield and High-quality Wilt-resistant Cotton Varieties

In the 1980s, verticillium wilt spread became more and more serious and hindered the development of cotton production by causing 10–20% yield loss and failures due to varieties that were not disease-resistant. Breeding wilt-resistant varieties increased, and by 1992 the number of disease-resistant high-yield varieties had expanded to more than 600,000 ha.

Introduction and Genetic Improvement of Bt Transgenic Cotton Varieties

Following the extremely serious outbreak of cotton bollworm in the Yellow River Basin in 1992/93, insect resistance has become one of the main directions of breeding new varieties.

The first batch of domestic transgenic cotton varieties with monovalent insect-resistant gene, such as GK12, Jinmian 26 and GKZ1 (National Anti-hybrid No. 1), were developed, and in 1999, China approved the first national insect-resistant cotton variety, Cotton Institute 38, which has led to cultivating and promoting transgenic cotton varieties in China. By 2013, China had approved 134 transgenic insect-resistant cotton varieties, 96 approved in the Yellow River Basin, 36 approved in the Yangtze River Basin, and 2 approved in Xinjiang. Among them, medium-mature varieties with the largest production demand were the main ones, seven early-maturing varieties with a growth period of less than 110 days, and one coloured cotton variety.

In 2000, 467,000 ha of insect-resistant cotton varieties were planted, mainly Lumianyan 15, CCRI NO. 29 and Xinmian 33B, accounting for 90% of the cotton-planting area in Shandong province. This basically eliminated the conventional non-insect-resistant cotton varieties. Since the popularization and application of transgenic insect-resistant cotton, 60 varieties including transgenic insect-resistant hybrid cotton and conventional insect-resistant flower Lumian No. 15, 16, 17, 21, 22 and 23 have been co-bred by Shandong Cotton Research Centre.

Lumian Yan 28, a conventional insect-resistant cotton variety bred by Shandong Cotton Research Centre, has achieved the coordinated improvement of insect resistance, high yield, stable yield and different management, and was approved by the National Crop Variety Certification Committee in 2006. It has been rated as the leading variety in the Yellow River Basin for ten consecutive years from 2007 to 2016. It is the domestic insect-resistant cotton variety with the largest cumulative promotion area and plays an irreplaceable role in China's cotton production.

Achievements in the Cultivation of New Long-staple Cotton Varieties

Over the past 40 years, 60 new long-staple cotton varieties of New Sea cotton series have been developed and sown in Xinjiang, providing variety guarantee for Xinjiang's cotton production in different periods. The per-mu yield of cotton has increased from about 60 kg in the late 1970s to more than 120 kg now.

Cotton region in the Yellow River Basin

In this area, cotton generally has a growing period of only 180 days, and the upland cotton cultivar which matures early is suitable for cultivation. It is the largest cotton-producing area in China and also an important commodity cotton base. Heat in the area is sufficient, the frost-free period is comfortable and sunshine is well recorded in the Yangtze River Basin. However, in early summer, Fuyu is relatively concentrated, and precipitation variability is great. It is not easy to solve this problem in production. In addition, the temporal and spatial distribution of meteorological elements in the cotton region of the Yellow River Basin is uneven, the stability of precipitation is poor, and natural disasters such as early rainfall, waterlogging, wind, freezing, and so on, occur. All these factors will have adverse effects on the yield and quality of cotton.

Cotton area in the Yangtze River Basin

In a subtropical humid climate zone, the frost-free period is 240–300 days, there is a 4–10-month average temperature of 21–24°C, more than 10°C accumulated temperature of 4600–6000°C, annual sunshine hours of 1700–2400 hours and annual precipitation of 800–1600 mm. Light condition is poor, there is often waterlogging, flooding, drought and typhoon weather. The soil in the plain area is mainly tidal soil and paddy soil with good fertility. Hilly cotton fields are mostly red soil and yellow-brown soil with poor fertility. There are large areas of saline-alkali soil along the coast. More than 90% of the cultivation system in this area implements two crops of grain (oil) and cotton in a year, and mainly adopts seedling transplanting cotton. The layout of cotton fields is concentrated, the yield per unit area is high, the uniformity of cotton fibre is high, the maturity is good but strong, the micronaire value is not enough, the textile industry is developed, and the transportation cost is low.

North-west cotton region

This area mainly includes the Hexi Corridor area of Xinjiang Uygur Autonomous Region and Gansu province. Compared with other cotton areas, the cotton quality produced in this area is better, the cotton grade is high and the quality is good. It is an important export cotton base, promoting development of the cotton industry. There is a dry climate, sufficient sunlight and a rich quantity of heat, which are very beneficial to form a high-quality cotton with rich lustre. While the land has little rain, the snow mountain area is large and provides a stable rich water resource for irrigation. Furthermore, the four seasons with a dry and cold winter climate reduce the cotton pest species and with a single planting structure the bollworm population has a limited time when plants provide habitat. As China's largest cotton-producing area and the largest high-quality cotton-production base (and the only long-staple cotton-production area in China), cotton production has unique climatic conditions.

Based on 38 years of meteorological data, regarding temperature, frost period and the average temperature in July as a zoning index, the cotton region of Xinjiang was divided into four sub-regions, namely, the eastern, southern, northern and special regions.

Cotton-growing Methods Applied in North-western China

Sowing

Seeds can be sown at 5 cm depth three days after the soil temperature is between 9.5 and 17.6°C, usually from 5–20 April, in Xinjiang region, following the laying of a pipe to

Fig. 5.12. Cotton mechanical seeding. (Photo: Zhang Wangfeng)

provide drip irrigation and plastic film mulch, in which holes are perforated to sow 1–2 seeds per hole, and subsequently covered in a semi-precision process (Fig. 5.12).

Row spacing

A 'wide and narrow' row spacing configuration in Xinjiang has one film mulch with three cotton rows spaced 76 cm–76 cm or six rows spaced 66 cm–10 cm–66 cm–10 cm. Within a row, plants are 9–12 cm apart in a zigzag pattern so theoretical density is generally 22.0–22.8 million plants/ha. 2.05 m of ultra-wide sheets of plastic film are used and drip-irrigation capillaries are arranged above the ground with two or three tubes to each film depending on the local situation.

Irrigation and fertilization of cotton

All cotton in the region is irrigated, as shown in Table 5.8, with fertilizer information. Drip-irrigation is the key approach with a pipe laid parallel to the direction of cotton planting and in the middle of the narrow inter-row. The length of the thin pipe arrangement is generally 50–100 m based on the size of field.

The maximum evapotranspiration of cotton is during flowering and the boll stage, in which the water consumption is 240–543.2 mm during the period from full flowering to boll-opening stage, while the maximum water consumption period from bud to early-flowering stages ranged from an average daily water consumption of 3.29 ~ 4.15 mm, initially, 16.5 ~ 20.8 mm in 5 days, and 33.0 ~ 41.6 mm in 10 days (Li and Zhao, 2014; Han *et al.*, 2015).

To produce 1000 kg of lint cotton, about 133.5 kg of nitrogen (N), 46.5 kg of phosphorus (P_2O_5) and 133.5 kg of potassium (K_2O) need to be absorbed. For seed cotton, about 50 kg of nitrogen (N), 18 kg of phosphorus (P_2O_5) and 40 g of potassium (K_2O) need to be absorbed (Tian *et al.*, 2019). According to the team test of drip irrigation, nitrogen uptake at the flowering stage accounted for 54% of the total uptake at the whole-growth stage, phosphorus uptake at the

Fig. 5.13. (a) 12 Row-spacing configuration of machine-harvested cotton in Xinjiang region; (b) wider row spacing. (Courtesy of Zhang Wangfeng)

flowering and boll stage accounted for 75% of the total uptake at the whole-growth stage, and potassium uptake accounted for 76% (Wang and Fok 2018). The machine serviced fertilizer with drip irrigation in field (see Fig. 5.14).

Chemical Regulation of Plant Growth

Hand topping and chemical topping is required to achieve three controls: fruit number

Table 5.8. Comparison of irrigation and fertilization in different cotton-producing areas in Xinjiang. (Table courtesy of Tian Jingshan)

Sub-region	East Xinjiang	South Xinjiang	North Xinjiang
Irrigation method	Drip irrigation	Flood irrigation in winter and drip irrigation in summer	Drip irrigation
Irrigation at sowing time	Dry seeding followed by irrigation	Seedling emergence with soil moisture	Dry seeding followed by irrigation
Number of irrigations	8–11	8–10	7–9
Irrigation amount (m^3/hm^2)	5250–6750	7500–9000	4500–5500
Fertilization applied by drip irrigation	Main fertilizer, plus small amount of phosphorus and potassium-based fertilizer	Phosphorus and potassium fertilizer are the main base fertilizer, and nitrogen fertilizer	Fertilizer and base fertilizer is applied in autumn in some farmland
Fertilizer application (kg/hm^2)	1200–1350	1500–1800	1050–1300

Fig. 5.14. Two common fertilizer units. (Photos: Lu Zhaozhi)

Table 5.9. The application of chemical topping in Xinjiang region. (Table courtesy of Zhaozhi Lu)

	The first time		The second time	
Chemical topping	Date	Amount (g/hm^2)	Date	Amount (g/hm^2)
Flumetra in chemical topping	20 –25 June	1500	5–10 July	2250
Chlormequat chloride	The same as hand topping	900		

of 7–9 per plant, plant height of 65–70 cm and topping time terminated strictly. After artificial topping, DPC must be sprayed with chlormequat chloride (120 g/hm^2). The cotton field with prosperous growth can be sprayed with 90–120 g of chlormequat chloride 7 days after the first time. Spraying a plant-growth regulator, such as flumetra (see Table 5.9) refers to inhibition of the plant growth to achieve the same technical measures as hand topping technique to regulate vegetative growth and reproductive growth.

References

An, J.J., Dang, Z.H., Gao, Z.L. *et al.* (2018) Sensitivity baseline and resistance of Hebei *Aphis gossypii* Glover population to neonicotinoid insecticides. *Journal of Hebei Agricultural University* 41, 112–116.

Dang, Z.H., An J.J., Gao, Z.L. *et al.* (2019) Fitness and resistance of different *Aphis gossypii* populations in Hebei to six pesticides. *Plant Protection* 45, 111–114.

Han, M., Zhao, C.Y., Jirka, S. and Gary, F. (2015) Evaluating the impact of groundwater on cotton growth and root zone water balance using Hydrus-1D coupled with a crop growth model [J]. *Agricultural Water Management* 160, 64–75.

Hu, H.Y., Ren, X.L., Ma, X.Y. *et al.* (2016) Monitoring of bollworm resistance to three insecticides in the cotton region of the yellow river basin. *Journal of Huazhong Agricultural University* 37, 65–69.

Jin, L., Wang, J., Guan, F., Zhang, J.P., Yu, S. *et al.* (2018) Dominant point mutation in a tetraspanin gene associated with field-evolved resistance of cotton bollworm to transgenic Bt cotton [J]. *Proceedings of the National Academy of Sciences of the United States of America* 115, 11760–11765.

Li, Y. and Zhao, M. (2014) Projections of water requirements of cotton and sugar beet in Xinjiang based on statistical downscaling model. *Transactions of the Chinese Society of Agricultural Engineering* 22, 70–79.

Lin, R., Liang, H., Zhang, R., Tian, C. and Ma, Y. (2010) Impact of alfalfa/cotton intercropping and management on some aphid predators in China [J]. *Journal of Applied Entomology* 127, 33–36.

Liu, S. and Liu, D. (1997) Chemical weed control effect and benefit of plastic mulching cotton. *Journal of Shaanxi Agricultural Science* 5, 26–27.

Liu, Y.D., Wu, K.M. and Gou, Y.Y. (2009) Population structure and introduction history of the pink bollworm, *Pectinophora gossypiella*, in China. *Entomologia Experimentalis et Applicata* 130, 160–172.

Lu, Y.H., Wu, K.M., Jiang, Y.Y., Xia, B., Li, P. *et al.* (2010) Mirid bug outbreaks in multiple crops correlated with wide-scale adoption of Bt cotton in China. *Science* 328, 1151–1154.

Lu, Y.H., Wu, K.M., Jiang, Y.Y., Guo, Y.Y. and Desneux, N. (2012) Widespread adoption of Bt cotton and insecticide decrease promotes biocontrol services. *Nature* 487, 362–365.

Ma, X.Y., Yan, M.A., Peng, J., Xi, J.P., Ma, Y.J. *et al.* (2010) Current situation and developing tendency of the weed researches in cotton field of China [J]. *Cotton Science* 22, 372–380. [In Chinese]

Ma, X., Ma, Y., Xi, J., Jiang, W., Ma, Y. and Li, X. (2012) Mixed weeds and competition with directly seeded cotton, north Henan Province. *Cotton Science* 24, 91–96.

Patima, W., Jiang, Y. and Ma, D. (2020) Effects of exposure to imidacloprid direct and poisoned cotton aphids *Aphis gossypii* on ladybird *Hippodamia variegata* feeding behaviour. *Journal of Pesticide Science* 45(1).

Sui, L.L., Tian, J.S., Yao, H.S., Zhang, P.P., Liang, F.B. *et al.* (2018) Effects of different sowing dates on emergence rates and seedling growth of cotton under mulched drip irrigation in Xinjiang. *Scientia Agricultura Sinica* 51, 4040–4051. [In Chinese]

Tang, Z.J. Wu, L. and Xie, D.Y. (2011) The research of genetic improvement effectiveness on cotton varieties of Henan Province from 1978 to 2007: improvement of disease resistance and agronomic traits. *Journal of Henan Agricultural Sciences* 40, 57–60.

Tian, L., Tian, C., Cui, J. and Lin, T. (2019) Analysis on green development strategy of Xinjiang cotton. *Meteorological and Environmental Research* 10, 19–25.

Wan, P., Huang, Y., Tabashnik, B.E., Huang, M. and Wu, K. (2012) The halo effect: suppression of pink bollworm on non-Bt cotton by Bt cotton in China. *PLOS ONE* 7.

Wang, F., Ma, Q., Li, H.B., Akedan, W., Xu, Y. *et al.* (1999) Study on insect resistance and utilization of transgenic cotton varieties in Xinjiang. *Xinjiang Agricultural Science* 4, 34–36. [In Chinese]

Wang, G. and Fok, M. (2018) Managing pests after 15 years of Bt cotton: farmers' practices, performance and opinions in northern China. *Crop Protection* 110, 251–260.

Wang, X., Ma, X., Jiang, W., Ren, X., Ma, Y. and Ma, Y. (2015) Competition between weeds and cotton. *Cotton Science* 27, 474–480.

Wu, K.M., Lu, Y.H., Feng, H.Q., Jiang, Y.Y. and Zhao, J.Z. (2008) Suppression of cotton bollworm in multiple crops in China in areas with Bt toxin-containing cotton [J]. *China Basic Science* 21, 1676–1678. [In Chinese]

Zhang, X.Q. and Chen, P.R. (1991) Spring populations of *Aphis gossypii* (Homoptera: Aphididae) in cotton fields: to spray or not to spray? *Agriculture, Ecosystems & Environment [J],* 35, 349–351.

Zhang, S., Yan, Ma, Y., Min, H., Yu, X.-Q., Li, N., Rui, C.-H. and Gao, X.-W. (2016) Insecticide resistance monitoring and management demonstration of major insect pests in the main cotton-growing areas of northern China. *Acta Entomologica Sinica* 59, 1238–1245.

Zhang, X.M., Lovei, G.L., Ferrante, M., Yang, N.W. and Wan, F.H. (2019) The potential of trap and barrier cropping to decrease densities of the whitefly *Bemisia tabaci* MED on cotton in China. *Pest Management Science* 76, 366–374.

6 Uzbekistan and Turkmenistan

Bahodir Eschanov* and Shadmon E. Namazov

According to literary sources, in the 3rd–4th centuries AD, the culture of *G. herbaceum* L. became widespread in Central Asia. Early-maturity forms adapted to the harsh arid climate were developed by folk selection. This continued into the 19th century although the bolls are small (2–4 g), the fibre is coarse and short (20–25 mm), and the yield is also low. With the integration of Central Asia into Russia in the 18th–19th centuries, this cotton was exported to Russia, but did not meet their requirements. Thus, cotton had been grown on a small scale in Uzbekistan, associated with the areas around the ancient cities, such as Khiva on the Amu Darya river from 2000 years earlier, according to Chinese records. But it was the 'Great Game' between the British and Russian Empires (Spoor, 1993), that resulted in railways being constructed between main cities and commercial centres in Central Asia and Russia.

Before 1860, the tsarist Russia's textile industry had relied on the USA to supply cotton fibre through the Baltic Sea ports, but the US Civil War hampered this, and Russia, envious of the UK's links with India, decided to venture into Central Asia. This inaugurated an era of regional specialization, and the small independent states covering modern Uzbekistan – the Kokand and Khiva khanates and the Bukhara Emirate – were forced to become the main suppliers of cotton to Russia (Djanibekov *et al.*, 2010).

Under the Russian influence, *Gossypium hirsutum* varieties were imported from Central America to increase cotton yields and improve its quality to meet the requirements of the Russian textile industry. Then, in1884, seeds of the more early-maturing cultivars King, Russels and Cleveland of the species *G.hirsutum* L., were brought. As a result of spontaneous reproduction, mechanical mixing and sowing of seeds formed 'factory mixtures'. Between 1860 and 1913, the cotton area was expanded from 35,000 ha to 441,600 ha, and the yield of raw cotton increased from 0.7 tons/ha to 1.2 tons/ha (Spoor, 1993). In 1870, for the first time, seeds of late-maturing *G. barbadense* L. were introduced, but this ended in failure. The Turkestan Agricultural Experimental Station was established in 1898, followed in 1905 by the Mirzachul Agricultural Experimental Station, where Bushuev was the first scientist to initiate cotton-breeding research.

The Andijan Experimental Field was set up in 1909 and in 1914, G.S. Zaytsev established a

*Corresponding author: bahodire@yahoo.com

DOI: 10.1079/9781800620216.0006

breeding department in Ulkan-Salik, near the Mirzachul Experimental Station. In 1918, G.S. Zaytsev was appointed as head of the Fergana breeding station on Pakhtali-Kul near Namangan province. This station was closed in 1919, and G.S. Zaytsev brought some of the seeds to the village of Qoplonbek near Tashkent. In 1922, he moved to a plot near the Bozsuv River. Later, by the initiative of G.S. Zaytsev, the plot was transformed into a Turkestan breeding station. At these stations, systematic breeding studies were carried out and the first domestic cultivars of cotton Navrotsky, 169 and 182, were developed.

The Cotton Research Institute was established in 1929, as a result of the incorporation of several research and experimental stations, including the Turkestan Selection Station. Breeding began at the Pakhtalik-Kul and Khorezm branches of this station near Namangan and expanded at the experimental stations of the All Union Cotton Research Institute. The Institute of Experimental Plant Biology was established in the system of the Academy of Sciences of Uzbekistan. In 1997 this was transformed into the Institute of Genetics and Experimental Biology of Plants. Later, a Decree of the President of the Republic of Uzbekistan #2125 of 10 February 2014 established the Institute of Cotton Breeding, Seed Production and Agrotechnologies, so now cotton-breeding research is conducted by this and its Experimental Stations, the Institute of Genetics and Experimental Biology, the Centre for Genomics and Bioinformatics of Uzbek Academy of Sciences, the Karakalpak Agricultural Research Institute, the Research Institute of Plant Genetic Resources, the Tashkent State Agrarian University, the Andijan Agricultural Institute and others.

The Russian and, from 1924, the Soviet influence resulted in Uzbekistan farmers being forced to grow cotton and, as less land was then available for cereals, Kazakhstan became the cereal-producing country. In the late 19th and early 20th centuries, one third of total irrigated land with more fertile soils in Central Asia was devoted to cotton production. Alongside the cotton were areas with lucerne, which would have been a source of natural enemies of cotton pests. The government strictly controlled the industry and introduced quotas to ensure efficient production at collective farms (*kolkhozes*). During the Soviet era, water from the Amu Darya River was diverted into Turkmenistan to increase the irrigated area of cotton and other crops, but this resulted in the drying of a large proportion of the Aral Sea. Following independence, Uzbekistan has decreased the area of cotton to grow more food crops (Fig. 6.1). Cotton is a universal fibre crop that supplies the textile industry and a number of other sectors of the economy by raw materials, as well as being the object of import–export (fibre, yarn, ready-made garments) in the international market.

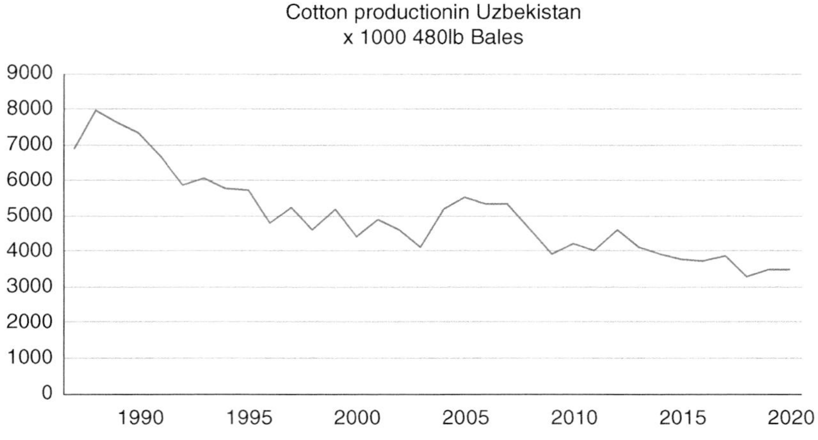

Fig. 6.1. Decline in production as crop diversity increased.

Cotton-growing Conditions

Uzbekistan is one of the most northern cotton-growing countries, but is the sixth largest cotton producer in the world, with cotton lint fibre yield of 750 kg/ha, annual production of 0.85–1.0 million metric tons of fibre (FAS/USDA, Global Market Analysis, 2020). Until 2018, Uzbekistan ranked fifth in the world in terms of cotton fibre exports, and today it processes almost 100% of its fibre in the country. Cultivars of tetraploid cotton species of *G. hirsutum* L. and *G. barbadense* L are now grown.

The cotton-growing regions of Uzbekistan are characterized by aridity, and an abundance of heat and light. Low air humidity in combination with high temperature and intense solar radiation determines high evaporation rate: from 900 mm in the north up to 1500 mm per annum in the south of the republic. The sum of effective temperatures during the cotton-growing season is 2300–3100° in the south in the Surkhandarya and Kashkadarya valleys and 1900–2300° in the north in the Khorezm region and the Republic of Karakalpakstan.

Long-term studies by scientists have established that sowing of fuzzy cottonseeds is recommended to start with a stable spring transition of average daily air temperatures of +10–12° and delinted seeds at +12–14° of average daily soil temperature at a depth of 10–15 cm.

The cotton-growing regions of Uzbekistan are divided into the following five zones according to soil and climatic conditions:

1. Northern zone: the Republic of Karakalpakstan and Khorezm province.
2. Central zone – includes two areas:

Table 6.1. Temperature and rainfall in cotton-growing areas.

Duration of the frost-free period	155–245 days
Number of days above 10 degrees	202–251 days
Number of days above 15 degrees	161–201 days
Precipitation	80–200 mm in desert areas and up to 500 mm in foothills

a) Territory of Bukhara region: Navoi and Bukhara provinces
b) Jizzakh, Samarkand, Syrdarya and Tashkent provinces
3. Fergana Valley zone: Andijan, Fergana and Namangan provinces
4. Southern zone: Surkhandarya and Kashkadarya provinces

Cotton Varieties

Changes in variety were sought over a series of seven periods to improve yields taking into account the level of soil salinity, stress factors (heat, water scarcity, extreme temperatures etc.), the degree of resistance to diseases and pests, as well as their sensitivity to agro-technical factors to optimize cultivar placement.

The first systematic cultivar replacement (1922–1931) was to replace low-yielding 'factory mixes' with cultivars that increased yields by up to 15–20%, fibre turnout up by 2–3% and earliness by 10 days.

The second cultivar replacement (1932–1941) introduced new high-yielding cultivars of the species *G. hirsutum* mainly by analytical breeding among the populations of American cultivars. The most widespread in the second cultivar replacement was the cultivar 8517. The long-staple cotton *G. barbadense* L. was finally mastered, and cultivars 35-1 and 35-2 were the most widespread.

The third cultivar replacement (1942–1946) was facilitated by the introduction of cultivars more resistant to *Verticillium dahliae* Kleb. And an increased yield up to 12–15%, fibre length up to 32.5 mm and fibre turnout up to 35.5%.

The fourth cultivar replacement (1947–1970) introduced cultivars that increased yields by 10–12% and improving early maturity up to 7–10 days. Long-staple cultivars S-6002 of *G. barbadense* L., with a fibre quality of 2–3 types, were also introduced.

In the fifth cultivar replacement (1971–1981), there was increased commercialization of *G. barbadense* cultivars, which were early-maturing and resistant to fusarium wilt. Also, *G.hirsutum* cultivars that were relatively early-maturing and resistant to the more aggressive race 2 of the *Verticillium dahliae* Kleb were introduced.

The sixth cultivar replacement (1982–2011) showed an expansion of the number of cultivars of *G. hirsutum* L. that occupied 80% of the sown area. New cultivars of *G. barbadense* L. were also introduced and were distinguished by early maturity (115–120 days), productivity (4.0 tons/ha and more) and high fibre quality (II-III types) (Ibragimov *et al.*, 2008).

The seventh cultivar replacement started in 2012 and is currently ongoing. According to the Resolution #47 of 30 January 2020 of the Cabinet of Ministers of the Republic of Uzbekistan, 'An effective organization of long staple cotton growing, production of new varieties and introduction of mechanisms of encouragement, the resumption of cultivation of *G. barbadense* L. varieties in Uzbekistan has begun. Now long-staple cotton varieties, such as Surkhan-14, Surkhan-16, Surkhan-103, CP-1607 and CT-1651 are commercialized.'

Cottonseed Production

In Uzbekistan, cotton breeding and seed production has a planned and nationwide programme carried out on the basis of a centralized state system. It includes breeding of new varieties, tested at the State Variety Testing Centre with new varieties allotted to zones. Seed production is at the elite seed farms of the Seed Development Centre, and seed-production farms in relation to cotton textile clusters, variety and seed control, which are engaged in mass-reproduction of the varieties while maintaining their biological and productivity qualities.

Thus, the seed system of the republic aims to provide all farms with the variety and sowing quality of seeds using five closely related sections:

1. preliminary propagation of seeds of new varieties;
2. state variety testing;
3. renewal of commercialized cotton varieties;
4. preparation of seed fund;
5. preparation of seeds for sowing and providing farms with seeds; and seed control.

In the first year, preliminary propagation of seeds of new varieties is organized at the first-year seed nursery on an area of at least 0.5 ha and thence the nursery-propagated seed is grown on an area of at least 1.0 ha. In subsequent years, elite seeds are sown and the area of reproduction is determined by the prospects of the variety. In elite seed farms, the total area of the seed propagation should not exceed 40 ha. Cotton on elite-seed farms is not defoliated.

Preliminary breeding of new varieties of cotton is methodically guided by research institutions with participation of the authors of the variety. The state variety testing is headed by the Crop Variety Testing Commission under the Ministry of Agriculture. Each variety testing site serves a specific soil-climatic zone, so it is organized under conditions specific to a particular area, and cotton varieties are tested using the agronomic techniques recommended for local conditions.

The purpose of the state varietal test is to evaluate all the main characteristics of new varieties in comparison with the commercialized variety for each region, the varieties with higher performance than the previously registered varieties are recommended for commercialization and suitability for a new zoning area. Cotton varieties submitted for state varietal testing are evaluated on a comparative basis for three years, before which this variety must be tested in a competitive varietal test of breeding institutions for the next two years. The varietal purity of cotton submitted to the State Variety Test of Cotton shall be not less than 96%: if the varietal purity of the new variety is less than 96%, it shall not be accepted for testing and seeds shall not be sown on farms where the first seed is propagated. Variety testing stations operate on the basis of special laws and methodologies for state varietal testing.

According to the state varietal test and local branches, a number of new varieties, which are superior to the regionalized varieties in terms of complexity, are presented for zoning in specific planting areas. Once the zoning of a new variety is confirmed, elite cultivation on primary seed farms will be discontinued and this work will be organized on elite farms in the regions where a particular variety is intended to be planted. Elite farms must clean the elite seeds variety up to 100% purity.

The plan for planting cotton varieties is organized on the basis of generalized and approved proposals of farms, regional and district agricultural departments of the Republican Centre for Cotton Seed Production in the Ministry of

Agriculture. This takes into account the zoning of varieties to each region and the quality, yield and export requirements of the fibre type.

Currently, seed is treated prophylactically with 65% liquid of P-4 and Bronopol 12% powder fungicides to control *Xanthomonas malvacearum* and *Rhizoctonia solani*.

The State Unitary Enterprise Center for the Provision of Services in the Agricultural Sector provides a methodological guide in ensuring the reproduction of the required amount of high-quality seeds, assessing the quality of the seed stock and selecting the best batch of seeds for sowing, monitoring compliance with state discipline in all areas of cottonseed. It controls the work of the republican cottonseed laboratories; develops seed analysis methodology; assesses the quality of the seed stock; monitors the accuracy of seed crop approbation; and implements other special measures on seed. It resolves disputes between agencies on seed quality in arbitration.

According to the five-year scheme of seed production, it is carried out on special elite seed farms and generation seed farms. A system of tender selection and training of farms specializing in seed production has been created. As a result, cottonseed production has radically improved, and the average seed consumption per ha decreased.

Pest Management

The severe winter conditions have resulted in relatively low populations of insect pests, but following the development of DDT and other organochlorine and organophosphate insecticides, these were widely applied from the air. Antanov aircraft were fitted with hydraulic nozzles. Aerial application of a defoliant was also practised to aid harvesting the crop by hand.

Some sprays were applied with a tractor-mounted air-assisted sprayer on which the fan oscillated from side to side while traversing a crop (Figs 6.2, 6.3 and 6.4). Known as the 'Sprayer ventilatory cotton OVH-600', it applied a relatively high spray volume of 250–300 litres per hectare, but spray distribution was very uneven (Matthews, 2004). A boom mounted unit had also been developed and was used in some trials during projects funded since independence (Figs 6.5 and 6.6). A modernised version of this boom with shielded 'drop-legs' was later developed as part of a World Bank project

However, the aerial sprays had an adverse impact on the rural population due to pollution and poisoning, so it had been decided to introduce a new policy of biological control. Technology was developed for large-scale rearing of the parasitic wasps *Trichogramma pintoi*, aimed at bollworm eggs and *Bracon hebetor* to control

Fig. 6.2. Russian air blast sprayer.

Fig. 6.3. Close-up of nozzles.

larvae. More recently, the predator *Chrysopa* spp., a predator of aphids, thrips and mites, was also bred in biofactories, so these could be released throughout the cotton area to control the major pests.

In 1970, the government of the former Soviet Union resolved to create a number of biolaboratories for mass-producing entomophages to release on crops against pest insects (Shepetilnikova and Fedorinchik, 1968; Chenkin, 1997). With this resolution, about 70 biological laboratories and over 300 farm product line biolaboratories were established for mass-rearing the parasitoid *Trichogramma* (Matthews, 1993). Subsequently, the area of entomophage releases in Central Asia increased every year. Mass-rearing of *Trichogramma* required, initially, the breeding of angoumis grain moth, *Sitotroga cereallela*, on a large scale, to have sufficient eggs in which *Trichogramma* could develop. This industrial-scale system had been developed in the USSR and 500 units were manufactured in Tashkent in the period 1979–85. *Bracon* was not reared using a large-scale mechanized system and relied on rearing the wax moth *Galleria melonella* as the host. *Chrysoperla* is also manually reared on *Sitotroga* eggs in glass jars. Most of the

Fig. 6.4. Sprayer being used in a cotton field.

Fig. 6.5. Sprayer fitted with vertical booms.

Fig. 6.6. Sprayer treating young cotton plants. (Figs 6.2–6.6 photos: Graham Matthews)

biofactories were set up on state farms (now co-operatives).

According to Tashpulatova (2007) and Tashpulatova and Zalom (2007) biological control was the leading method to control leaf-chewing pests in Uzbekistan. Its annual application on cotton crops involved releasing *Trichogramma* at the pest threshold of 20–25 bollworm eggs/100 plants, assuming 50% hatch, to provide control of pest eggs of three generations, repeating their release 3–4 times every 15–20 days (Figs 6.7–6.9) (Adashkevich *et al.*, 1991).

By 1990, at the time of the break-up of the Soviet Union, more than 700 bio-units (including 550 mechanized lines) had been established in Uzbekistan. Operational and maintenance costs of the bio-units had fallen sharply, so the World Bank was asked for funding for upgrading the biofactories. Efforts were made to establish the costs of rearing the natural enemies, but field trials indicated that the *Trichogramma* being released from one of the biofactories had little or no impact on bollworms, attributed to poor-quality control in production of the entomophage. However, under the Soviet system, there were

Fig. 6.7. Rearing *Trichogramma*.

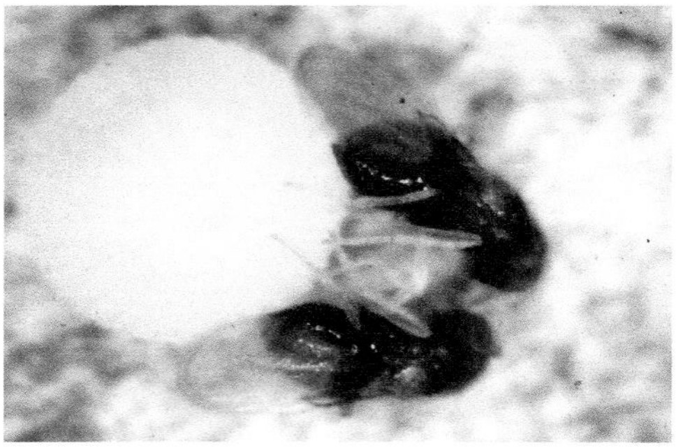

Fig. 6.8. *Trichogramma* on a Helicoverpa egg.

Fig. 6.9. Releasing *Trichogramma*. (Figs 6.7–6.9 photos: Graham Matthews)

large areas of alfalfa which provided a reservoir of natural enemies, which were very effective in reducing populations of early-season sucking pests. Chemical sprays against bollworm reduced the overall pest load (i.e. pests other than bollworm) and resulted in increased yields. With an increased availability of pesticides as the market was liberalized, there was greater interest in chemical control, by applying more modern insecticides, using tractor-mounted boom sprayers.

Since Uzbekistan's independence was declared on 31 August 1991, consideration for locally produced cereals resulted in the area sown to cotton being reduced from 1.8 million ha in 1990 to 1.4 million ha in 2006, while the area under cereals increased from 1.0 million to 1.6 million ha. The area of lucerne also decreased.

The cotton industry had a bad reputation, as it was claimed that the harvesting of the crop was largely with forced labour. The Uzbekistan government decided that mechanization was needed on 70% or more of the crop by 2016 to shorten the harvest season by several weeks, although this would reduce the demand for labour in rural areas, and their income. Child labour was eliminated in 2012 and the government is collaborating with the World Bank and International Labour Organization to ensure full compliance with the implementation of the Program of Decent Labour.

Due to very cold winters, the cotton crop is sown during April–early May and harvested in September. The crop is concentrated in the periphery of Aydar Lake (near Bukhara), along the Amu Darya in the border area with Turkmenistan and also near Tashkent along the Syr Darya. It is also grown in the Fergana Valley. During 2010/11 the yield of lint reported was 752 kg/ha, with new varieties adopted since 2009; these varieties are: Bukhara 102, Bukhara 8, Andijan 35 and Khoresm 150. As 720,000 tonnes of cotton a year are now used by the Uzbekistan textile industry, its output rose by 80% between 2014 and 2018, while exports of raw cotton decreased.

Yields of cotton were very low in 2003, which may have been due to a number of factors, including poor rainfall at the time of sowing, poor seed quality, more pests following a mild winter and insufficient natural enemies for release. There was also a slow response to the need for insecticides, which were not available when needed. Subsequent research has indicated that rearing of wasps on alternative hosts means that their effectiveness in the field may not be sustained, so a policy of integrated pest management is now followed which can involve some sprays being applied where and when required. The main insecticides then registered were pyrethroids, such as lambda-cyhalothrin and deltamethrin, the latter also available mixed with triazophos. However, when these were applied with a sprayer with an oscillating air blast, the gaps in spray coverage allowed significant damage due to bollworm. In worst-case scenarios of relatively heavy pest pressure, less than half of the cotton area in Uzbekistan is sprayed one or two times with insecticides during a season. Herbicides are rarely used because weather tends to be cold and wet in the spring, resulting in weed death prior to cotton planting. There is therefore a need for more information and guidance to be available as illustrated booklets suitable for farmers to read. One booklet was issued by the Asian Development Bank (Fig. 6.10).

Future of Cotton Production

The cluster system introduced by the initiative of the president, Sh. Mirziyoyev, is one of the most important changes of the last three years. As a result of the establishment of cotton textile clusters in the country, the main cotton crop is grown by clusters. For example, in 2019, on 1,034,000 ha in the country cotton is cultivated by 88 cotton-textile clusters in 88 districts. A total of 73% of the total yield of raw cotton was produced on 677,500 ha (66% of the cotton area). The average yield of row cotton across the clusters was 2.9 t/ha, and compared to 2018 increased by 0.55 t/ha. The yield of row cotton of 65 clusters was more than 3.0 t/ha, while 11 clusters achieved 3.5 t/ha.

More than 3 million tons of cotton was harvested and processed by 97 cotton textile companies and 11 cooperative cotton textile organizations in 2020. Voluntary co-operatives of farmers were established on the basis of ginneries on 173,000 ha of cotton.

According to the decision #4633 of the President of Uzbekistan, 'On measures to widely introduce market principles in the cotton sector', signed on 6 March 2020, the prospects for the development of cotton in a market economy are set out. The resolution identified new mechanisms for the introduction of market principles in the country, ensuring free competition in the cultivation, purchase and sale of raw cotton, as well as increasing the interest of farms in growing cotton.

In particular, from the harvest of 2020, the state will abolish the practice of setting plans for the production and sale of raw cotton. The practice of setting the purchase price of raw cotton will be abandoned. Growers of raw cotton were given the right to freely choose regionalized cotton varieties. The supply of certified seeds has been maintained and will be gradually transferred to the seed clusters of the Seed Development Centre under the Ministry of Agriculture and cotton-textile clusters. Fifty-two elite seed farms established in 2019 will provide farmers, co-operatives and cotton-textile clusters with quality-guaranteed seeds.

These clusters installed drip-irrigation systems on 9400 ha. Also, 11,600 units of agricultural machinery and 191 high-efficiency cotton-picking machines were purchased. As a result of modernization of ginneries by clusters, the output of cotton fibre increased from 34% to 35–37%.

Great attention is paid to the manufacturing of textile products in the country. For example, in the 1990s, the volume of cotton fibre processing was only 7%, but in 2020, as a result of measures taken in the industry, it had reached 100%.

Cotton Growing in Turkmenistan

Agriculture accounts for almost half the value of Turkmenistan's economic production, with cotton being a priority crop. All land-use rights are allocated by the state to farmers and leaseholders. Farms depend on melting snows of the Pamir mountains feeding the Amu Darya and

Fig. 6.10. An advisory booklet published by the Asian Development Bank.

Syr Darya rivers. Part of the flow along the Amu Darya has been diverted to the Karakum Canal, which stretches 750 miles across southern Turkmenistan, towards the Caspian Sea. Although the building of the canal has been severely criticized, due to the consequent ecological disaster of the shrinking of the Aral Sea, its development was planned to vitalize farming and living conditions in Turkmenistan. Crops in Turkmenistan can only be grown by irrigation and hence most of the farming regions in Turkmenistan lie along the Amu Darya river and the Karakum Canal. The situation began to change rapidly after 1990, when the government started to stimulate wheat production in order to achieve a higher degree of national food self-sufficiency. However, despite this relative decline of cotton production, Turkmenistan remains a significant cotton producer in the region (Fig. 6.11).

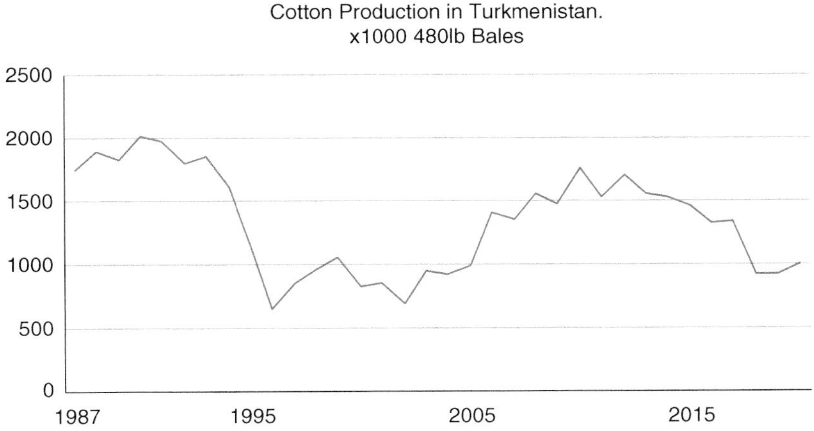

Fig. 6.11. Changes in production since 1987. (Figure courtesy of Graham Matthews)

The long-staple cotton (*G. barbadense*) is generally sown in the Mary region in April/May with harvesting in September, so plants are exposed to insect pests over a period of five months when temperatures are high with low relative humidity. Medium-staple *G. hirsutum* cotton is also grown. Insecticides are not widely used today in Turkmenistan. Some sprays have been applied at the end of a crop in April to reduce the spread of whiteflies from glasshouses into early cotton. The extremely cold winters undoubtedly reduce the severity of most potential insect pests in the large areas of wheat, cotton and lucerne grown in the desert under irrigation. Integrated pest management includes a closed season, crop rotation and conservation of natural enemies by growing lucerne. Choosing a smooth-leaf variety gives some resistance to bollworm, while the barbadense cotton has gossypol as a natural control.

Despite the cold winter, important beneficial insects thrive there in the absence of large-scale insecticide use. Biological control is being used in many areas. For example, golden eye (*Chrysopa*) is a common lacewing keeping aphid and thrips populations in check. Coccinellids also play a key role in biological control. If the cotton bollworm (*Helicoverpa armigera*) population increases, farmers can get the egg parasitoid *Trichogramma pintoi* reared on *Sitotroga* and/or *Bracon* parasitoids from biolaboratories for release in cotton; however, the equipment at these biolaboratories had declined over recent years. Scouting has been used to determine if an insecticide spray is needed.

The main market for pesticides is fungicidal seed treatments as cotton seedlings are prone to disease, especially when growth is slowed by low night-time temperatures. Bronocol is the main ingredient used. Frequent passage of tractors with harrows for mechanical weeding leads to soil compaction affecting crop development, so there is a potential for increased herbicide use. So far, the small quantity of herbicides imported is applied on wheat, but many fields have large infestations of grass weeds (e.g. *Cynodon*, *Echinocloa*) as well as *Amaranthis* and *Convolvulus*.

After harvesting, the stalks are removed for fuel, or shredded and ploughed in. Winter ploughing is considered to reduce the overwintering pupae of *Helicoverpa*. Yields in 2019 were poor, mainly due to problems with water supply.

References

Adashkevich, B.P., Adilov, Z.K. and Rasulev, F.K. (1991) Biological control in action. *Journal of Plant Protection (Zashita Rasteniy)* 7, 9–10.

Chenkin, A.F. (1997) *History of Development and Problems of Plant Protection*. Rossel′khozakademii͡a, Moscow.

Djanibekov, N., Rudenko, I., Lamers, J.P.A. and Bobojonov, I. (2010) Pros and cons of cotton production in Uzbekistan. In: *Case Study #7-9 of the Program: Food Policy for Developing Countries: The Role of Government in the Global Food System*. Cornell University, Ithaca, New York.

Ibragimov, P. Sh., Avtonomov, V.A., Amanturdiev, A.B., Namazov, S.E., Zaurov, D.E. *et al.* (2008) Breeding and genetics, Uzbek Scientific Research Institute of Cotton Breeding and Seed Production: breeding and germplasm resources. *Journal of Cotton Science* 12, 62–72.

Matthews, G.A. (1993) Biological control takes precedence in Uzbekistan cotton. *Pesticide Outlook* 4, 36–38.

Matthews, G. (2004) IPM at Ak Altin. *Outlooks on Pest Management* 15, 238–239.

Shepetilnikova, V.A. and Fedorinchik, N.S. (1968) *Biological Control of Crop Pests*. Colos, Moscow.

Spoor, M. (1993) Transition to market economies in former Soviet Central Asia: dependency, cotton, and water. *European Journal of Development Research* 5, 142–158.

Tashpulatova, B. (2007) Integrated pest management in Uzbekistan. In: *Integrated Pest Management in Central Asia.* Proceedings of the Central Asia Region Integrated Pest Management Stakeholders Forum sponsored by the USAID IPM-Collaborative Research Support Program (CRSP). Dushanbe, Tajikistan, 41–46.

Tashpulatova, B. and Zalom, F. (2007) Enhancing the efficiency and product lines of biolaboratories in Central Asia. In: *Integrated Pest Management in Central Asia.* Proceedings of the Central Asia Region Integrated Pest Management Stakeholders Forum sponsored by the USAID IPM-Collaborative Research Support Program (CRSP). Dushanbe, Tajikistan. 21–26.

7 Cotton Growing Along the Nile

Graham Matthews*

Egypt

Cotton was widely used in the ancient civilizations of Mesopotamia, Egypt and the Indian subcontinent, but it was not until the beginning of the 19th century, during the reign of Muhammad Ali Pasha, that Louis Alexis Jumel, a French textile engineer, spotted a cotton bush in a Cairo garden with unusually long and strong fibre. Muhammad Ali was keen to support industrialization in Egypt and encouraged Jumel to develop the seeds of the plant in Cairo, which may have come via Christopher Columbus from Barbados. Muhammad Ali appointed Jumel plantation manager, who increased his cotton crops all over the Delta region of the River Nile. Muhammad Ali continued to dominate the industry in Egypt and marketed cotton profitably in Europe. Later, the heads of state, using the success of exporting cotton, started to get loans from some banks in Europe, as they wanted to modernize the production of cotton and thus increase the exports. Egyptian cotton started to make a name for itself, so demand for the high-quality lint skyrocketed. Exports rose from 50.1 million pounds of cotton in 1860 to 250.7 million pounds by 1865.

Unfortunately, when the American Civil War stopped their cotton exports to Europe, the price of Egyptian cotton rose, so Muhammad Ali's grandson, Ismail, decided to develop a part of Cairo into a city like Paris, and supported the construction of the Suez Canal. The latter had an immediate effect of easing the trade in cotton from India to Europe. Once the war in America ended and exporting cotton to Europe resumed, it had a disastrous impact on the cotton industry in Egypt, presumably as the price of American cotton was lower.

Government intervention and other factors resulted in a redefinition of property rights towards the end of the 19th century, which meant that there was a massive redistribution of land away from villages and nomadic peoples to the well-connected owners of large estates. By the 1930s it was the universal practice to include cotton and a leguminous crop in all rotations, so few, if any, farmers grew the same crop two years in succession upon the same land. All the rotations were built around cotton and berseem, the latter providing a source of natural enemies that helped to reduce insect pests in the cotton. The rotation could also include corn, wheat, barley, grain sorghum, rice, onions, beans and vegetables. Thus, in the canal-irrigated lands, cotton was followed by wheat in year 1 and in year 2 by corn and berseem, the latter being sown in the standing corn and carried over until the following year. When such a rotation is used,

*g.matthews@imperial.ac.uk

 Pest Management in Cotton: A Global Perspective
(eds G.A. Matthews and T. Miller)
DOI: 10.1079/9781800620216.0007

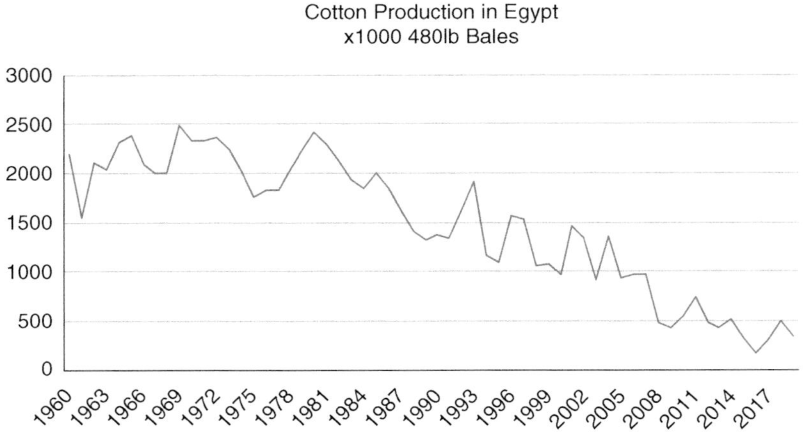

Fig. 7.1. Decline in cotton production, 1960–2018.

Fig. 7.2. Children looking for leafworms on cotton.

the land is occupied by cotton from March to October (Norris, 1934).

Norris reported that the most common and destructive insect was the pink bollworm (*Pectinophora gossypiella* Saund.), which appeared in 1913, probably by bringing seeds from India, and subsequently spread over the entire country. The method of control by law was to treat seed by heat at the gin as this heating gave a fair degree of protection. Storey (1921) reported that the percentage of green bolls attacked by pink bollworm had been between 22–29%. Other insects common in the cotton were the spring cotton bollworm (*Earias insulana* Boifd.) and the Egyptian cotton worm (then called *Prodenia liture* Fab. but now *Spodoptera littoralis*). Norris commented that although these insects are destructive and often do a great deal of damage, they are not general over the entire country. It is commercially estimated that the average annual damage by all insects varied between 15% and 50% of the crop.

The principal plant disease was cotton wilt (*Fusarium*) and was checked by rotation as attempts to develop a wilt-resisting variety have been met with only some degree of success. The short-staple varieties appear to be somewhat resistant. The principal districts suffering from wilt are in lower Egypt where drainage is poor and other soil conditions are unfavourable. Otherwise, the country is remarkably free of cotton diseases.

According to Norris, after Muhammed Ali became interested in cotton production, the seed of Sea Island, Brazilian, Peruvian and American upland cottons were introduced, so the early Egyptian varieties were the result of selection and natural crossing. They received little or no attention from the plant breeders. These varieties apparently rose to general favour by the spinners, only to break down biologically and disappear from the market in a few years. The single exception is the variety known as Ashmouni, which appeared about 1860 and is still one of the leading varieties of the country. The long-staple group includes Sakellaridis varieties, largely grown on the high-water-table lands of the northern Delta, with fibres light cream in colour and very fine and silky, with a staple length 13/8th inch and longer. By 1934, the trend was towards a shorter staple as the Sakellaridis and other long-staple varieties occupied about 75% of the total cotton area in 1922, with Ashmouni and Zagora and other shorter-staple varieties on about 25%; but by 1932, Sakellaridis had decreased to 43%, whereas the shorter varieties had increased to about 50% of the total cotton area.

Post- Second World War, the leafworm, *Spodoptera littoralis*, was controlled by arranging for children to search for larvae on the cotton leaves (Hosny *et al.*, 1981; Fig. 7.2). With the availability of the pheromones of cotton pests in Egypt (Hall *et al.*, 1982), the Natural Resources Institute, UK, undertook several projects. Microencapsulated formulations were easily manufactured by known methods of interfacial polymerization, so they could be applied to large areas with conventional spray equipment or used to control the release characteristics when placed in traps. In one project the pheromone of *Spodotera littoralis* was used in traps in an attempt to mass-trap the moths (Critchley and El-Dieb, 1981) following studies by Campion *et al.* (1980). Subsequently, the possibility of combining the pheromone with an insecticide λ-cyhalothrin, in 500 point-sources per ha at low rates of application of both components was investigated. It was envisaged that male moths would be attracted to contact the sources and subsequently suffer lethal or sub-lethal effects which would prevent mating. Following various experiments, it was concluded that none of the attracticide treatments represented a viable control technique (Downham *et al.*, 1995). According to Hosny *et al.* (1986), the minimum economic damage threshold practical for the cotton leafworm, *Spodoptera littoralis*, on cotton was 10,000 egg masses per ha for most years in Fayoum, Egypt.

The pheromone of the pink bollworm was used in large-scale field trials conducted in three governorates in Egypt during the 1987 cotton season. These trials showed that two or three applications of pink bollworm pheromone (formulated as either microcapsules or hollow fibres) integrated with one or two applications of conventional insecticides gave as good a control of bollworm as did four or five applications of the insecticides alone (Moawad *et al.*, 1991). Their results also underlined the importance of early-season pheromone treatments, and by avoiding insecticides, cross-pollinating by bees was considered to improve the yields in spite of a general increase in pink bollworm infestation throughout the governorate (Fig. 7.3).

Where hollow-fibre, laminate-flake and microencapsulated formulations of the synthetic sex pheromone *(Z,Z)* and (Z,E)-7–11-hexadecadienyl acetate of *Pectinophora gossypiella* (Saunders) were applied aerially in large-scale trials in the Delta region of Egypt in 1985, yields of seed cotton showed that adequate levels of control were achieved with all three pheromone formulations. The trials with 100-ha blocks of cotton showed these were at least as effective as the insecticide sprays and that the costs of the pheromone formulations and their aerial application also compared favourably with the insecticide programme. Greater numbers of beneficial insects were recorded in the pheromone-treated areas than in the insecticide-treated area (Hall *et al.*, 1982).

Insecticide spraying began in the 1950s using aerial application with contract Antanov aircraft from Eastern Europe and there was less emphasis on crop rotation (Figs 7.5, 7.6 and 7.7). To simplify the aerial application of insecticides and other farm operations, all cotton in a village is planted together, rather than in small patches, as hitherto. This simplification of the ecosystem may be partly responsible for the general increase in bollworm numbers. An average of three consecutive sprays at 15-day intervals using a different dual-purpose pesticide each time, thus 3–4 sprays of triclorphon, toxaphene, carbaryl or chlorpyrifos were applied, was current common practice (Salama, 1983) to protect the cotton crop against bollworm and leafworm infestations. Spray drift also affected adjacent crops. Some farmers used a tractor sprayer with tank and long hose carried through the crop by several helpers with the spray operator directing a lance into the crop (Fig. 7.4). This resulted in both the operator and helpers being highly exposed to the spray. Later, the first nationwide cotton leaf worm and cotton bollworm resistance-management programme was adopted by Egypt in 1979 to prevent or delay the resistance to pyrethroids as well as other insecticides. In this programme, based on alternate different classes of chemicals (organophosphates (OP) + insect growth regulators (IGR), pyrethroids (PY) and carbamates (CAR)), the pyrethroid application was used only once per season and solely on cotton by the ground spray. From 1993 to 1995 a rotation programme and the use of biocontrol agents were encouraged while conventional insecticides were used below the recommended doses. Disruption pheromones for pink bollworm were used in small areas initially, reaching to 50% of the total cotton areas by 1995, using conventional insecticides when the infestation reached a threshold (El-Wakeil and Abdallah, 2012).

Fig. 7.3. Twist-tie pheromone on cotton plant.

One potential biopesticide being tried was the nuclear polyhedrosis virus of the cotton leafworm, *Spodoptera littoralis* (Boisd.) (SLNPV). When combined with a low concentration of diflubenzuron, the second instar larvae were more susceptible than later instars (Moawed and Elnabrawy (1987). A locally produced wettable powder formulation of the nuclear polyhedrosis virus (NPV) was sprayed at three different application rates, 5×10^{11}, 1×10^{12} and 5×10^{12} polyhedral inclusion bodies (p.i.b.) ha^{-1}, using a

Fig. 7.4. Spraying from end of hose carried by helpers.

Fig. 7.5. Aircraft from eastern Europe to spray cotton.

Fig. 7.6. Aircraft spraying in the Nile Delta.

knapsack sprayer fitted with a cotton tailboom (Fig. 7.8). The level of control, assessed using crop-damage levels, was compared with the traditional method of controlling leafworm, the hand collection of egg masses. Control with the NPV was dose-dependent and an application rate of 1×10^{12} p.i.b. ha^{-1} resulted in a level of control equivalent to the hand collection of egg masses, which represented a five-fold reduction in the virus dose previously used (Jones *et al.* 1994).

The cultivated area with the cotton in the season of 2009 was only of 287,961 Feddan, as the area of cotton in Egypt has been decreased over recent years to allow a larger area of wheat to

Fig. 7.7. Planning aerial sprays near Zigazag.

Fig. 7.8. Spraying baculovirus using knapsack with tailboom to direct sprays within the crop canopy.

be grown (El-Wakeil and Abdallah, 2012). According to El-Adi *et al.* (2008), mating disruption of pink bollworm has continued to be used, but there is now interest in the first commercially grown, genetically modified crop in Egypt. This new cotton crop contains a gene purchased from Monsanto that makes the plants resistant to certain insects, but it will retain its unique Egyptian characteristics in every other respect. The aim is to maintain the long-staple fibres for which Egypt is well known. In May 2007, and for the first time, field trials of the new long-staple cotton varieties, namely Giza 86, 89 ,90, 91 and 96, have been cultivated (Mansour, 2009).

A question now is, Should the cotton crop be grown organically? Farmers will grow cotton if it is profitable, and there is a clear demand for high-quality long fibres, which was the key factor in the Egyptian cotton industry. The organic agriculture movement was started in Egypt more than 20 years ago (Fig. 7.9), chiefly to alleviate the increasing threat of pesticide poisoning to Egyptian farmers as cotton cultivation is one of the most pesticide-intensive crops.

The Centre for Organic Agriculture in Egypt operates an inspection and certification scheme according to the EU Regulation 2092/91. The optional release of biological control agents gives farmers the ability to react rapidly when pest populations increase. But the limited number of commercially available agents points to the need for research to successfully combine inundation and inoculation biological control agents with other organic pest-management practices. Effect of organic sources of nutrients on insect pests in cotton was studied in many experiments. Such technology aimed at meeting high nutrient demand of crops by organic sources alone to reduce the reliance on chemical fertilizers, thereby reducing the possible weak symptoms of using excess fertilizers, made the plants refractory against infestation of cotton insect pests (Naik *et al.*, 2009). Despite the growth of organic agriculture, there has been a lack of research-based information to address the need for a greater understanding of the mechanisms operating in organic systems, including plant–pest interactions (El-Wakeil *et al.*, 2015).

Sudan

Since Pliny the Elder described the 'wool-bearing trees' of Aethiopia in his *Natural History*, the cultivation of cotton has been a well-known aspect of the textile production of ancient Sudan (Yvanez and Wozniak, 2019). Cotton production is now the subject of a renewed interest in the archaeology of the ancient world, from India to Africa, and general trends are starting to emerge (Wild *et al.* 2007; Bouchaud *et al.* 2018). Archaeological studies discovered many cotton textiles

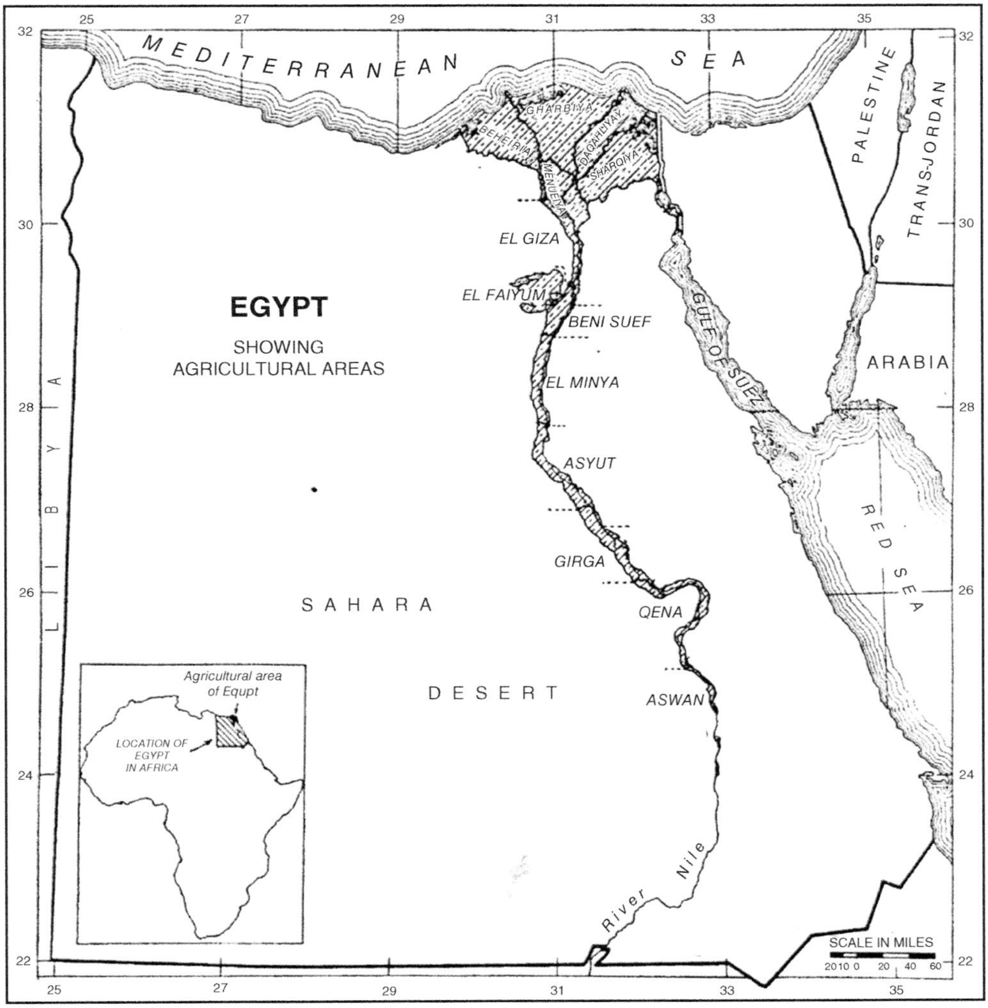

Fig. 7.9. Production area is confined to the Nile valley (Upper Egypt) and Delta (Lower Egypt).

in Meroe and Lower Nubia, and more recently several other archaeobotanical studies have been completed in the same regions. The Meroitic kingdom developed from Central Sudan at the very beginning of the 3rd century BCE. Over the next five centuries, their power covered a vast territory including 1500 km of the Nile valley, from northern Nubia to the fertile plains of the south in the Gezira region. The area included very diverse populations composed of sedentary, semi-nomads and nomadic groups, settling along the rivers and in the adjacent deserts. Meroitic graves and a few settlements have delivered a remarkably rich assemblage of archaeological textiles, increasing dramatically from the 1st century CE onward, overwhelmingly made of cotton fibre, in a clear demarcation from earlier textiles made of linen and wool. Meroitic textiles and dress practices are therefore closely intertwined with cotton production, placing cotton at the heart of the economy and culture of this ancient African kingdom.

It seems that cotton production occurred at first as small-scale experiments before scaling up during the 3rd century AD, in conjunction with the spread of the water-wheel in the Nile valley. Cotton in Nubia, and possibly in other neighbouring areas, probably belonged to the African species *G. herbaceum*, which was in all likelihood domesticated in southern regions, perhaps Ethiopia. We suggest that the increase of exchanges across the Indian Ocean during Antiquity created a favourable context for the emergence of cotton production and its relative expansion before the Islamic period (Bouchaud *et al.*, 2018).

Since the 1860s, cotton was grown in an area in the delta of the Baraka river near the Red Sea in north-eastern Sudan and close to Tokar, a town of 40,000 people and in another delta where the Gash river flows from the Eritrean foothills, near Kassala. In both these areas, the vegetation is Acacia-desert grass in a savannah, with annual rainfall of about 20–32 cm. Rainfall results in an annual flood in the delta so cotton is sown in September–October, depending on the flood, and is harvested from mid-December to April. The crop was grown on a receding water table in both places where the deep-rooted cotton plant grew well on the better-flooded areas. Cotton is not grown there now. This Egyptian-type cotton had similar pests to the Gezira, but the red bollworm (*Diparopsis watersi*) was also a major pest (Tunstall, 1958).

Early in the 20th century, cotton was grown in an area close to the White Nile in 1904. In studies on the ways of irrigating the land, in 1914, cultivation of 24 sq. km (9.3 sq. mi) was initiated as it was discovered that Egyptian-type long-staple cotton could be grown and was welcomed as a better raw material for the British textile industry. Then, taking advantage of the lowest Nile flood for 200 years, the Sennar Dam was constructed on the Blue Nile south-east of Khartoum to provide a reservoir of water. The dam, completed in 1925, enabled the development of the Gezira Scheme, initially financed by the Sudan Plantations Syndicate in London, aided by the British government, while the area was governed as part of Anglo-Egyptian Sudan. Water from the Blue Nile was distributed through canals and ditches to tenant farms, situated between the Blue and White Nile rivers.

The Gezira (which means 'island') is particularly suited to irrigation. This is because the soil slopes away from the Blue Nile, so water flows naturally through the irrigation canals by gravity. As the soil has a high clay content, seepage of water into the desert was minimized. At first it was thought that farmers would grow wheat, but when it was found that Egyptian-type long-staple cotton could be grown, it was considered to be a better cash crop (Fig. 7.10). The original Gezira scheme had 4300 km (2700 mi) of canals and ditches and this was increased by the Managil Extension to cover 8800 sq. km (3400 sq. mi), about half the country's total land under irrigation.

Cowland (1947) recognized that jassids were an important pest in the Gezira that led to initiating insecticide trials. Snow and Taylor (1952) reported that large-scale spraying to control jassids started in the Gezira in the crop season 1945/46; yield increases from about 1250 feddans (acres) sprayed with 0.1%. DDT emulsion was very satisfactory. In the following seasons the area sprayed increased and in 1947/48 spraying was extended to the cotton farms along the White Nile between Khartoum and Kosti. In 1949/50, the total area sprayed in the Sudan was over 150,000 feddans.

When the use of aerial sprays was initiated in the Gezira (Joyce, 1975), experiments using large plots were required because of the movement of jassids (Joyce and Roberts, 1959). Jassids (*Empoasca lybica* de Berg) were a key pest, the life cycle of which from egg to gravid adult took 16–24 days, and the adults could live for up to 40 days, with no diapause. During the 100 days from late August to early December, when breeding on cotton is of economic importance, a single male and female could give rise to some 50,000 progeny. During May–July, when crops are confined to irrigated gardens and river banks, jassids were widely distributed and could be found also on tree hosts, which are numerous, especially in the southern Gezira and along river banks. They could also be carried by air movement over long distances, as the great majority of catches of *E. lybica* in sticky traps were made before the increase in population on cotton that occurs from September onwards. The rate of increase of infestations of *E. lybica* on cotton was found to be positively correlated with the concentration of nitrogen recorded 2–4 weeks previously in the cotton leaf. This concentration affected not only the rate of increase of the initial colonizers, but also the rate of recovery of populations during November and December after spray-applications of DDT (Joyce, 1961).

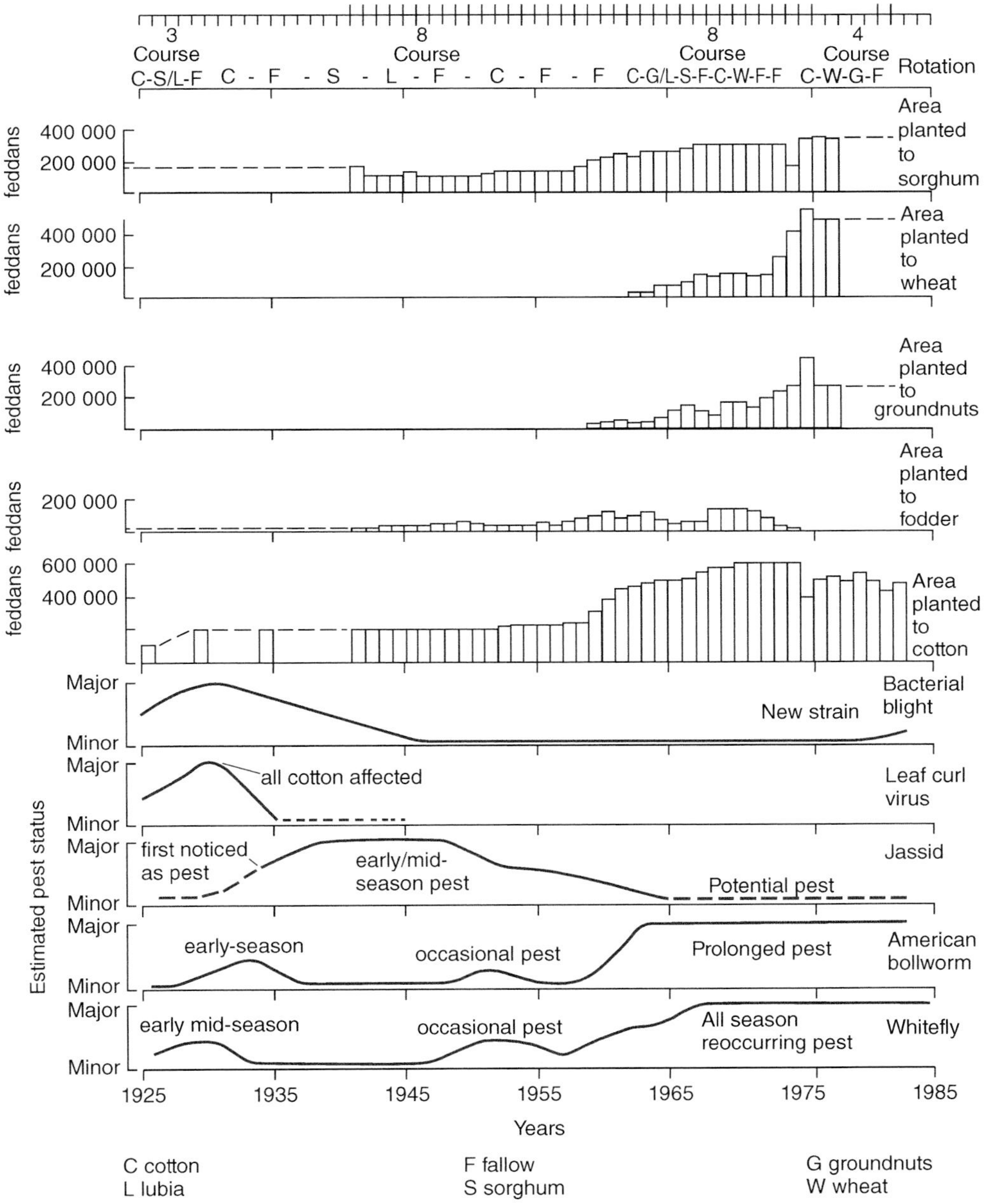

Fig. 7.10. Time profile of Sudan Gezira with particular reference to cotton insect pests. (From Griffiths, 1984)

The yield response to DDT sprays of Gezira cotton, estimated over the period 1949/50–1956/57, varied from season to season, related to the amount of rain falling in July, some six weeks before the crop was sown. A series of experiments in 1956/57 showed that Domains Sakel cotton, grown in the drier northern part of the Gezira, tended to give an increased response to DDT spray, especially at higher levels of nitrogenous fertilizer. A very highly significant part of this response was due to the control of jassid, *E. lybica* and thrips (*Caliothrips* spp.). In contrast, the variety X1730A grown in the wetter, southern part of the Gezira, gave no overall response to spraying, nor was the response to nitrogenous fertilizer affected by spraying (Joyce, 1959). Looking back at data prior to spraying, Joyce (1959) concluded that at least half of the deleterious effect of poor pre-sowing rains on Gezira yields is a 'pest effect', which can be eliminated by DDT spray.

During the course of later experiments, the commercial crop practice changed towards earlier sowing, heavier nitrogen dressing and more frequent spraying, with apparently beneficial effects on yield and consistency of cropping (Burhan and Jackson, 1973). Later, two ULV spray spectra were examined, namely a coarse (120 μm VMD) and fine (80 μm VMD) spray, which were released in the early morning (0600 h), mid-morning (0900 h), midday (1130 h) and mid-afternoon (1530 h). Early-morning sprays produced a rather narrow useful swath from both coarse and fine sprays but there was a high risk of drift from fine droplets. Both droplet spectra in the mid-morning spray provided a good deposit pattern within the cotton canopy and had the least amount in the air at 100 m downwind. Furthermore, the fine spray gave a swath almost 50% wider than the coarse spray. The physical structure of the crop canopy was the factor determining penetration of spray as turbulence generated by wind or the aircraft had no detectable effect (Uk, 1987).

Fortunately, when sowing the *Barbadense* cotton, the intensity of bollworm (*Helicoverpa armigera*) attack has generally been less than when *Hirsutum* cotton is grown in the area. The main difference seems to be the presence of more glands with gossypol in the *Barbadense* varieties, which gives them some protection. As a result of insecticide applications against larvae on cotton in the Sudan Gezira, a very substantial mortality of males, and virgin and inseminated females, was detected (Topper, 1987a). It was considered that the *H. armigera* adults attacking cotton within the Gezira were locally produced, with local populations building up in groundnut fields between July and September, which colonized sorghum before infesting cotton later. Further studies showed that feeding was important for adult survival and moths were strongly attracted to crops with plentiful honeydew or nectar (Topper, 1987b)

Another major pest is the whitefly, *Bemisia tabaci*, which has been recorded worldwide on cotton and in the Sudan *Trialeurodes lubia* has also been associated with cotton (Butler and Hennebury, 1990). Economic damage is a result of direct feeding and yield loss, but also lint contamination with honeydew and associated fungi (Gerling *et al.*, 1980). *Bemisia tabaci* is also an important vector of plant diseases, such as cotton leaf crumple and cotton leaf curl. The increase in whitefly infestations was considered to be due to the control strategies which were based on intensive aerial spraying and adoption of a system of 'package deal' contracts in which aspects of insect control were delegated to commercial companies (Eveleens, 1983). One of the insecticides being used was dimethoate (Rogor), which may not have been very effective due to much of the spray failing to penetrate the crop canopy or whiteflies being resistant due to the continued use of organophosphates. The lack of control of whitefly by dimethoate, monocrotophos or other organophosphorus insecticides resulted in a tremendous flare-up of whitefly by 1980/81 (Dittrich *et al.* 1985; Ahmed *et al.*, 1987).

Possibly because of this, another method was adopted which involved applying aldicarb as a soil treatment on the furrow at planting. This was successful but it was discontinued mainly because of international criticism of using the highly hazardous insecticide, which, it was claimed, had adversely affected animals in the area (Mann and Danauskas,1984). Where used, it did result in much higher yields.

Concern about the use of insecticides led to a World Bank request to look at the management of the chemicals. Unfortunately, the insecticides were being imported in 200-litre drums which were manhandled from lorries and then stored in the open without shade or adequate fencing. Later, proper stores were erected, but as the chemicals were imported there was an accumulation of empty drums. This was despite a refinery in the country where it was probable that the insecticide active ingredients could have been formulated and made use of recycled drums. There was also the question of whether degradation of insecticides stored outdoors had occurred. It was suggested that if the level of active ingredient had decreased, it would be possible to adjust the amount and avoid having to dispose of old stocks.

Bindra (1985) pointed out that, in relation to pest control and growing irrigated cotton in the Sudan, the 'crisis phase' around 1967/68 and the 'disaster phase' in 1980/81 were the results of uncontrollable whitefly-induced lint stickiness and the American bollworm (ABW) on newer varieties. He suggested that the higher pest incidence on newer varieties compared with Sakel-type cottons; the Lambert-type cottons had a greater build-up of the whitefly, while Acala cottons would have not only larger populations of both whitefly and ABW but also increased losses from the latter. Varieties of *G. hirsutum* generally contain less gossypol than in *G. barbadense* (Yu *et al.*, 2012) and would therefore increase the risk of damage to both these pests. Varietal changes in the Gezira, therefore, had profound ecological effects which favoured the whitefly and ABW, which made sprays less effective, leading to the cotton-pest-control predicament. He hoped that switching to cultivars less favourable to the pests and more suited to efficient pesticide application would ease the cotton-protection problem.

According to Salman (2013), the Gezira scheme began to collapse in the 1970s (Fig. 7.11). This was due to several factors including:

- poor maintenance of the irrigation canal network which led to serious water management and distribution problems;
- the use of river water with a large sediment load and extensive weed growth which clogged up the canals, all of which reduced irrigation efficiency;
- the reservoirs of Sennar and Roseires dams lost half of their storage capacities as a result of siltation, leading to the lack of water for irrigation and lower yields and low profits;
- during the 1973 oil crisis, the government failed to provide the required agricultural inputs, which resulted in low yields;
- low cotton prices because of increased supply on the international market from Asia, China and India;
- costly pest control for cotton; inadequate financing and marketing arrangements for most crops; inefficient agricultural processing; disillusioned farmers and low-cost recovery for irrigation water deliveries; and
- tenants often found it impossible to make ends meet from farm income, despite the fact that the average farm size was 20 feddans.

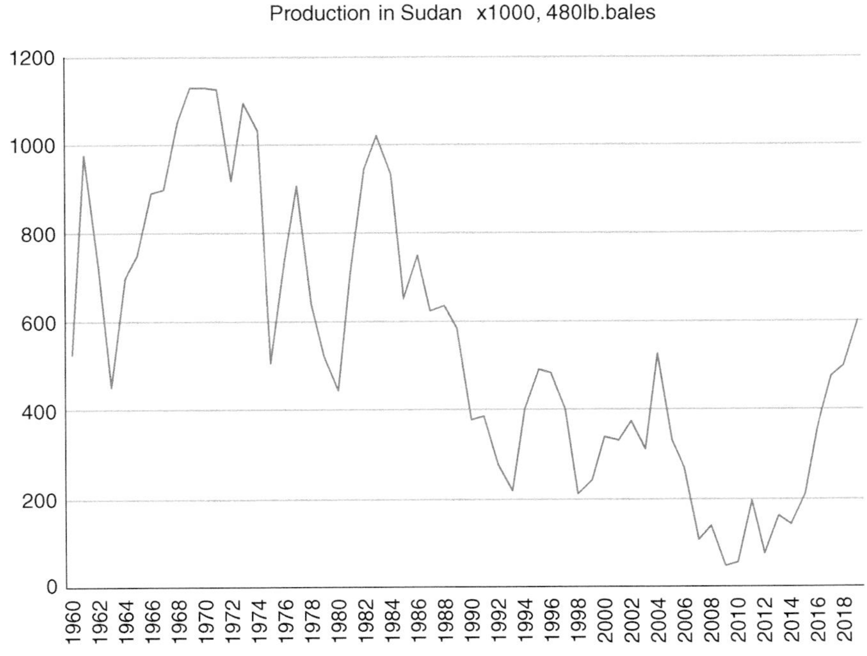

Fig. 7.11. Changes in production in Sudan, 1960–2018.

When the Electrodyn sprayer was initially in development, a prototype (Fig. 7.12) was taken to the Gezira to assess its potential use on large cotton plants and indicated penetration within the crop canopy was limited.

In June 2005, the National Council passed the controversial Gezira Scheme Act of 2005, but this was followed by a change in government policy in 2010 that led to further problems between tenants, the government and others, while wheat was grown on a larger area of the Gezira. In consequence, cotton production had declined from 1984 through to 2009.

Fig. 7.12. Examining the possibility of using the 'Electrodyn' Sprayer.

However, genetically modified cotton was trialled in the Gezira for the first time, and according to Abdallah (2014), Sudanese farmers growing Bt cotton said that the crop yield had increased to four tonnes per ha, which encouraged them to increase the cotton-cultivation area. The seed had been donated by the Chinese Ministry of Agriculture (CN1), but farmers observed that if there were a Sudanese company that would produce the seed, the cost of production would be less. However, they were satisfied with the Bt cotton production using the Chinese strain. From a decade ago, when it was difficult to obtain 3000 kg/ha, it is now very easy to obtain 4000 kg/ha. This has led to an increase in area from 45,000 ha in 2011 to 185,000 ha in 2020 (Fig. 7.13).

According to an ICAC report (2018), the release of Bt cotton varieties has provided a concrete platform for multi-disciplinary research. Management of cotton requires implementation of the agricultural packages integrated with other elements of production inputs, financing and agricultural policies as a mechanism for sustainable cotton farming. More emphasis has been given to extra-fine cotton research as this type of cotton is of immense importance in the marketing policy of Sudan. New extra-fine cotton lines have been developed with intermediate reaction to bacterial blight disease.

Some of these studies at the University of Gezira, but conducted in the Gezira Research

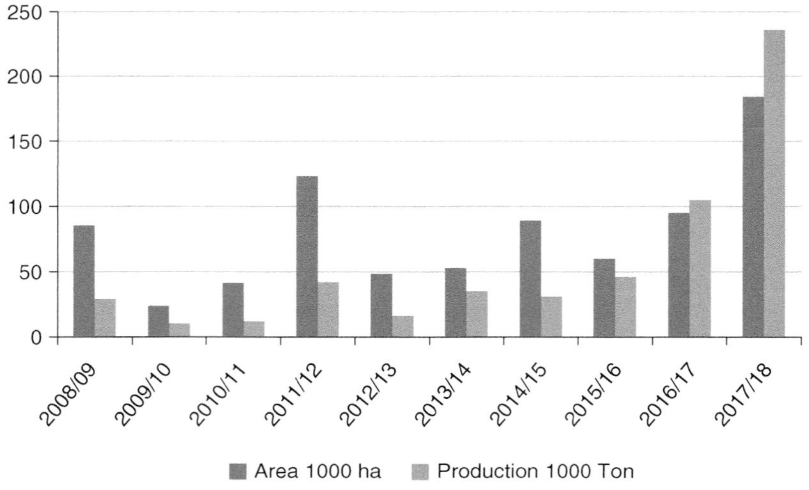

Fig. 7.13. Sudan cotton area (ha) and production (1000 tons) during the last ten seasons.

Station during the season 2014/15, evaluated eight Egyptian cotton experimental lines (extra-long-staple cotton) along with Barakat-90 for yield, quality and bacterial blight resistance. One promising line was better than the other tested genotypes in most of the parameters of lint quality measured. It was expected that the good traits found in this genotype could be used and transferred to new genotypes in a breeding programme (Mohamed, 2017). The advantage of seed cotton yield of these new lines over Barakat-90 was in the range of 4–28%. They had longer, stronger fibre compared to Barakat-90. The lines were earlier cropping and gave

Fig. 7.14. Unloading drums of insecticide.

Fig. 7.15. Drums stored in the sun.

45.6–61.2% of their yield in the first pick and finer fibres compared to 43.5% for Barakat-90. Hence, these lines signify improvement in seed-cotton yield, fibre quality, earliness of maturity and reaction to bacterial blight in Sudan extra-fine cotton. In addition to the commercial Chinese Bt cotton variety, one more Chinese (China2) variety and two Indian hybrids have been released for commercial production. The two Indian hybrids are characterized by good fibre quality to improve Sudan cotton marketability. Research is being conducted to optimize the Bt cotton varieties' needs, irrigation water, fertilizers and other protection measures specifically for sucking pests, like jassid.

The textile industry has suffered many problems, including a scarcity of electricity, but during the last five years the textile and clothing sector has witnessed resurgence due to direct foreign investment (FDI) with a large company such as Sur International, with Turkish Qatari and Sudanese shares establishing a number of factories in Khartoum and Hasahisa targeting 300,000 yards annually.

One problem in the Sudan was the storage of drums of insecticide imported from Europe, rather than getting the active ingredient imported and formulated locally. Movement of drums was done manually from the lorries. Drums were left unshaded and protected only by a wire fence (Figs 7.14–7.16). Subsequently, the drums could not be recycled and were often left in the open. Later, better storage facilities were introduced.

Fig. 7.16. Discarded drums.

References

Abdallah, N.A. (2014) The story behind Bt cotton: where does Sudan stand? *GM Crops & Food* 5, 241–224.

Ahmed, A.H.M., Elhag, E.A. and Bashir, N.H.H. (1987) Insecticide resistance in the cotton whitefly (*Bemisia tabaci* Genn.) in the Sudan Gezira. *Tropical Pest Management* 33, 67–72.

Bindra, O.S. (1985) Relation of cotton cultivars to the cotton-pest problem in the Sudan Gezira. *Euphytica* 34, 849–856.

Bouchaud, C., Clapham, A., Newton, C., Tallet, G. and Thanheiser, U. (2018) Cottoning on to cotton (*Gossypium* spp.) in Arabia and Africa during antiquity. *Plants and People in the African Past*, 380–426.

Burhan, H.O. and Jackson, J.E. (1973) Effects of sowing date, nitrogenous fertilizer and insect pest control on cotton yield and its year-to-year variation in the Sudan Gezira. *The Journal of Agricultural Science* 81, 481–489.

Butler, G.D. and Hennebury, T.J. (1990) *Bemisia* and *Trialeurodes* (Hemiptera: Aleyrodidae). In: Matthews, G.A. and Tunstall, J.P. (eds) *Insect Pests of Cotton*. CAB International, Wallingford, UK.

Campion, D.G., Hunter-Jones, P., McVeigh, L.J., Hall, D.R., Lester, R. and Nesbitt, B.F. (1980) Modification of the attractiveness of the primary pheromone component of the Egyptian cotton leafworm, *Spodoptera littoralis* (Boisduval) (Lepidoptera: Noctuidae), by secondary pheromone components and related chemicals. *Bulletin of Entomological Research* 70, 417–434.

Cowland, J.W. (1947) The cotton jassid (*Empoasca libyca*, Berg.) in the Anglo-Egyptian Sudan, and experiments on its control. *Bulletin of Entomological Research* 38, 99–115.

Critchley, B.R. and El-Dieb, Y.A. (1981) An improved pheromone trap for the mass-trapping of cotton leafworm moths in Egypt. *Appropriate Technology for Egyptian Agriculture* 4, 6–10.

Dittrich, V., Hassan, S.O. and Ernst, G.H. (1985) Sudanese cotton and the whitefly: a case study of the emergence of a new primary pest. *Crop Protection* 4, 161–176.

Downham, M.C.A., McVeigh, L.J. and Moawad, G.M. (1995) Field investigation of an attracticide control technique using the sex pheromone of the Egyptian cotton leafworm, *Spodoptera littoralis* (Lepidoptera: Noctuidae). *Bulletin of Entomological Research* 85, 463–472.

El-Adi, E.M., Hosny, M.M. and Campion, D.G. (2008) Mating disruption for the control of pink bollworm *Pectinophora gossypiella* (Saunders) in the delta cotton growing area of Egypt. *Tropical Pest Management* 34, 210–214.

El-Wakeil, N. and Abdallah, A. (2012) Cotton pests and the actual strategies for their management control. In: Giuliano, B. and Vinci, E.J. (eds) *Cotton: Cultivation, Varieties and Uses*. Nova Science Publishers, Hauppauge, New York, pp. 1–59.

El-Wakeil, N., Abd-Alla, A., El Sebai, T.N. and Gaafar, N.M.F. (2015) Effect of organic sources of insect pest management strategies and nutrients on cotton insect pests. In: Gorawala, P. and Prathamesh, S.M. (eds) Agricultural Research Updates. Vol. 10. Nova Science Publishers, Hauppauge, New York.

Eveleens, K.G. (1983) Cotton-insect control in the Sudan Gezira: analysis of a crisis. *Crop Protection* 2, 273–287.

Gerling, D., Motro, U. and Horowitz, R. (1980) Dynamics of *Bemisia tabaci* (Gennadius) (Homoptera: Aleyrodidae) attacking cotton in the coastal plain of Israel. *Bulletin of Entomological Research* 70, 213–219.

Griffiths, W.T. (1984) A review of the development of cotton pest problems in Sudan Gezira. MSc thesis, University of London.

Hall, D.R., Nesbitt, B.F., Marrs, G.J., Green, St J., A., Campion, D.G. and Critchley, B.R. (1982) Development of microencapsulated pheromone formulations. In: *Insect Pheromone Technology: Chemistry and Applications*. ACS Symposium Series 190. American Chemical Society, Washington, DC, pp. 131–143.

Hosny, M.M., Iss-hak, R.R., Nasr, El-S.A. and El-Shafei, S.A. (1981) The efficiency of hand picking egg-masses of the cotton leafworm *Spodoptera littoralis* (Boisd.) in relation to the rate of infestation. *Research Bulletin* 1529. Ain Shams University, Cairo.

Hosny, M.M., Topper, C.P., Moawad, G.M. and El-Saadany, G.B. (1986) Economic damage thresholds of *Spodoptera littoralis* (Boisd.) (Lepidoptera: Noctuidae) on cotton in Egypt. *Crop Protection* 5, 100–104.

ICAC Report (2018) *Sudan Cotton Crop Developments During the Last 10 Years*. Country Status Report, 13th Meeting of the Inter-Regional Cooperative Research Network on Cotton for the Mediterranean and Middle East Regions. Luxor, Egypt.

Jones, K.A., Irving, N.S., Grzywacz, D., Moawad, G.M. and Frargahly, A. (1994) Application rate trials with a nuclear polyhedrosis virus to control *Spodoptera littoralis* (Boisd.) on cotton in Egypt. *Crop Protection* 13, 337–340.

Joyce, R.J.V. (1959) The yield response of cotton in the Sudan Gezira to DDT spray. *Bulletin of Entomological Research* 50, 567–594.

Joyce, R.J.V. (1961) Some factors affecting numbers of *Empoasca lybica* (de Berg) infesting cotton in the Sudan. *Bulletin of Entomological Research* 52, 191–230.

Joyce, R.J.V. (1975) Sequential aerial spraying of cotton at ULV rates in the Sudan Gezira as a contribution to synchronised chemical application over the area occupied by the pest. *Proceedings of the 5th International Congress on Agricultural Aviation*. Cranfield, IAAC, 47–54.

Joyce, R.J.V. and Roberts, P. (1959) The determination of the size of plot suitable for cotton spraying experiments in the Sudan Gezira. *Annals of Applied Biology* 47, 287–305.

Mann, J.B. and Danauskas, J.X. (1984) Human effects associated with the use of aldicarb on cotton in Sudan, Africa. *Studies in Environmental Science* 25, 571–578.

Mansour, S. (2009) Egypt: Biotechnology. GAIN Report Number EG9012. Global Agriculture Information Network, USDA Foreign Agriculture Service. Available at: http://www.sciencedev.net/Docs/Biotechnology_cairo.pdf (accessed 29 July 2021).

Moawad, G., Khidr, A.A., Zaki, M., Critchley, B.R., McVeigh, L.J. and Campion, D.G. (1991) Large-scale use of hollow fibre and microencapsulated pink bollworm pheromone formulations integrated with conventional insecticides for the control of the cotton pest complex in Egypt. *International Journal of Pest Management* 37, 10–16.

Moawed, S.M. and Elnabrawy, I.M. (1987) Effectiveness of nuclear polyhedrosis virus and insecticides against the cotton leafworm, *Spodoptera littoralis* (Boisd.). *International Journal of Tropical Insect Science* 8, 89–93.

Mohamed, I.E.S. (2017) Evaluation of eight Egyptian cotton genotypes for yield, quality and bacterial blight resistance, Gezira State, Sudan. MSc thesis, University of Gezira. Available at: http://repo.uofg.edu.sd/bitstream/handle/123456789/611/Ikram%20Elsadig%20Suliman%20Mohamed%2C2017.pdf (accessed 29 July 2021).

Naik, M.I., Prasanna, S.O., Manjunatha, M., Shivanna, B.K. and Pradeep, S. (2009) Effect of organic sources of nutrients on major sucking pests in Bt cotton and their natural enemies. *Karnataka Journal of Agricultural Sciences* 22, 648–650.

Norris, P.K. (1934) *Cotton Production in Egypt*. Technical Bulletin No. 451. United States Department of Agriculture, Washington, DC.

Salama, H.S. (1983) Cotton-pest management in Egypt. *Crop Protection* 2, 183–191.

Salman, S. (2013) The Nile basin cooperative framework agreement: a peacefully unfolding African spring? *Water International* 38, 17–29.

Snow, O.W. and Taylor, J. (1952) The large-scale control of the cotton jassid in the Gezira and white Nile areas of the Sudan. *Bulletin of Entomological Research* 43, 479–502.

Storey, G. (1921) The present situation with regard to the control of the pink bollworm in Egypt. *Ministry of Agriculture, Egypt, Technical and Scientific Service. Bulletin* 16.

Topper, C.P. (1987a) The dynamics of the adult population of *Heliothis armigera* (Hübner) (Lepidoptera: Noctuidae) within the Sudan Gezira in relation to cropping pattern and pest control on cotton. *Bulletin of Entomological Research* 77, 525–539.

Topper, C.P. (1987b) Nocturnal behaviour of adults of *Heliothis armigera* (Hübner) (Lepidoptera: Noctuidae) in the Sudan Gezira and pest control implications. *Bulletin of Entomological Research* 77, 541–554.

Tunstall, J.P. (1958) The biology of the Sudan bollworm *Diparopsis watersi* (Roths.) in the Gash Delta, Sudan. *Bulletin of Entomological Research* 49, 1–23.

Uk, S. (1987) Distribution patterns of aerially applied ULV sprays by aircraft over and within the cotton canopy in the Sudan Gezira. *Crop Protection* 6, 43–48.

Wild, J.P., Wild, F. and Clapham, A. (2007) Irrigation and the spread of cotton growing in Roman times. *Archaeological Textiles Newsletter* 44, 16–18.

Yu, J., Yu, S., Fan, S., Song, M., Zhai, H., Li, X. and Zhang, J. (2012) Mapping quantitative trait loci for cottonseed oil, protein and gossypol content in a *Gossypium hirsutum* x *Gossypium barbadense* backcross inbred line population. *Euphytica* 187,191–201.

Yvanez, E. and Wozniak, M.M. (2019) Cotton in ancient Sudan and Nubia: archaeological sources and historical implications. *Cotton in the Old World: Revue d'ethnoécologie* 15.

8 Cotton in Southern Africa

Graham Matthews* and John Tunstall

Colonists in the USA began to grow cotton along the James River in Virginia in 1616, and with African slaves brought in to provide the labour needed to harvest the crop, cotton growing expanded in the southern states. Following slave emancipation at the end of the American Civil War, this labour under paid employment enabled the USA to continue as a primary source of cotton. The introduction and growth of cotton elsewhere in the world was stimulated by a shortage of cotton availability during the period of civil war in America. In southern Africa, some cotton was planted in the Western Cape as early as 1690, with a further attempt to grow cotton there in 1846, but it was in the 1920s that European settlers in South Africa and Swaziland (now Eswatini) (Sikhondze, 1989) took an interest in growing cotton. This coincided with the UK government deciding that it should examine the potential to grow cotton in Africa, as the impact of the First World War had questioned relying entirely on obtaining cotton from the USA. Portuguese settlers also ventured into cotton production in Angola and Mozambique.

In the 1920s, the Empire Cotton-Growing Corporation (ECGC) was established by the British government and set up a colonial research programme with plant breeders, agronomists and entomologists to develop a cotton industry with the small-scale African farmers. One of the earliest programmes was set up at Barberton in South Africa, where extensive research was carried out to select cotton varieties suitable for southern Africa and at Gatooma (now Kadoma) in Southern Rhodesia (now Zimbabwe), but very soon Parnell (1925; Parnell *et al.*, 1949) had utilized the hairiness of MU8 from India to select the U4 varieties, which proved to be resistant to the jassids (Fig. 8.1), but yields remained low due to bollworm and the lint was often stained due to damage caused by *Nematospora gossypii*, the spores of which were transmitted by cotton stainers, *Dysdercus* spp. (Figs 8.28 and 8.29) (Pearson, 1934). Biological control was studied to determine if *Trichogramma lutea* would be effective against bollworm (Parsons and Ullyett, 1936). In Rhodesia, seeds from the USA, notably Coker Wilds variety, failed to survive the infestations of jassids (*Empoasca*, now *Jacobiasca fascialis*) (Figs 8.2 and 8.3).

Pearson later moved to Nyasaland (now Malawi) and the studies on the insect pest complex, especially into the problem of red bollworm, *Diparopsis castanea*, in Malawi resulted in a major report on the control of insect pests in the Shire Valley (Pearson and Mitchell, 1945). This led him to recommending to the UK government to establish a Cotton Pest Research Scheme administered by the Nyasaland government in 1956,

*Corresponding author: g.matthews@imperial.ac.uk

(eds G.A. Matthews and T. Miller)
DOI: 10.1079/9781800620216.0008

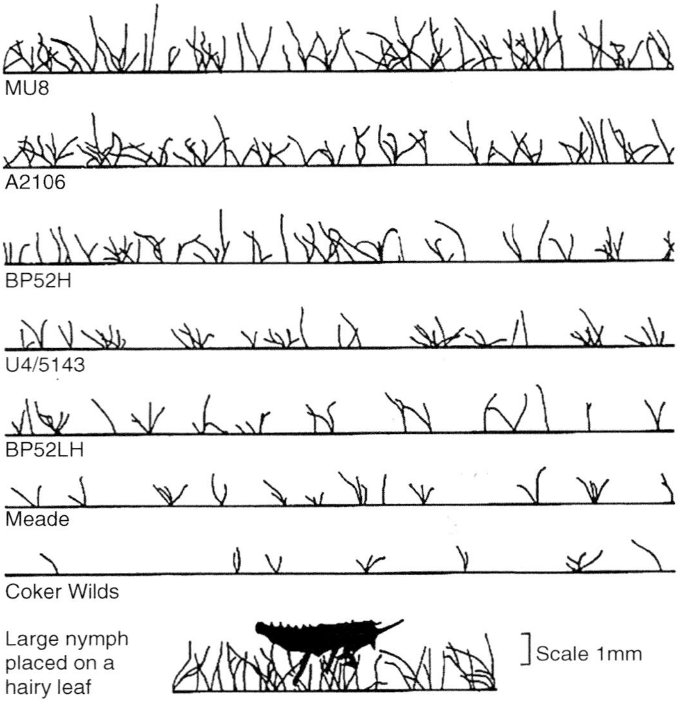

Fig. 8.1. Range of hairiness of cotton leaves in contrast to the hairiness of the variety MU8 from India resistant to jassids and the Coker Wilds cotton from USA which was susceptible. (From Parnell *et al*., 1949)

Fig. 8.2. Jassid adult *Jacobiasca*.

Fig 8.3. Jassid damage on suceptible variety, with row top left, resistant to damage.

which covered each of the cotton-production areas of the Federation of Rhodesia and Nyasaland. This project was particularly aimed at the small-scale cotton growers in Nyasaland, but benefitted these farmers throughout the Federation, who had not been adequately helped previously, while farming was largely governed by the settlers. The Department of Research and Specialists in Southern Rhodesia assisted the starting up of the project by seconding an entomologist and later provided an extension officer attached to the project. The project became a

Figs 8.4 and 8.5. *Diparopsis* moth emerging from soil, and drying wings during the night at about 3 am.

model research programme for the setting up of the Agricultural Research Council of Rhodesia and on the break-up of the Federation in 1967, the scheme became a unit of a newly formed Agricultural Research Council of Malawi. Initially, detailed studies were on the red bollworm to determine when moths emerged (Tunstall, 1968) and revealed that the first instar larvae were the most exposed stage to insecticide sprays, before entering a bud or boll.

The scheme fulfilled its purpose in the development of greatly improved integrated pest management with pesticide application and demonstrated the importance of operating research on a regional base where there is a range of farming levels from smallholder to estate cotton production. The region has since seen further progress being made in combating red spider resistance (Duncombe, 1973) and introduction of newer pesticides of sufficient safety in use (Brettell, 1986).

In 1957/58, there was a viewpoint in Nyasaland that someone should go to South America to find the natural enemies of *Sacadodes*

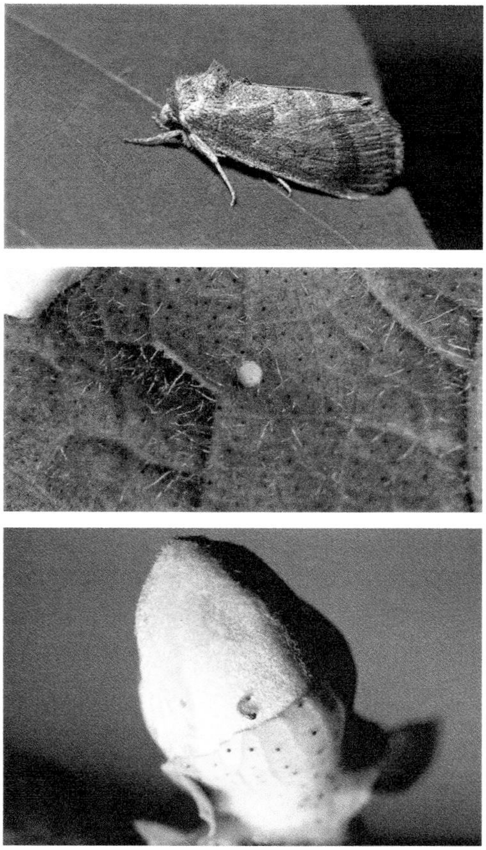

Figs 8.6, 8.7 and 8.8. *Diparopsis* moth, egg on cotton leaf and first instar larva on a bud.

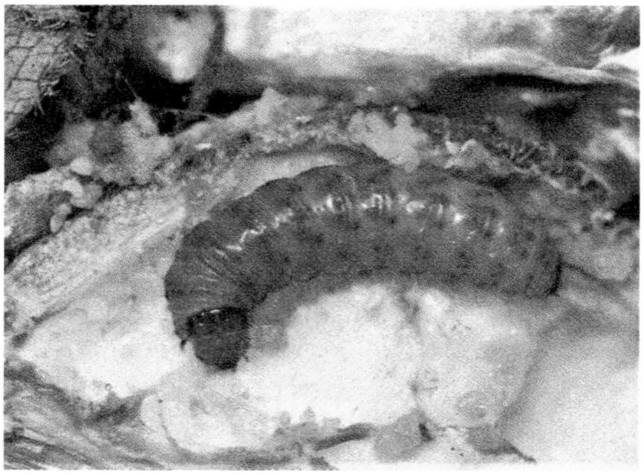

Fig. 8.9. Fully grown larva inside a cotton boll.

pyralis, a bollworm closely related to *Diparopsis* (Pearson, 1954), but this was not accepted. Pearson had wanted a chemist in the team and to show farmers how yields had been increased in Texas using DDT, so the third entomologist to join the team had studied some chemistry at university.

Detailed studies on the biology of *Diparopsis* on the emergence of moths following a pupal diapause (Tunstall, 1968) later included 24-hour, 7-day observations of the life cycle of the bollworm, which confirmed that the larvae were most exposed to a spray deposit only between egg hatching and penetration of the first instar larvae into a bud or boll. The insecticide endrin had been tried in some initial trials in 1955/56 at 280 g a.i./ha applied at weekly or fortnightly intervals, and increased yield by 450–900 kg/ha (Gledhill, 1979), but the toxicity of endrin was a concern, so after laboratory bioassays (Matthews, 1966), a programme using carbaryl against red bollworm and changing to DDT to control *Helicoverpa armigera* was developed (Tunstall and Matthews, 1966). Particular attention was given to the method of application. Using endrin in trials, the operator had a lance fitted with two nozzles pointing laterally into the plants either side of the spray operator, so the person applying the insecticide was walking towards the spray (Fig. 8.10).

Fig. 8.10. Lance with two nozzles spraying laterally.

Subsequent research showed that as cotton plants grew in height during the period when bollworm caused damage, coverage of the canopy had to be better than can be achieved with only one or two nozzles. Thinking about mechanized sprayers spraying from a boom, an attempt was made to try having a horizontal boom carried by two operators (Fig. 8.11), but it was soon realized that this was not appropriate as both operators did not walk at the same speed. The other early decision was that the operator should not walk forward into the spray. This led to the idea of mounting nozzles on a vertical boom fitted to the tank on a lever-operated knapsack sprayer. The first sprays on small plants needed only one pair of nozzles, then for each 30 cm increase in plant height an additional nozzle pair was used. The number of nozzles used increased during the season in relation to plant height and the nozzles were angled slightly upwards to improve deposition on the underside of leaves (Fig. 8.12). The spray volume applied was 50 litres/ha for each pair of nozzles, so on the largest plants, the volume of spray had increased to 200 litres/ha. To reduce exposure of the operator to insecticides, it was decided to mount a series of nozzles on a 'vertical' boom mounted on the knapsack sprayer tank and referred to as a 'tailboom' (Figs 8.13–8.15) (Tunstall *et al.*, 1961). A cone nozzle Y3 was selected with a 60° angle and obtained from a US company, and the operator needed to pump to maintain a pressure of 3 bar at the nozzle.

The improved spraying technique and timing of sprays related to crop monitoring for bollworm eggs was examined over two seasons from 1960–1962, and in both Southern Rhodesia and Nyasaland in a large trial comparing sprayed and unsprayed cotton in adjacent plots on farms. See increase in yield in 1962 (Fig. 8.25) and production (Fig. 8.26).

Subsequently the adoption of the technique by farmers was facilitated by active collaboration with the extension services and enabled cotton production to increase. The *Cotton Handbook* was first published by the Rhodesia Cotton Growers' Association in 1964 (Fig. 8.27) with complete revisions in 1971, 1977, 1985 and 1995, by which time it was the *Cotton Handbook Zimbabwe*. Other minor revisions were circulated to registered growers. Information for the small-scale farmers was provided in

Fig. 8.11. A Two-man boom sprayer was tried.

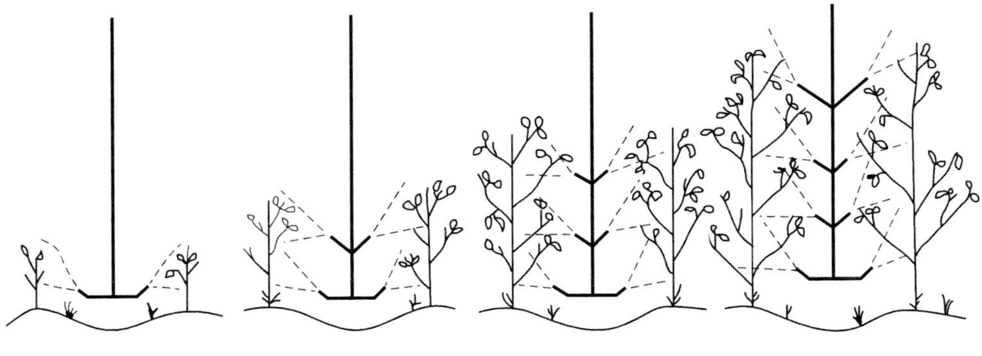

Fig. 8.12. Increase in number of nozzles on tailboom with plant height.

booklets translated into a local language. A poster was also provided that showed pictures of the pests and recommended equipment with changes in the number of nozzles needed on the tailboom as the plants gained height. A training school was also introduced (Burgess, 1983). Later, a pegboard was developed to facilitate scouting (Figs 8.16a and b). The spray programme was also adopted by cotton farmers in Northern Rhodesia (now Zambia).

An important decision was to avoid any highly toxic insecticide, as its use would require the operator to wear protective clothing. However, by mounting the spray nozzles behind the operator to avoid the body touching sprayed foliage, the operator was advised to wear shoes, long trousers, a long-sleeved shirt and a hat. Studies confirmed that the operator was less exposed to the spray when using the sprayer fitted with a tailboom (Figs 8.19 and 8.20). The hat was mainly to shade the head from tropical sun. Gloves were not included as the insecticide was supplied in sachets (Figs 8.17 and 8.18) and any of the powder could be washed off the hands immediately. Operators were advised to always have a separate supply of water for washing and drinking.

Fig. 8.13. Nozzles mounted on rear of knapsack sprayer.

Later, with ULV sprays, gloves and eye shield were recommended but operators were trained to keep the spray downwind of the body.

Farmers with a larger area of cotton soon wanted to use a tractor sprayer, so a boom fitted with shielded tailbooms was designed (Fig. 8.21). The shield acted like the person operating the knapsack sprayer by preventing branches being caught by the nozzles and hosing to each nozzle (Tunstall *et al.*, 1965). Similarly, an ox-drawn sprayer was also developed (Fig. 8.22). Later, some farmers wanted aerial sprays, and further research with a helicopter showed that good yields could be obtained (Fig. 8.23) (Matthews and Tunstall, 1965). Fixed-wing aircraft were used commercially as operational costs were lower (Fig. 8.24).

The programme of spraying insecticides was considered to be part of an integrated pest-management programme (IPM). Plant breeders had utilized the resistance to jassids by having pubescent leaves and added resistance to *Xanthomonas malvacearum*, the cause of Blackarm disease, so in 1960 a new variety, Albar 637, was available and increased the yield potential. Detailed studies of the activity of first instar larvae between hatching from the egg to

Fig. 8.14. Original 'tailboom' with Dexion support for nozzles and a later version with sachets of insecticide.

finding a bud showed that the larvae took longer to find a bud on plants with pubescent leaves compared to glabrous leaves, and on sprayed plants could pick up the spray deposit and fail to reach a bud (Matthews, 1966). Sowing acid-delinted seed reduced the spread of certain seed-borne diseases. Cultural control was achieved by making sure farmers followed the policy of a closed season of at least two months, after harvesting and sowing the next crop, which had been introduced in 1938 (McKinstry, 1938). There were unexpected outbreaks of pink bollworm (*Pectinophora gossypiella*), where cotton plants had not been removed or ploughed after

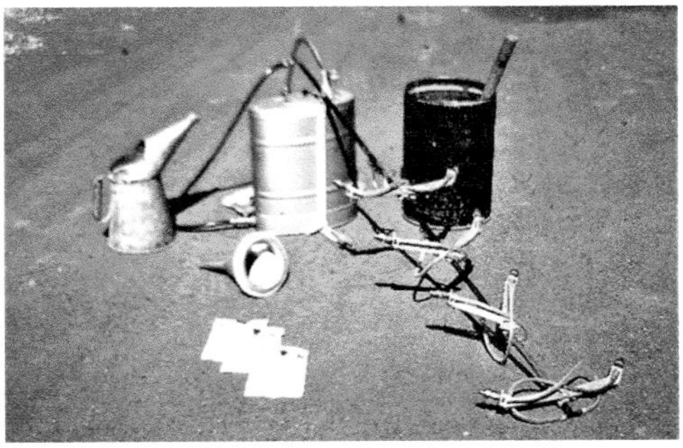

Fig. 8.15. The jug, mixing drum and filter were later combined into a single 3 gallon container.

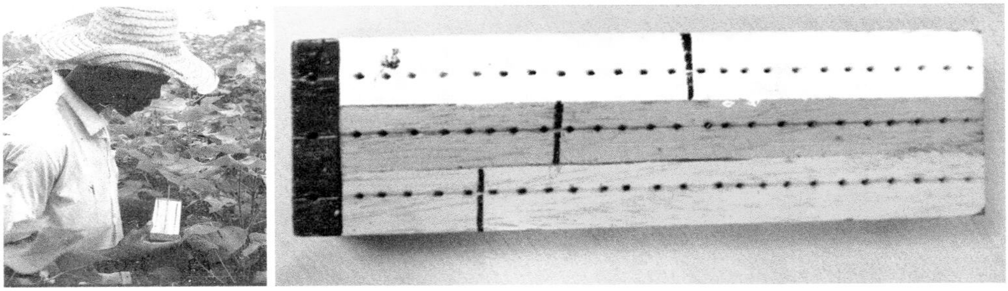

Fig. 8.16. (a) Farmer scouting with pegboard; (b) pegboard.

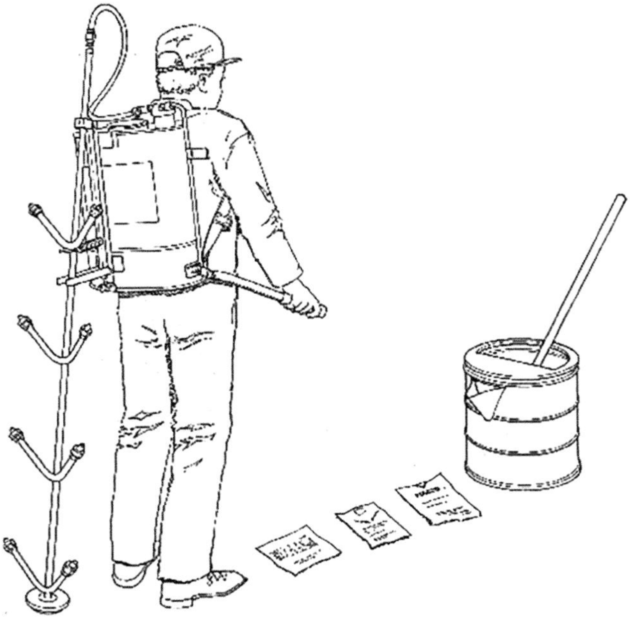

Fig. 8.17. Sprayer with tailboom and specially made mixing container with filter within the spout to facilitate pouring into the sprayer.

harvest near the Mozambique border (one in Malawi and also in the Sabi-Tanganda valley area of Zimbabwe). In the latter area, pink bollworm was also found on wild *Hibiscus dongolensis* alongside the cotton area (Matthews *et al.*, 1965).

These developments in the 1960s were also taken up elsewhere in Southern Africa. An entomologist from Mozambique visited the Cotton Pest Research team and subsequent return visits were made to see cotton in the Limpopo Valley and further north. The CPR team collaborated

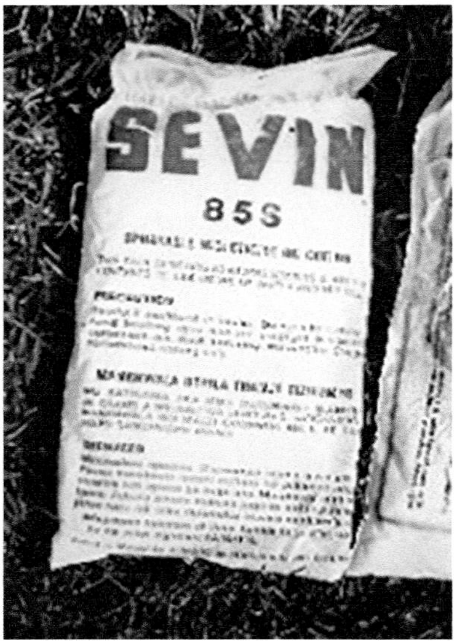

Fig. 8.18. Sachet of carbaryl pesticide.

Fig. 8.19. Operator with 35 mm film as sampling surface for insecticide deposit, positioned in different parts of the body.

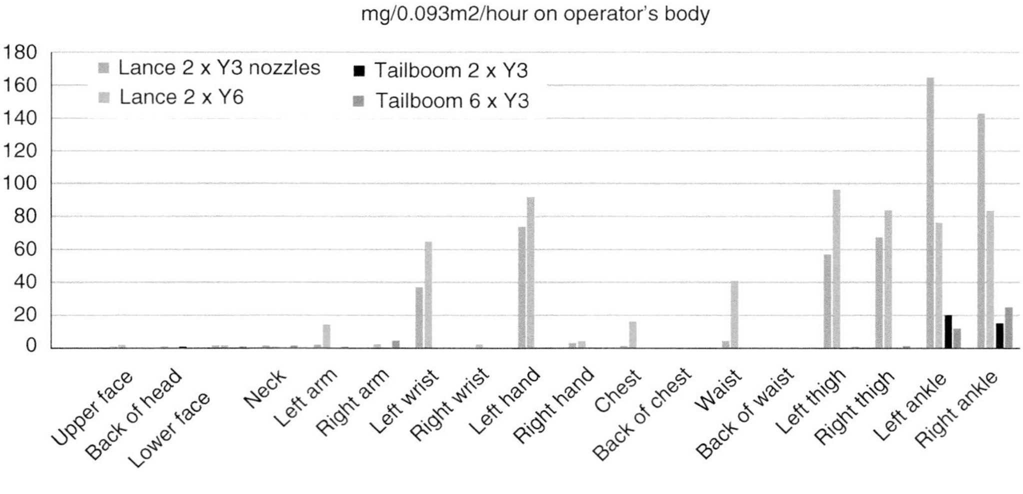

Fig. 8.20. Operator exposure with single lance versus tailboom.

Fig. 8.21. Tractor-spraying in Zimbabwe.

Fig. 8.22. Ox-drawn sprayer in Zimbabwe.

Fig. 8.23. Helicopter spraying.

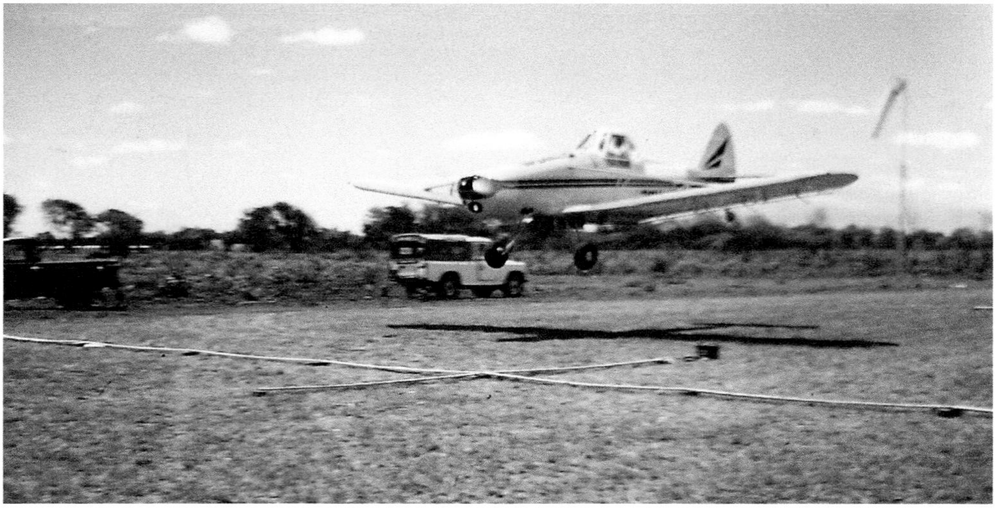

Fig. 8.24. Checking swath width with fixed-wing aircraft.

with trials in Northern Rhodesia (covered by Federation), Swaziland and South Africa, as the insect pest complex was similar throughout the region.

The ability to control bollworm vastly increased yields, but production was low in some seasons due to drought conditions and insecurity due to the campaigns seeking independence

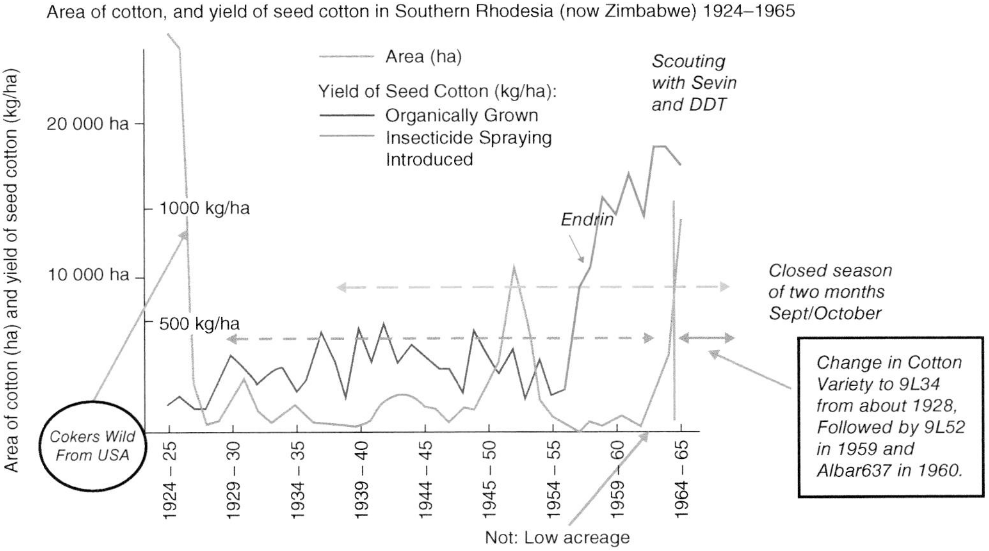

Fig. 8.25. Changes in area and yields obtained in Southern Rhodesia between 1925 and 1966 (Matthews, 2014).

from European control (Fig. 8.26). Unfortunately, land reforms in Zimbabwe in 2000 resulted in the loss of cotton production on large farms (Fig. 8.26).

Other studies on the red bollworm included using traps with a moth to see if the pheromone would attract wild moths. Pupae were sent to London to determine the chemical structure of the pheromone released (Nesbitt *et al.*, 1973). Subsequently, detailed trials with dicastalure and combined with gossyplure were carried out in Malawi to determine if traps could be used to monitor the moths, but traps were not used to assess oviposition on farmers' cotton (Marks, 1976a; Marks, 1976b; Marks, 1978).

South Africa

Cotton had been introduced into South Africa by the explorers that used Cape Town as a key port *en route* to and from Asia. It was grown on a small scale in the Western Cape in 1690, and in 1909 William Seherffius was attracted to move from his position as the head of the Kentucky Experiment Station and expand South African cotton cultivation, which was stagnating at a lowly 30 bales. He initiated a research programme and encouraged cotton growing in Zululand, where 80,000 ha of land were designated for settlement in 1917 to provide farming opportunities for soldiers returning from the battlefields of the First World War. The area of cotton jumped from 250 acres in 1917 to 500 acres in 1918 to 4000 acres in 1919 (Schnurr, 2011). The area of cotton more than doubled and by 1925, Zululand cotton growers thought that they were on the cusp of the greatest yield southern Africa had ever seen. Production increased from just over 800,000 pounds in 1922 to over 8 million pounds in 1925 from over 30,500 acres of cotton, accounting for just under half of the 67,500 acres devoted to cotton throughout South Africa. Subsequent poor weather conditions and other factors resulted in low yields so cotton production declined. The next phase was dependent upon research at Barberton, where apart from the efforts in plant breeding referred to earlier and efforts to control jassids, the life cycle of red and American bollworms was investigated by Parsons and Ullyett (1934), while the cotton stainers were also studied (Pearson, 1934).

Following the research in Southern Rhodesia under the CPR Scheme, cotton farmers

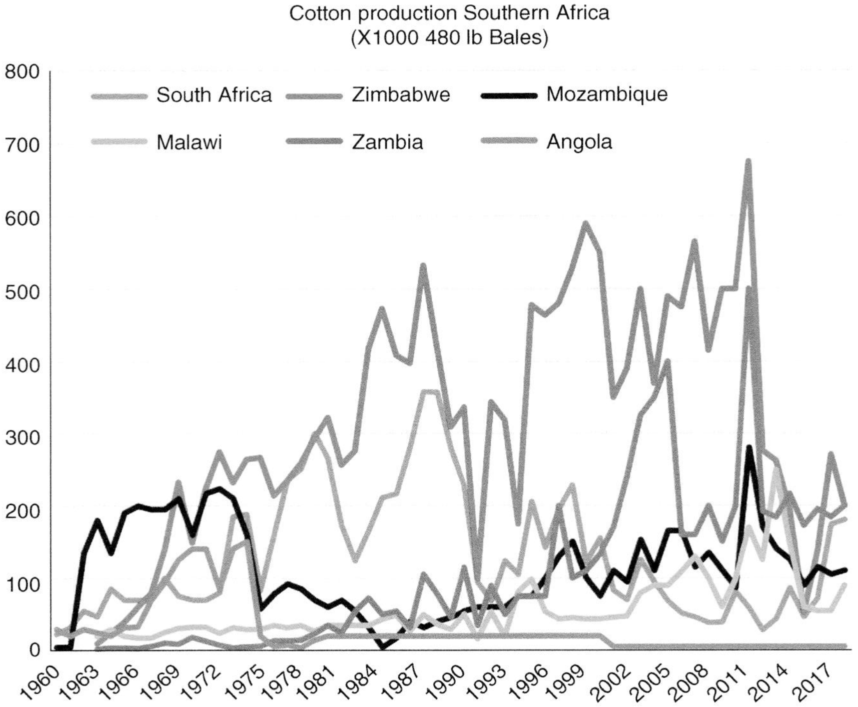

Fig. 8.26. Production of bales of cotton from six countries in southern Africa 1960–2018. (Source: https://www.indexmundi.com/agriculture)

adopted similar recommendations and production increased in the 1960s before decreasing in the 1990s (Fig. 8. 26). In South Africa, cotton is now grown in the north-west, Limpopo and Mpumalanga regions, in Northern Cape and to the east in KwaZulu-Natal. The crop depends on adequate rainfall, but some areas have irrigation. Seventy-five per cent of local production is harvested by hand.

Unlike other parts of southern Africa, genetically modified cotton has been grown by both commercial and small-scale farmers and has contributed significantly to improving the yields and the profitability of cotton production, but the area planted has not increased. Smallholders in Makhathini Flats, KwaZulu-Natal, showed that poor-resource farmers who adopted Bt cotton in 1999–2000 benefitted by obtaining higher yields and lower chemical costs, although seeds were more expensive, giving higher gross margins. In protecting the Bt crops from sucking pests, a highly hazardous insecticide, monocrotophos, was unfortunately used (Bennett *et al.*, 2004). Adopters of GM cotton were shown to be more efficient than non-adopters (Beyers *et al.*, 2003). However, where multiple sources of uncertainty abound, such as rainfall and pest density, depending on the relative importance of each random variable, Shankar *et al.* (2008) concluded that a plausible interpretation of the results is that Bt technology best produces its effects in Makhathini, when the crop-growth conditions, such as rainfall, are good. In 2007, the Mkahathini ginnery was closed under liquidation by the Land Bank, but in 2010 the Department of Agriculture and Fisheries purchased a ginnery intending to recommission it to process locally produced cotton. It opened in 2012 and the plan was to transfer the ginnery to the Ubongwa Farmers' Cooperative with over 1500 farmer members.

The textile industry is relatively small with a large number of small and microenterprises and only 20 large companies, five of which account for 40% of the income. The industry relies on the export market as the local market is dominated by imports, particularly from China.

Fig. 8.27. Cotton Handbook for the Federation of Rhodesia and Nyasaland and later the Cotton Handbook in Zimbabwe.

Eswatini (Formerly Swaziland)

A few settlers in Swaziland began to grow cotton during the First World War with some continuing during the 1920s and 1930s although the prices paid for cotton were low. The Department of Agricultural Research and Specialists Services (DARSS) was initiated by the Ministry of Agriculture in 1959, with a central research station at Malkerns, and sub-stations at Big Bend and Nhlangano. In the late 1950s, external brokers and farmers owning cattle invested in an increase in cotton growing (Sikhondze, 1989). Cotton was grown mostly in the south-east of the country near Big Bend, where wild cotton, *Gossypium herbaceum* subsp. *africanum*, could be found. The production of Swaziland cotton has until now been processed through South Africa.

Zimbabwe

Indigenous cotton was grown in some areas of what was then Southern Rhodesia at the end of the 19th century. During the early 1900s, the first research trials were conducted by the British South Africa Company using seed from Egypt, Brazil, the United States and Peru (Poulton and Hanyani-Mlambo, 2008).

The early settlers in Southern Rhodesia aided by the British Cotton Growing Association (BCGA) endeavoured to develop a viable cotton-growing industry to reduce the British textile industry's dependence on American cotton, but progress was slow during the period of British South Africa Company Administration up to 1923. Then with the formation of the Empire Cotton Growing Corporation (ECGC) the situation was transformed when the colony's first cotton research station was started at Gatooma (now Kadoma) in 1924. In 1925, Major Cameron was sent to head the cotton station and played a key role in the development of the cotton industry until he died in 1958.

In 1936, the Cotton Research Industry Board (CRIB) was established to oversee the development of the cotton industry, with a cotton growers' representative on the board to keep cotton farmers informed about the board's functions. Three ginneries in Gatooma were funded by the British Cotton Growers Association (BCGA) and in 1941 the first cotton-spinning mill was built with a capacity of 1000 spindles. This was later increased to 35,000 spindles with the construction of the No. 2 mill in 1951. A small factory for the manufacture of absorbent cotton wool was also located in the Cotton Research Institute. CRIB was reconstituted in order to broaden the scope of its activities and was responsible for establishing and developing 'within the Colony, textile and allied industries', as well as supervising all research work on cotton (Nyambara, 2000; Mlambo and Phimister, 2006; Nyambara, 2014).

During the Second World War, A.H. McKinstrey and A.N Prentice had been relocated from Gatooma to East Africa as the ECGC considered it was more important that they supported an area where there was a larger number of small-scale growers of cotton. McKinstrey returned later to become in charge of the Gatooma Research Station where he closely collaborated with the Cotton Pest Control Research programme when it was established covering the Federation of Rhodesia and Nyasaland in 1956 and, as indicated above, resulted in the ability to control bollworm. The area of both rain-grown and irrigated cotton came under large scale-management immediately. This increase was achieved by close collaboration between the CMB, the Commercial Cotton Growers' Association and the Cotton Research Institute, notably with regard to cotton seed-breeding programmes.

With the successful introduction of IPM in 1963, having a cotton advisory officer stationed at the Cotton Research Institute (now Kadoma) was pivotal, facilitating liaison between researchers and extension workers in the consolidation of research and development in pest management and the very high-quality seed used in Zimbabwe (Mariga, 1994). As pointed out later by Hillocks (1995), IPM in cotton included consideration of disease and weed control as well as insect pest control.

After the Rhodesian government declared unilateral independence (UDI) in 1965, it developed an industrial sector designed primarily to service the domestic economy supplemented by an export market in neighbouring South Africa. Only about 11% of Zimbabwean firms were engaged in export production, with few producing large-volume runs of fabrics for export; only two in the spinning and weaving industry provided around 75% of the fabric supplied to the domestic clothing industry (Jackson, 2004).

Relatively little attention was given in Southern Rhodesia to the small-scale indigenous farmers, although production by smallholders expanded following independence. On some small irrigation projects, such as the one at Nyanyadzi, there were very high yields of seed cotton from improved methods of pest control. This led the World Bank to put pressure on Zimbabwe to deregulate and liberalize the cotton trade in 1994 (Larsen, 2002). Ginning and marketing companies Cargill (a transnational corporation) and Cotpro (an arm of the Commercial Cotton Growers Association) competed with the newly named Cottco (formerly CMB – Cotton Marketing Board) by offering different contract packages to growers, although Cottco led the pricing system by offering competitive packages to growers. With lower prices in the 1990s, communal farmers began to face mounting constraints. Inflation increased and economic stagnation made it difficult to secure input credit. As in Malawi, there have been reports that the import of second-hand clothing adversely affected the local textile industry with more of the cotton then being exported. However, in 2000, Zimbabwe produced 350,000 tonnes of seed cotton in comparison with 1987/88. By offering different packages to growers, communal farmers produced 323,000 tonnes (Larsen, 2002).

Production then fell (Fig. 8. 26) when land reform resulted in the loss of the large-scale cotton farms, although the number of small-scale growers had increased (Poulton and Hanyani-Mlambo, 2008; Matthews and Tunstall, 2019). At this time, a school to train farmers in scouting and other factors for growing cotton was closed. Unfortunately, although an effort was made to provide inputs for the small-scale farmers, continued political problems and unfavourable weather resulted in a further decline in production in 2018/19, to some extent due to farmers no longer using the 'tailboom' to spray their crops, so control of pests was not as effective.

Red spider mites were observed to have resistance to dimethoate sprayed not only on cotton but also on many other crops, so an acaricide rotation scheme was introduced to restrict the use of one mode of action for no more than two seasons, before its use was transferred to a new part of the country (Duncombe, 1973).

Zambia

Some cotton was grown in Zambia as early as 1911, but as in Zimbabwe between the two World Wars, yields were low due to insect infestations.

Progress was made after 1956 when, following trials, cotton was grown by small-scale and large farms, production reaching 2.6 million kg in 1966 and 12.7 million kg of seed cotton by 1971 (Harkema, 1972). Some aerial spraying and mechanical harvesting were tried. During the 1960s, boron deficiency was observed and a soluble boron was added to insecticide sprays prior to flowering (Rothwell *et al.*, 1967). Subsequently, cotton has been an agricultural success story led by private cotton ginneries and smallholder production. Zambia's cotton purchasing, processing, and marketing was controlled by the state-owned Lint Company of Zambia (LINTCO), which purchased seed cotton from an estimated 140,000 small farmers at a fixed price and extended services such as the provision of certified seeds, pesticides, sprayers, bags, and advice on growing techniques. LINTCO was the principal buyer of seed cotton, the sole provider of extension services, and the sole distributor of inputs on credit. However, by the early 1990s, LINTCO was operating at a loss, having accumulated substantial unpaid debts.

In 1994, Zambia's president, Frederick Chiluba, implemented a wide-ranging restructuring of Zambia's economy. This resulted in the sale of LINTCO to Lonrho Cotton and Clark Cotton, two private ginners and exporters. Subsequently, the cotton industry initially grew rapidly and attracted new ginning companies and independent outgrower agents to the market. However, by 1998, ginning capacity exceeded production capacity and there were too many outgrower agents competing for a limited number of farmers. Competition between agents led to growing distrust and a lack of transparency in price-setting, with agents and ginners vying to outbid each other for cotton. The entire value chain became volatile as ginners and outgrowers experienced very high incidences of defaulted loans. Production has varied, but not so well since 2007 (Goeb, 2011), partly due to lack of support for cotton growers in contrast to the support farmers received to increase maize production.

Instead of employing a large number of extension agents with high overhead cost to the ginning company, in the new model, ginners used almost no direct-hire employees to deliver services. Instead, distributors were mobilized via formal written contracts to identify farmers. The distributors would acquire the inputs from a ginner company, on credit, deliver them along with technical advice to the farmers, and ensure that farmers sold their cotton back to Dunavant to recover the input credit. In this scheme, the distributor's compensation was directly tied to the amount of credit recovered.

With the public-sector reforms implemented by the Zambian government in the mid-1990s, the Cotton Development Trust (CDT) was created in 1999 after exhaustive consultations by the Ministry of Agriculture and Cooperatives (MACO, currently the Ministry of Agriculture and Livestock, MAL) and other stakeholders. The primary aim of forming the CDT was to facilitate the development of cotton in Zambia through improved service delivery to the industry and enhanced participation of the private sector. All the assets of the Magoye Regional Research Station located in Southern Province were transferred to the CDT Board of Trustees in addition to a capital and recurrent grant from the World Bank – International Development Association (IDA) under the Agricultural Sector Investment Programme (ASIP). These reforms included commercialization of research activities, which became the key guiding principle for the creation of the trusts that were involved in agricultural research, including the CDT.

CDT, despite ongoing financial and operational challenges continued to make positive contributions to the development of cotton in Zambia. New varieties with higher yield and lint quality were adapted to meet the national demands and challenges of the market and for good agronomic performance in the various cotton-growing environments of Zambia. Provision of foundation seed to the industry improved both the lint quality and quantity marketed.

The CDT set up an IPM unit at Magoye primarily to reduce pesticide use on insect control through the introduction of alternative pest-management strategies that are less hazardous to the farmers and more environmentally friendly. The unit has expanded its IPM training activities to pilot areas in: Magoye, Sinazeze, Chikanta, Mumbwa/Nangome and Chipata. Other contributions aimed to increase arable farm land available for cropping activities, ameliorate the low soil productivity of degraded farmlands by improving soil acidity and organic matter content and introduce conservation tillage and farming methods for the CDT commercial and research farms, improve the skills and technical knowledge of CDT staff in research and farm management, including proper operation and maintenance of tractors and farm equipment, and facilitation of knowledge exchange through participation in regional and international conferences and exchange visits. At one stage, CDT became the biggest employer in the Magoye area with over 500 people employed seasonally.

Cotton is grown mainly in the eastern part of the country near Chipata and the Luangwa Valley. In this area, kidney cotton, *Gossypium brasiliense*, has been observed, presumably as a result of Arab traders travelling inland from Dar es Salaam much earlier. Cotton is grown also in areas north-west and south of Lusaka. More recently, the China-Africa Cotton Company has invested in helping farmers with better quality acid-delinted seed and by purchasing seed cotton at harvest with cash (Tang, 2019).

Malawi

In 1889, the area, explored by David Livingstone from 1858 onwards was established as the British Central Africa Protectorate (BCA), ratified in 1891 and later renamed Nyasaland in 1907. It became independent as Malawi in 1967. Dawe (1993) describes how shortages of cotton during the American Civil War stimulated interest in getting supplies from other countries and describes early interest in Central and East Africa. The British Cotton Growing Association (BCGA) provided limited amounts of Egyptian cottonseed to African farmers in the Upper Shire Valley in 1903, but in 1905 the Association sent an expert to Nyasaland, who advocated wider seed distribution, building ginneries and providing markets in lowland areas suitable for growing long-staple Egyptian cotton by African farmers, although until 1914 they were discouraged from growing cotton in upland areas where it might compete with existing estate farming. From 1902 to 1922, almost all cotton was grown on European-owned estates, but prices of seed cotton slumped and the number of Europeans growing the crop declined drastically with a majority switching to growing tobacco. In 1910, the BCGA built a ginnery and agreed to purchase part of the cotton crop. Production by Africans flourished along the Shire Valley, where river, and later rail, transport was available, but it was often abandoned in areas where transport costs were high.

After a rapid rise from virtually nothing before 1908, cotton production for the whole Protectorate was around 1000 short tons in the years prior to 1918, before declining until 1925 when it rose to around 3000 short tons per annum before another slump in the early 1930s. Part of the problem was the distribution of seed to white planters who had no experience of growing cotton. From 1925 to 1936, on average 80% of production was from the Lower Shire districts of Nsanje and Chikwawa, but production there later declined, as parts of the area became permanently inundated by the Shire River. In 1935, the Zambezi Bridge was constructed, creating an uninterrupted rail link from Blantyre to the sea at Beira, and in the same year a northern railway extension to Lake Nyasa was completed. The close border with Mozambique led to the BCGA and the government being concerned that growers had an alternative market for their cotton. By 1936, the UK Colonial Office considered that, except for tea, European agriculture in Nyasaland had failed in economic terms and that indigenous farmers would be the cotton producers of the future. Under the government ordinance, cotton plants had to be uprooted and burned after harvest to prevent disease and the growers had to have their seed cotton ginned within Nyasaland and not sell it to Mozambique (Robins, 2016).

In a subsequent report (1937), Professor J.W. Munro from Imperial College, London, impressed by the work of the ECGC, considered that an addition to the staff in Nyasaland may be necessary. Subsequently, Eric Pearson, who had been studying cotton stainers in South Africa (Pearson, 1934, 1958), moved to Nyasaland and research was established at Makanga, near Chiromo, in Shire Valley and Chitala for the Lake Shore area. According to Ducker (1929), yields were low due to jassids. During the Second World War, cotton was exported to South Africa, Australia and India, but by 1951, 80% was being sent to Britain. As discussed earlier, Pearson and Mitchell

Fig. 8.28. Young nymphs of *Dysdercus* on open boll.

Fig. 8.29. Cotton stainer, *Dysdercus intermedius*.

(1945) reported on the situation and the need to control insect pests in the Shire Valley.

The Cotton Pest Research Scheme, initiated in 1956, began trials on farms in 1960 and demonstrated that the introduction of a spray programme could increase their cotton yields significantly (Tunstall and Matthews, 1966). A monthly newsletter was distributed to the extension staff so it was read with the senior staff and any questions could be answered or other advice given before the staff returned to their base and disseminated the information to farmers in their area (Fig. 8.30). A *Cotton Handbook of Malawi* was published in 1971 and a second edition in 1976 (Fig. 8.31). As in Zimbabwe, posters were distributed to farmers and advice was also provided in a weekly short radio programme.

All the farmers growing cotton were registered when they obtained the seeds and were obliged to return the harvested seed cotton to the same Farmers' Marketing Board office. Sachets of insecticide were stocked at the FMB depot and were sold for cash or a slightly higher price if the farmer arranged to repay at harvest. When each load of seed cotton was taken to the FMB depot, the farmer received at least 50% of the value and

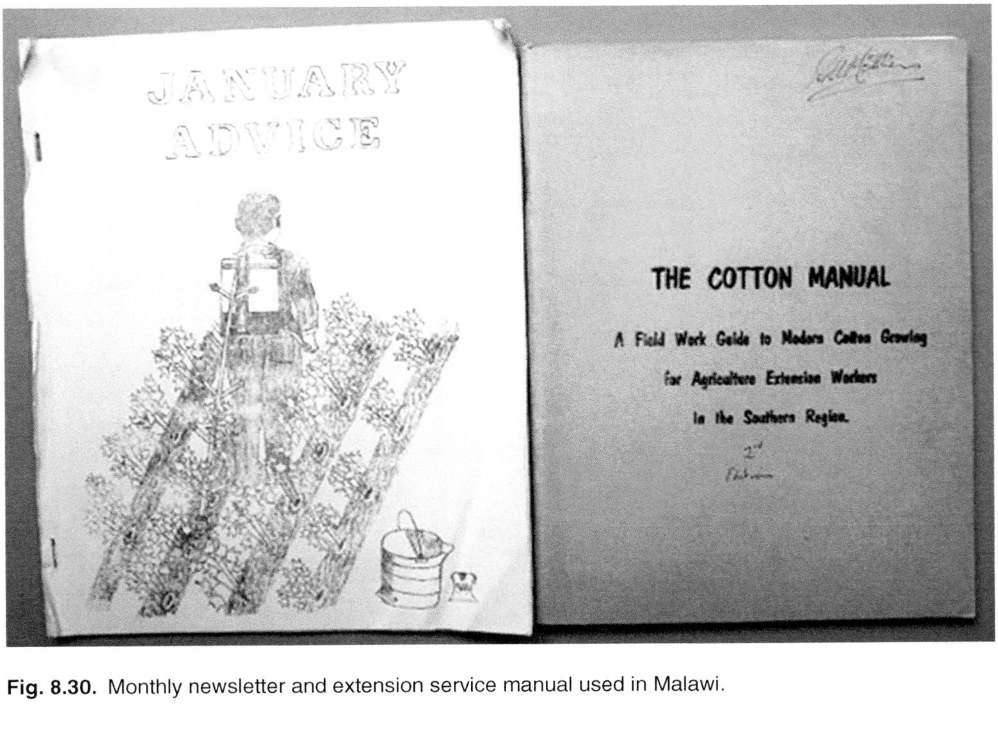

Fig. 8.30. Monthly newsletter and extension service manual used in Malawi.

any loans to purchase insecticide were progressively repaid, so that by the end of the season over 90% of the loans had been recovered. The area of cotton grown and sprayed expanded more in areas which had no traditions or inbuilt resistance to new methods of growing the crop (Beeden, 1971), but collecting water for spraying the tall plants (over a metre high) later in the season was identified as a problem. The feasibility of using aerial spraying, using a Piper Pawnee, was examined on 250 acres of cotton in three blocks involving 96 farmers and expanded a year later with 316 farmers (Beeden, 1971). Although it saved farmers the need to carry water to their fields, it was not continued.

At the time the Cotton Pest Research Scheme was relocated in Malawi, the government of Malawi applied for an IDA credit (World Bank) to help finance a programme to induce existing farmers in the Shire Valley to adopt improved agricultural practices and to settle others on as yet uncleared land (Anon, 1968). A similar scheme financed by aid from Germany was set up in the Lake Shore area. The immediate impact of the programme proposed was to increase yields, especially of cotton, aiming to more than double cotton production with a modest increase in area. Improved production of the staple food maize was also expected. Although the supply of sprayers was facilitated by a 40% subsidy, by 1970 the adoption of the use of insecticides was limited to only 28% of the cotton farmers (Farrington, 1977), which was due to the majority of farmers being unable to get sufficient water to spray using a knapsack sprayer. Research using ultra-low-volume sprays enabled similar yields to be obtained by spraying an oil-based formulation of the same insecticides at up to 2.5 litres/ha (Matthews, 1971; Matthews, 1973) (Figs 8.31 and 8.34). This technique was not pursued in Malawi, although further research examined a very low-volume application technique (Mowlam *et al.*, 1975; Nyirenda, 1991), so production never was as high as in Zimbabwe.

Some farmers had suffered loss of seedlings due to the elegant grasshopper and wondered if spraying would be profitable. To assess the problem a trial was arranged where the field was sown and then divided into plots so that in some the number of plants was decreased at random

Fig. 8.31. Cotton Handbook and Ultra-low Volume Spraying of Cotton – A New Application Technique booklet.

to simulate grasshopper damage (Fig. 8.33b). Treatments which ranged from only 10% to 100% of the original plant population were protected from bollworm damage by the recommended spray programme. The low-density plants had more sympodial branches and each plant had more bolls at harvest compared with the plots from which no plants had been removed. Although it showed that plants could compensate for early-season damage, poor stands would be undesirable in a pest-management programme as bollworm infestation continued later in the season (Matthews *et al.* 1972).

The closure of the Cotton Research Project in 1975 and other factors resulted in inadequate assistance to cotton farmers. The situation might have been different if ULV spraying had been adopted as in West Africa, where increased yields significantly improved production from 1975 until 1995, when the oil-based formulations were withdrawn by the agrochemical companies. The introduction in West Africa was regarded as revolutionary as the farmers no longer needed to take large volumes of water to fields to spray their crops, and the ULV sprayer was lightweight in contrast to the knapsack sprayers used previously. ULV sprayers continued to be used but the insecticides were diluted and applied in ten litres of water per ha.

The World Bank set up the Shire Valley Agricultural Development Project, also referred to as the Chikwawa Cotton Project, in 1968

Fig. 8.32. Very late weeding of cotton in Malawi.

Fig. 8.33. (a) *Zonocerus elegans*. A grasshopper that can damage young seedling cotton plants; (b) part of a trial to simulate loss of plants due to grasshoppers.

Fig. 8.34. Spinning disc sprayer applying ULV spray in Malawi. The nozzle is held downwind so that the operator does not walk where plants have just been sprayed.

(Anon, 1968), which was designed to help farmers achieve higher yields prior to investment in development of an irrigation scheme to provide water from the Shire river along a contour that would allow water to be fed into smaller channels routing the water through the farms between the level of the main canal and the river. In the first phase of the project (1968–1972), the extension staff successfully persuaded a considerable proportion of the growers to adopt spray technology, alongside improved infrastructure of credit services, markets, all-weather roads and boreholes, but Phase II failed. According to Garbett (1984), this was due to cotton requiring a much greater labour input, compared with other crops, such as sorghum and maize, as well as the lack of new crop varieties suitable for the valley's various ecological zones, which were never developed. A survey of 20 cotton farmers in the Lower Shire in 1969 showed the extent of the spraying they did in relation to scouting for insect pests (Fig. 8.35).

Colman (1984) pointed out that the adoption of spraying was not as good as expected because income from a higher yield of sprayed cotton was reduced by the higher cost of labour to obtain and transport water to fields to use in the sprayers. If ULV spraying had been introduced in 1975, as in Francophone Africa, the costings would have been significantly different. However, a key factor that affected yields was the unreliability of rainfall and too much dependence on cotton, without rotating it with other crops, including a legume crop to improve soil fertility. Perhaps with irrigation, the situation would have been significantly better.

Much later, the plans for the development of the Shire Valley Irrigation Project were disclosed in 2017. The project is a 14-year programme (2018–2031) aimed to provide access to reliable gravity-fed irrigation services, secure land tenure for smallholder farmers, and strengthening the management of wetlands and protected areas in the Shire Valley. The first project under the programme will initiate the process of transformation of the Shire Valley and pave the way for agricultural commercialization. The indicative objectives for the second and third phases would be to increase agricultural productivity in targeted smallholder-owned commercial farm enterprises; support value chain and value addition; extend the area supported with irrigation and farm development; and continue and expand efforts to address land degradation and sustainable management of forests, wetlands and protected areas (Anon, 2017).

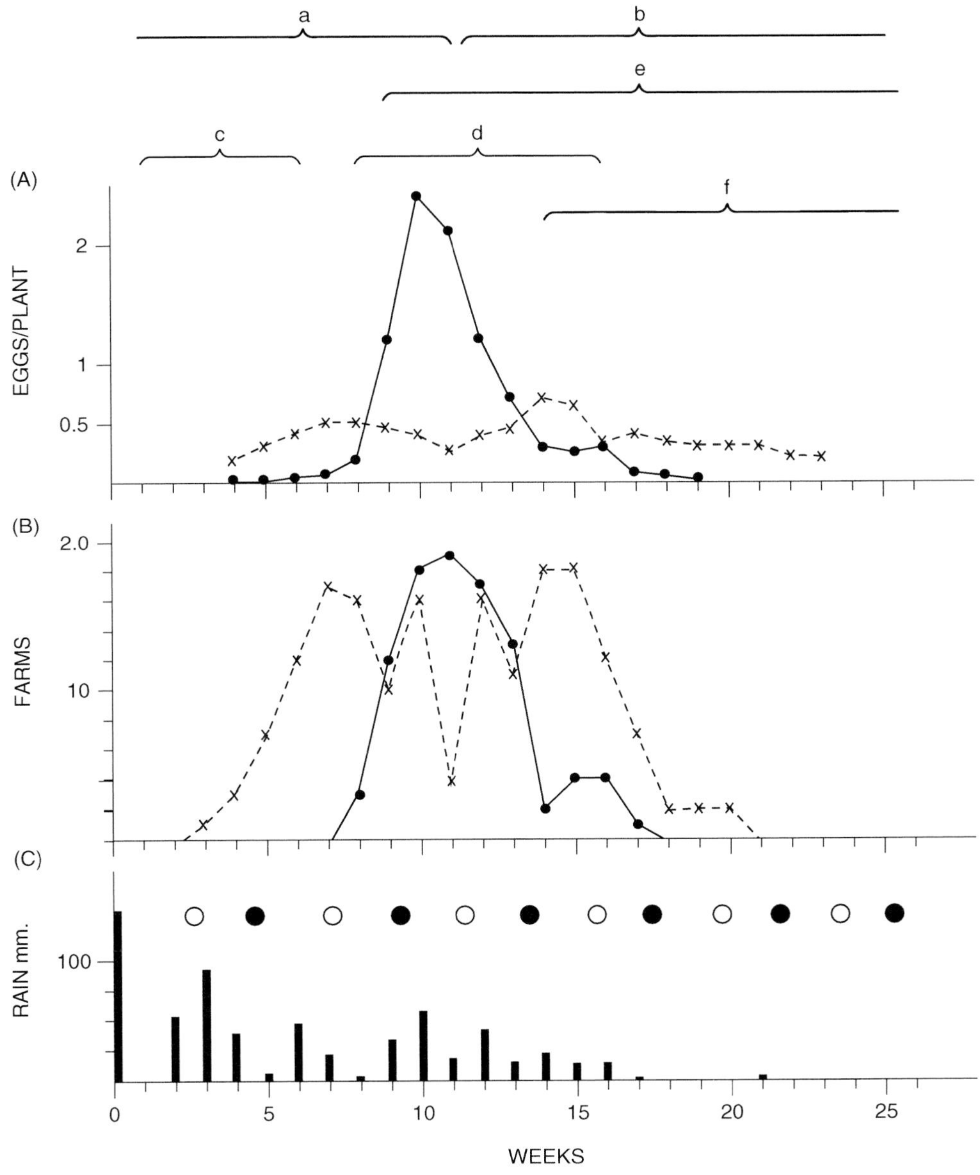

Fig. 8.35. Data from a survey of 20 farmers in the Shire Valley during 1969.
Key:
A. Bollworm eggs per plant – dotted line, *D. castanea*; solid line, *H. armigera*
B. Number of farms spraying each week – dotted line, carbaryl; solid line, DDT
C. Weekly rainfall totals at Ngabu with indication of lunar cycle
(a) Period of emergence of *D. castanea* from overwintering pupae; (b) continuation of *D. castanea* infestation within present season's crop, especially unsprayed cotton; (c) period of jassid and grasshopper infestation; (d) period of *H. armigera* infestation and flowering period; (e) period of infestation of *Earias* spp. and cotton stainers; (f) period of crop maturation and harvesting

In practice, young extension staff were not sufficiently trained to help farmers adopt new technology and relied on basic instructions about early garden preparation, early planting and weeding (Fig. 8.32), crop spacing and ridging (Garbett, 1984.

Further studies on oviposition by bollworm in cotton showed that more red bollworm eggs were present on plants near the field edge, whereas the *H. armigera* eggs were more uniformly distributed (Beeden, 1974).

In 1967, the increase in cotton production encouraged David Whitehead & Sons, operating in Gatooma, Southern Rhodesia, to set up a company (as a subsidiary of Lonrho Malawi Limited). The company employed over 2000 and contributed largely to the welfare of Malawians. However, beginning in the 1980s, Lonrho decided to pull out, as textile production had shifted from Europe to east and south-east Asia, and presented challenges to the company, accompanied by the import of cheap second-hand cotton goods by aid agencies. The company was eventually taken over by Mapeto Wholesalers to create Mapeto DWS, which has since added a ginnery in Salima and sought Chinese investment to expand employment and production of cotton fabrics.

The Agricultural Production and Marketing Board was replaced by a Farmers Marketing Board (FMB) in 1962, with growers' representatives as members, to buy, sell and process farm products, to promote price stability and to distribute subsidized seed and fertilizer sales. After 1994, a reformed Agricultural Development and Marketing Corporation (ADMARC) acted as the main provider for smallholders of fertilizer and seed at subsidized prices, and of local markets to sell their crops, as few private traders emerged. After further efforts to promote private trading, three large private companies and a few smaller ones entered the cotton-buying market, but these colluded to keep prices low. Most cotton was still exported, as the domestic clothing industry was very small, but by 2010, most cotton markets were again operated by ADMARC. However, lack of support for the cotton growers resulted in less production of cotton and in ADMARC concentrating on other crops, including maize, beans, pigeon peas, rice and groundnuts.

From 2003 to 2005, the Cotton Development Assistance Model (2003–2005) was driven by private-sector players, especially ginners, who provided inputs to farmers at concessionary prices that ranged between 5% and 15% of prevailing commercial prices. Production increased by about 12% from 53,000 MT to about 60,000 MT during the initial years. Subsequently, a Cotton Production Up-scaling Model (2011–2014) enabled the government to support cotton farmers by injecting the equivalent of about $10 million, with production increasing to 100,000 MT of cotton. Subsequently, when a Contract Farming Model (2013–2015) was implemented by ginners, production increased on average by 7000 MT during the implementation period. However, this initiative was discontinued due to high default rate by farmers which amounted to MK 1.3 billion.

The government had, in the meantime, talks with China and set up the China-Africa Cotton Development Ltd registered in Hong Kong, which acquired Cargill Malawi and subsequently set up the Malawi Cotton Company Limited (MCC). It was reported that the China-Africa Cotton was committed to invest and develop the African agriculture industry and establish seven ginneries and two cottonseed oil-extracting mills and one special seed plant in Africa (Mtika, 2019). However, by 2015, cotton production had decreased significantly and only three ginners remained operational. The Chinese link has failed to transfer skills to Malawians as the whole processing chain is carried out by Chinese staff, while most farmers have blamed the Malawi government for failing to properly regulate the cotton industry in the interest of farmers and the economy. Thus, farmers are unable to recoup their investment in inputs.

As in Zambia, it is not yet clear whether this is primarily a means of buying African-grown cotton for export or a system to help farmers achieve higher yields with new technology, such as genetically modified seeds and using drones to spray crops when necessary.

Mozambique

In 1926, the Portuguese government opted for a labour-intensive system based on forced peasant production, requiring minimal investment. Legislation enabled the establishment of cotton zones, with concessions to local commercial interests, that received the exclusive right to purchase, at fixed prices, the cotton that small-scale farmers within the region were compelled to grow. The concessionary companies were responsible for ginning the cotton and selling it overseas, preferably to the Portuguese textile industry. However, the low prices paid resulted in the overall production to decrease by 50% by 1931. To help the textile industry in Portugal, concessions were made to increase cotton production, but higher-paid jobs in South African mines, as in Malawi, attracted 75–105,000 Mozambicans annually to work there rather than grow cotton, while others went to sugar estates (Isaacman *et al.*, 1980).

In the 1960s, production was initially high due to settlers at an irrigation scheme near the Limpopo river with a tractor-mounted mistblower being used (Fig. 8.36). Later, teams applying ultra-low volume sprays managed by an agrochemical company were used in the Zambezian province near the Zambezi river in the 1970s. Some 3000 ha on small-scale farms were treated with ULV sprays that increased yield from 300 kg/ha to at least 1000 kg/ha (Watson-Cook, 1971). Monocrotophos was applied, and when asked about safety for the spray operators, a bottle of 'antidote' (atropine) for organophosphate poisoning was produced with a teaspoon. After a long period of political conflict cotton production only recovered much later with the use of VLV sprays and for a short period, some sprays were applied with an 'Electrodyn' sprayer in the Nampula province (Javaid *et al.*, 2000).

A questionnaire survey of 100 small-scale cotton growers conducted in two cotton-growing provinces of Mozambique to assess the constraints faced by the farmers in relation to the timing of insecticide application revealed that the majority of the farmers started spraying too early (four weeks after germination) and were unaware that such sprays reduced the population of natural enemies of insect pests and might not lead to increased yields. The farmers applied an average of three sprays in the cotton-growing season, about half of them according to pest observance and 41% by following instructions of extension officers, but were constrained by the unavailability of insecticides and sprayers. Another constraint was lack of information on the timing of insecticide application. More than half of the farmers had suffered mild insecticide poisoning and needed more effective training on the judicious and safer use of insecticides, integrated with non-chemical pest-management practices. Seventy-one farmers used the ULV sprayers and 20 used the Electrodyn sprayers. The remaining had used both types of sprayers. The sprayers are owned by Joint Venture Companies (JVCs) and are lent to the smallholders for insecticide applications (Javaid *et al.*, 1998).

The first textile mill in Mozambique was established in the late 1940s, and a modest cotton-based textile and clothing industry developed steadily until 1974. Then, a mass exodus of Portuguese residents who had managed the

Fig. 8.36. Mistblower spraying cotton near Limpopo river at the wrong time of day (noon).

companies, and the effects of socialist policies and the long-lasting civil war in Mozambique, precipitated the demise of the industry. Cotton lint production plummeted from 50,000 MT in 1973 to 2000 MT in 1985 as many factories closed down or were reduced to operating at very low utilization rates. There are now four textile companies, despite the overwhelming importation of garments from China.

Angola

In Angola, there was a brief surge of production in the 1860s at the time of the Civil War in the USA, but production remained insignificant until the 1930s as a new law in 1926 required farmers in parts of the Luanda, Cuanza Norte, and Malange districts to plant cotton. With government technical aid and a tax reduction in return for developing the production and marketing of the cotton, production increased to a peak of 22,000 tons in 1944, and then remained at this level for the next 20 years, until compulsory planting was abolished in 1963. When cotton planting was no longer compulsory, the number of African cotton growers declined; only 41,596 of them planted cotton in 1964, compared with 54,842 in 1960. Nevertheless, cotton production increased rapidly in the years 1964–73 with an increasing number of commercial farms, but in 1975 production dropped off when commercial farms were abandoned by their Portuguese owners. Now, in 2018, there are efforts to increase production again. A major insect pest in Angola is *Diparopsis tephragramma* (Clements, 1951), whereas *Diparopsis castanea* is the red bollworm in southern Africa.

The internal demand for textiles has risen significantly, but the majority of the population rely on second-hand clothing from the USA and Europe. However, some new textile manufacturing companies have been established since 2014.

Cotton in Southern Africa – the Future

The growing of cotton has, in southern Africa, been dictated by recent political issues, so that it is now increasingly dependent on the small-scale growers. Large-scale cotton farms were very successful until 2000, but there has not been adequate research and extension supported by government to enable the small-scale farms to achieve the same level of production. Whether it will be possible, by introducing the newer genetically modified varieties, to minimize the need for insecticides and improve weed management has yet to be tried. The successful cultivation of GM cotton varieties will need the farmers to have adequate training so that they have the necessary skills to react to changes in pest and disease infestations brought about by these varieties and to possible measures protecting their bollworm resistance.

As pointed out by Hillocks (2005), Bt cotton is a technology with the potential to benefit African smallholders but successful deployment of the technology depends on several factors:

- implementation of biosafety regulations, particularly with respect to regulation of seed multiplication and distribution;
- farmers understanding that Bt cotton alone cannot deliver yield increases, if basic crop management is poor;
- that Bt cotton is promoted as a component of integrated crop management (ICM) systems;
- farmers understanding that insecticide spraying to control non-lepidopterous pests may still be required if the full benefit from Bt varieties is to be realized;
- there is proper regulation of resistance-management strategies; this will become more important the more widely Bt cotton and maize are adopted, and it may be advisable to implement them at the outset, even though they may not at first be necessary (due to natural refugia).

References

Anon. (1968) *Malawi Shire Valley Agricultural Development Project*. Unpublished report.
Anon (2017) Shire Valley Irrigation Project. Ministry of Agriculture, Irrigation and Water Development, Government of Malawi. COWI A/S Parallelvej 2 DK-2800 Kongens Lyngby, Denmark.

Beeden, P. (1971) Agricultural development in the lower shire valley of Malawi. Available at: https://agris.fao.org/agris-search/search.do?recordID=US201302384496 (accessed 2 August 2021).
Beeden, P. (1974) Bollworm oviposition on cotton in Malawi. *Cotton Growing Review* 51, 52–61.
Bennett, R., Ismael, Y., Morse, S. and Shankar, B. (2004) Reductions in insecticide use from adoption of Bt cotton in South Africa: impacts on economic performance and toxic load to the environment. *Journal of Agricultural Science* 142, 665–674.
Beyers, C.T.L., Ismael, Y. and Piesse, J. (2003) Can GM-technologies help the poor? The impact of Bt cotton in Makhathini Flats, KwaZulu-Natal. World Development 31, 717–732.
Brettell, J.H. (1986) Some aspects of cotton pest management in Zimbabwe. *Zimbabwe Agricultural Journal* 83, 4146.
Burgess, M.W. (1983) Some aspects of cotton pest management in Zimbabwe. *Crop Protection* 2, 247–250.
Clements, A.N. (1951) A revision of Diparopsis Hmps. (Agrotidae Lepidoptera). *Bulletin of Entomology Research* 42, 491–497.
Colman, D. (1984) Smallholder agriculture in the lower shire valley in Malawi: analysis of the experience with cotton. *Agricultural Administration* 15, 25–43.
Dawe, J.A. (1993) *A History of Cotton Growing in East and Central Africa: British Demand, African Supply*. PhD thesis, Edinburgh University.
Ducker, H.C. (1929) Cotton in Nyasaland. *Empire Cotton Growing Review* 6, 326–382.
Duncombe, W.C. (1973) The acaricide spray rotation for cotton. *Rhodesian Journal of Agricultural Research* 70, 115–118.
Farrington, J. (1977) Research based recommendations versus farmer' practices: some lessons from cotton spraying in Malawi. *Experimental Agriculture* 13, 9–15.
Garbett, K. (1984) Labour and cotton in the lower shire valley, Malawi. *Social Analysis: Journal of Cultural and Social Practice* 16, 108–126.
Gledhill, J.A. (1979) The Cotton Research Institute of Gatooma. *Rhodesia Agricultural Journal* 76, 103–118.
Goeb, J.C. (2011) *Impacts of Government Maize Supports on Smallholder Cotton Production in Zambia*. MSc thesis, University of Michigan.
Harkema, R.C. (1972) Zambia cotton production and textile industry. *Geography* 57, 345–348.
Hillocks, R.J. (1995) Insect pests, diseases and weeds of cotton in Africa. *Integrated Pest Management Reviews* 1, 31–47.
Hillocks, R.J. (2005) Is there a role for Bt cotton in IPM for smallholders in Africa. *International Journal of Pest Management*, 131–141.
Isaacman, A., Stephen, M., Adam, Y., Homen, M.J., Macamo, E. and Pililao, A. (1980) 'Cotton is the mother of poverty': peasant resistance to forced cotton production in Mozambique, 1938–1961. *The International Journal of African Historical Studies* 13, 581–615. Boston University African Studies Center.
Jackson, P. (2004) What is the enabling state? The views of textiles and garments entrepreneurs in Zimbabwe. *Journal of International Development* 16, 769–783.
Javaid, I., Uaine, R.N. and Massua, J. (1998) Pest management constraints in smallscale cotton farms in Mozambique: timing and application of pesticides. *Insect Science and Its Application* 18, 251–255.
Javaid, I., Uaine, R.N. and Massua, J. (2000) Studies on very-low volume (VLV) water-based sprays for the control of cotton pests. *International Journal of Pest Management* 46, 81–83.
Larsen, M.N. (2002) Is oligopoly a condition of successful privatization? The case of cotton in Zimbabwe. *Journal of Agrarian Change* 2, 185–205.
Mariga, I. (1994) Cotton research and development. In: Rukuni, M. and Eicher, C. (eds) Zimbabwe's Agricultural Revolution. University of Zimbabwe Publications, Harare, pp. 219–233.
Marks, R.J. (1976a) Field studies with the synthetic sex pheromone and inhibitor of the red bollworm *Diparopsis castanea* Hmps (Lepidoptera: Noctuidae) in Malawi. *Bulletin of Entomological Research* 66, 243–265.
Marks, R.J. (1976b) Field evaluation of gossyplure, the synthetic sex pheromone of *Pectinophora gossypiella* (Saund.) (Lepidoptera: Galechiidae) in Malawi. *Bulletin of Entomological Research* 66, 267–278.
Marks, R.J. (1978) The influence of pheromone trap design and placement on the catch of the red bollworm Diparopsis castanea Hampson (Lepidoptera: Noctuidae). *Bulletin of Entomological Research* 68, 31–45.
Matthews, G.A. (1966) Investigations of the chemical control of insect pests of cotton in central Africa: Part I: laboratory rearing methods and tests of insecticides by application to bollworm eggs. *Bulletin of Entomological Research* 57, 69–76; Part II: tests with larvae and adults. *Bulletin of Entomological Research* 57, 77–91; Part III: field trials. *Bulletin of Entomological Research* 57, 193–197.
Matthews, G.A. (1971) Reduced volume of spray on cotton. *Proceedings of the Cotton Institute Conference*, Blantyre, Malawi, pp. 172–181.

Matthews, G.A. (1973) Ultra-low volume spray application on cotton in Malawi. *International Journal of Pest Management* 19, 48–53.
Matthews, G. (2014) A retrospective: the impact of research on cotton pest control in Central Africa and development of ultra-low volume spraying for small scale farmers between 1958–72. *Outlooks on Pest Management* 25, 25–28.
Matthews, G.A. and Tunstall, J.P. (1965) Aerial and ground spraying for cotton insect pest control in Rhodesia. *Empire Cotton Growing Review* 42, 180–192.
Matthews, G.A. and Tunstall, J.P. (2019) The changes in cotton production in Zimbabwe – 1924–2018. *Outlook on Pest Management* 30, 75–79.
Matthews, G.A., Tunstall, J.P. and McKinley, D.J. (1965) Outbreaks of pink bollworm (*Pectinophora gossypiella* Saund) in Rhodesia and Malawi. *Cotton Growing Review* 52, 197–208.
Matthews, G.A., Rowell, J.G. and Beeden, P. (1972) Yield and plant development of reduced cotton stands in Malawi. *Experimental Agriculture* 8, 33–48.
McKinstry, A.H. (1938) Major pests of cotton in Southern Africa. In: *Third Conference on Cotton Growing Problems*. Empire Cotton Growing Corporation, London.
Mlambo, A. and Phimister, I. (2006) Partly protected: origins and growth of colonial Zimbabwe's textile industry, 1890–1965. *Historia* 52, 145–175.
Mowlam, M.D., Nyirenda, G.K.C. and Tunstall, J.P. (1975) Ultra-low volume application of water-based formulations of insecticides on cotton. *Cotton Growers Review* 52, 360–370.
Mtika, C. (2019) Malawi bequeaths cotton production to Chinese firm. Centre for Investigative Journalism Malawi (CIJM).
Munro, J.W. (1937) *Cotton Pest Control Work in Southern and Central Africa and the Rhodesias*. Report of a tour to the Empire Cotton Growing Corporation.
Nesbitt, B.F., Beevor, P.S., Cole, R.A., Lester, R. and Poppi, R.G. (1973) Sex pheromones of two noctuid moths. *Nature London* 244, 208–209.
Nyambara, P.S. (2000) Colonial policy and peasant cotton agriculture in Southern Rhodesia, 1904–1953. *The International Journal of African Historical Studies* 33, 81–111.
Nyambara, P.S. (2014) The politics of locating the third spinning mill in Southern Rhodesia, 1951–1953. *Historia* 59.
Nyirenda, G.K.C. (1991) Effect of swath width, time of application and height on the efficacy of very-low-volume (VLV) water-based insecticides on cotton in Malawi. *Crop Protection* 10, 111–116.
Parnell, F.R. (1925) Breeding jassid resistant cottons. *Empire Cotton Growing Review* 2, 330–336.
Parnell, F.R., King, H.E. and Ruston, D.F. (1949) Insect resistance and hairiness of the cotton plant. *Bulletin of Entomological Research* 39, 539–575.
Parsons F.S. and Ullyett, G.C. (1934) Investigations on the control of the American and red bollworms of cotton in South Africa. *Bulletin of Entomological Research* 25, 349–381.
Parsons, F.S. and Ullyett, G.C. (1936) Investigations on Trichogramma lutea, Gir. as a parasite of the cotton bollworm, Heliothis obsolete. Fabr. *Bulletin of Entomological Research* 27, 219–235.
Pearson, E.O. (1934) Preliminary observations on cotton strainers and internal boll disease of cotton in South Africa. *Bulletin of Entomological Research* 25, 383–414.
Pearson, E.O. (1954) The relationship between the African and South American red bollworms of cotton, Diparopsis and Sacadodes. *Cotton Grower Review* 27, 1–6.
Pearson, E.O. (1958) *The Insect Pests of Cotton in Tropical Africa*. ECGC and Commonwealth Institute of Entomology, London.
Pearson, E.O. and Mitchell, B.L. (1945) *A Report on the Status and Control of Insect Pests of Cotton in the Lower River Districts of Nyasaland*. Zomba, Government Printer.
Poulton, C. and Hanyani-Mlambo, B. (2008) The cotton sector of Zimbabwe. *Africa Region Working Paper Series* No. 122 (2009). World Bank.
Robins, J.E. (2016) *Cotton and race across the Atlantic – Britain, Africa, and America, 1900–1920*. University of Rochester Press, New York.
Rothwell, A., Bryden, J.W., Knight, H. and Coxe, B.J. (1967) Boron deficiency of cotton in Zambia. *Cotton Growing Review* 44, 23–28.
Schnurr, M.A. (2011) The boom and bust of Zululand cotton, 1910–1933. *Journal of Southern African Studies* 37, 119–134.
Shankar, B., Bennett, R. and Morse, S. (2008) Production risk, pesticide use and GM crop technology in South Africa. *Applied Economics* 40, 2489–2500

Sikhondze, B.A.B. (1989) The Development of Swazi Cotton Cultivation 1904–85. PhD thesis, School of Oriental and African Studies. University of London.

Tang, X. (2019) The impact of Chinese investment on skill development and technology transfer in Zambia and Malawi's cotton sector. Working Paper No. 2019/23. China Africa Research Initiative, School of Advanced International Studies, Johns Hopkins University, Washington, DC.

Tunstall, J.P. (1968) Pupal development and moth emergence of the red bollworm (*Diparopsis castanea* Hmps.) in Malawi and Rhodesia. *Bulletin of Entomological Research* 58, 233–253.

Tunstall, J.P. and Matthews, G.A. (1966) Large-scale spraying trials for the control of cotton insect pests in Central Africa. *Cotton Growing Review* 43, 121–139.

Tunstall, J.P., Matthews, G.A. and Rhodes, A.A.K. (1961) A modified knapsack sprayer for the application of insecticides to cotton. *Cotton Growing Review* 38, 22–26.

Tunstall, J.P., Matthews, G.A. and Rhodes, A.A.K. (1965) Development of cotton spraying equipment in Central Africa. *Cotton Growing Review* 42, 131–145.

Watson-Cook, D.J. (1971) Ultra-low volume spraying from the ground and the air. *Proceedings of the Cotton Institute Conference*, Blantyre, Malawi, 165–171.

9 Cotton Growing in East Africa

Joe Kabissa*, Pius Elobu and Anthony Muriithi

Tanzania

Joe Kabissa

Introduction

Tanzania's economy continues to be agriculture-led with over 67% of Tanzania's workforce still engaged in agriculture, which contributes 30% of Tanzania's total exports. Cotton is one of five crops categorized as 'strategic crops' as it is widely regarded as the crop with perhaps the greatest potential for socioeconomic development. It is widely cultivated and has a more inclusive value chain than other crops. Its economic potential is, however, undermined by low productivity coupled with limited domestic processing of lint and other by-products into value-added items.

Historical Context

Cotton farming in Tanzania, then called Tanganyika, started in the 20th century at the initiative of German textile manufacturers. By 1907, a total of 14,826 ha had been sown to cotton in the Kilosa district of Morogoro region by a non-government organization called Kolonial-Wirtschaftliichos Komitee (KWK) (Beckert, 2015). Further expansion of cotton cultivation became problematic without the requisite experience on how cotton should be grown. Their push to grow cotton as a plantation crop, like tea and rubber, using peasants as low-cost rural workers to work on such cotton plantations failed.

When the British colonial administration took over the administration of Tanganyika in 1919 after the First World War, cotton production was entrusted to the British Cotton Growers Association (BCGA) which provided seeds and expertise on cotton. But just like KWK, BCGA put African farmers under considerable pressure to step up cotton production. However, attempts to grow cotton

*Corresponding author: joekabissa@gmail.com

 Pest Management in Cotton: A Global Perspective (eds G.A. Matthews and T. Miller)
DOI: 10.1079/9781800620216.0009

as a plantation crop proved unsuccessful again, because the indigenous population was not incentivized either to grow more cotton or adopt new production technologies and preferred to grow cotton on their own rather than to work for wages on cotton plantations owned by Europeans.

When the BCGA was replaced by the Empire Cotton Growing Corporation (ECGC), their approach was different and they initiated cotton research at two stations, Ukiriguru located in the Western Cotton Growing Area (WCGA) and Ilonga in the Eastern Cotton Growing Area (ECGA), in the 1930s and 1949, respectively. Up until 1940, the expansion of cotton production in both these areas was largely held back as the varieties were highly susceptible to leaf-sucking pests – jassids (*Jacobiasca* spp.) and two major cotton diseases – bacterial blight, *Xanthomonas malvacearum* and fusarium wilt, *Fusarium oxysporum*.

Based on studies in South Africa, the jassid problem was subsequently resolved by breeding for increased leaf and stem hairiness in all cotton varieties, coupled with successful breeding for resistance against the two major cotton diseases. This had a significant impact on programmes to expand cotton cultivation in the WCGA between 1947 and 1956 when the Sukumaland Development Scheme was in progress. These factors coupled with high global prices for cotton in the aftermath of the Second World War and the Korean War greatly contributed to increased cotton output between the late 1940s and mid-1960s (Kabissa, 2014).

New Institutions, Policies and Interventions

Cotton growing and marketing in Tanganyika up until the early 1950s was largely with the private sector operated by Asians of Indian origin, who were responsible not only for buying seed cotton from farmers but also ginning it in Uganda at that time. Due to cheating on farmers during the weighing of seed cotton at buying posts, the Agricultural Marketing Cooperative Societies (AMCOs) emerged to compete with them and eventually were responsible for the input supply, marketing and ginning. The AMCOs were followed up by the formation of the Lint and Seed Marketing Board (LSMB) in 1952 whose major functions under the colonial government were to sell lint overseas and oversee cotton quality.

After the launch of LSMB, the government involvement in the cotton industry increased significantly. In 1967, the post-independence administration sought to transform the agriculture-led economy using 'self-reliance' as its driving principle and 'Ujamaa villages' as its basic units of production needed for rural development and socialist transformation. Under the compulsory villagization programme, it was envisaged that the agricultural sector would become modernized as a result of people being moved to villages where the provision of key services, such as extension services and other social services, would become centralized and thence amenable to most rural dwellers.

However, the implementation of the villagization process resulted in the dismantling of co-operatives as voluntary and independent member-driven organizations confined to cotton farmers and their replacement by so-called multipurpose village co-operatives for commercial/political/administrative activities. These lacked the knowledge to purchase the cotton crop or administer inputs and input credit whereas the marketing co-operatives were business-focused. In consequence, annual cotton output decreased from 75,373 metric tons of lint during the 1966/67 marketing season to 42,170 tons by 1975/76, when the co-operatives were abolished and replaced by a state enterprise called Tanzania Cotton Authority (TCA), which also incorporated the functions undertaken by LSMB. The TCA was soon overwhelmed by the multiplicity of tasks conferred upon it and could not arrest the declining trend in cotton at a time when the economy was facing increasing hardships, due to the war against the Idi Amin regime in Uganda, a growing debt crisis and deteriorating terms of trade for exports in the late 1970s and early 1980s. By 1984, further changes resulted in cotton marketing reverting to a three-tier system involving primary co-operative societies (PS), regional co-operative unions (RCUs) and the parastatal Tanzania Cotton Marketing Board (TCMB) that had existed prior to 1976 and now replaced the TCA to function as the RCU's statutory agent for the sale of lint overseas.

The Ordeal of Market Liberalization

Policy and institutional changes that took place between 1967 and 1990 were, by and large, geared at fine-tuning the functioning of the single-chain marketing system. However, such changes did not result in the desired impacts because the declining trend in cotton output from a peak of 75,373 tons of lint obtained during 1966/67 continued until a low of 54,000 tons was reached by the 1989/90 marketing season. The failure of a marketing system based on marketing parastatals and co-operatives to deliver on increased commodity production has tended to be attributed to the propensity of such a single marketing system to keep the profitability of designated commodities low by way of substandard margins owing to both explicit and implicit taxation (Bates, 2015).

Under the structural adjustment programme (SAP), Tanzania was compelled to replace the single-channel marketing system by a more open and competitive market structure through a process of liberalization, defined as the move to market-determined prices, and increased the participation of the private sector in such activities (Akiyama *et al.*, 2001). The introduction of competition all along the production, marketing, processing and export chain was envisaged to improve efficiency, reduce marketing costs and hence raise farmers' share of the free on-board price (FOB) and hence motivate them to boost production and productivity.

However, the switch from a single- to a multichannel marketing system took effect in 1994 and resulted in an unregulated entry of newcomers into the cotton market, the majority of whom preferred to go into buying and ginning cotton rather than producing it. This made it difficult to provide public goods and services, such as quality control, extension, R & D as well as seasonal inputs and input credit. These unintended outcomes of liberalization went on to significantly impact both the output and quality of cotton, so the quality of Tanzanian lint deteriorated considerably (Kabissa, 2014).

The Bumpy Road to Recovery

In 2001, the Tanzania Cotton Board (TCB) became the *de facto* regulator of the cotton industry following the enactment of the Tanzania Cotton Industry Act No. 2 of 2001, and was empowered to co-ordinate and promote the cotton industry as well as oversee and manage a newly formed cotton development fund (CDF). This fund provided the financing for the procurement and distribution of seasonal inputs as well as R & D at Ukiriguru and Ilonga. The formation of the Tanzania Cotton Association (TCA) and Tanzania Cotton Growers Association (TACOGA) in 1997 and 2002, respectively, also helped to unite cotton buyers, ginners and farmers and to defend their interests, including lobbying government, as the absence of member-driven co-operatives meant farmers had become liable to manipulation by politicians.

These changes enabled cotton output to recover significantly and for the first time in over 100 years exceeded 100,000 tons of lint cotton on four occasions during the first decade of the 21st century. Buoyed by such signs of recovery, national lint output was projected to reach 270,000 tons by 2015 by merely trebling average seed cotton yield per ha from the level of 750 kg to 2250 kg. Such projection was partly motivated by the fact that some countries such as Benin, Turkey, Burkina Faso and Australia were already attaining considerably higher cotton outputs in spite of their areas under cotton cultivation being quite comparable to Tanzania's (Kabissa, 2014).

Hopes of further growth of the cotton industry were boosted during the 2008/09 cropping season following the launching of the Cotton and Textile Development Program (CTDP). The latter, as implemented by the Tanzania Gatsby Trust under the joint financing of the Gatsby Charitable Foundation (GCF) and Department for International Development (DFID) of the UK, sought to specifically address constraints relating to low production and productivity, and low quality of both seed cotton and lint, as well as to limit domestic processing of lint into textiles and apparel. To address the issues of low production and productivity, CTDP sought to institutionalize contract farming following its successes in other crops in addressing farmers' needs for inputs, credit, training needs and marketing outlets. Despite being piloted and refined over time, the promotion and further uptake of contract farming on conventional cotton was stopped by government order in 2018 although it continues to be used for the production of organic cotton and certified cottonseed for planting.

The release by Ukiriguru of a high-yielding variety called UKM08, in 2008, coupled with the introduction of the use of acid-delinted cottonseed for planting, further heightened the hopes of revamping productivity and production in the country. The introduction of seeds of an improved cotton variety promised to help the industry do away with burgeoning use of low-quality recycled seeds in the aftermath of liberalization. The introduction of delinted seed was anticipated to help improve agronomic practices by facilitating a switch from the age-old practice of broadcasting fuzzy seeds to planting in appropriately spaced rows. Nevertheless, the uptake of delinted seed as well as wider adoption of UKM08 took quite a while owing to weaknesses of the current cottonseed system as well as farmers' aversion to change.

Between 2008 and 2010, the government made two interventions aimed at the pest-control problem on cotton. After liberalization, the collapse of the input credit system meant that the vast majority of cotton farmers were unable to provide optimal control of insect pests due to lack of financial capacity to purchase adequate quantities of insecticides. During the 2008/09 cropping season, the government introduced a subsidy programme that enabled farmers to buy insecticides at prices that were up to 30% below market prices. Such intervention was followed up in the 2009/10 cropping season by the introduction of the so-called input voucher scheme under which farmers were to be provided with both seeds and insecticides at subsidized prices in quantities that would be enough to satisfy the needs of one acre as per recommendation packages from cotton R & D stations.

However, these interventions were temporary measures that needed more sustainable solutions. Due to weak systems for administering the subsidies, farmers, *per se*, do not often get the full benefits of the programmes. In the case of the input voucher scheme, farmers ended up paying more for the pesticides than in previous seasons because pesticides coming under the subsidy programme came from the open market rather than from the Cotton Development Trust Fund (CDTF). The latter had, for many years, been ordering such pesticides in bulk, a procedure that tended to make their unit prices more competitive than those for pesticides obtained from open markets. Perhaps more importantly, because the pesticides given to farmers under the voucher scheme came from the open market, most of the expired and fake products as well as those that had not been duly registered for use on cotton found their way onto the cotton market. Not surprisingly, cotton output in that season ended up being significantly lower than in the previous one.

In another attempt aimed at boosting cotton production, the government, during the 2014/15 cropping season, sought to put up an irrigation scheme involving the use of water drawn from bore wells to irrigate 300 ha of cotton farmland in the Kishapu district of Shinyanga region, but it did not happen.

Current Status of Cotton in Tanzania

In spite of the many challenges facing cotton production, farmers keep on growing the crop because it is their most viable source of livelihood in the semi-arid areas where it is grown. In such areas, cotton also serves as an important insurance crop in the event of food-crop failure due to drought and other hazards. Furthermore, unlike most other cash crops, cotton provides the flexibility to be grown either as a low-input-low-output crop (which is good for resource-poor farmers) or as a high-input-high-output crop (which is good for the well-to-do farmers). Finally, it just happens that the revenues from cotton sales quite often come at an optimal time in the annual cropping calendar when farmers face multiple needs for cash.

During the 2018/19 cropping season, the area sown to cotton reached an all-time high of 750,000 ha following the expansion of cotton farming into new areas of Katavi and Dodoma regions during the 2017/18 and 2018/19 cropping seasons, respectively. The crop is now grown in 17 out of 26 regions of mainland Tanzania by an estimated half a million farmers. The bulk of the crop is produced in the regions of Simiyu, Shinyanga, Mwanza, Mara and Geita Tabora, Singida, Kagera, Katavi, Kigoma and Dodoma, which collectively account for nearly 99% of total national output. These regions, with the exception of Dodoma and Katavi, used to constitute the so-called WCGA. The remainder of the crop comes from areas scattered in the regions of Manyara, Coast, Morogoro and Iringa. These regions, together with Kilimanjaro, Tanga and Iringa, collectively constitute the so-called

ECGA. Up until 1934, the ECGA accounted for more cotton produced in the country than the WCGA. However, production in the ECGA went into decline and by 1990 it accounted for just up to 10% of total national output. To date, the ECGA accounts for less than 1% of national output. Such dramatic decline in cotton output is, by and large, due to cotton growing in the ECGA being constrained by severe attacks from insect pests and intense competition from weeds. Due to soil and weather conditions in the ECGA favouring more crops to be grown successfully, crop competition tends to be more severe in the ECGA than in the WCGA. This fact, coupled with poor marketing arrangements for cotton in the ECGA, tends to make cotton farming less attractive than in the WCGA.

Cotton production is prohibited in all the south and south-western regions of Rukwa, Mbeya, Songwe, Njombe, Ruvuma, Mtwara, Lindi and parts of the new region of Katavi. The ban was instituted during the early 1940s in order to prevent the northward spread of the red bollworm, *Diparopsis castanea*, into these regions from Mozambique, Malawi and Zambia where it is endemic (Kabissa and Nyambo, 1989). In these regions there exists a 200 km-wide quarantine area between them and the Tanzanian borders with the said countries where no cotton is to be grown. In spite of the existence of such a quarantine area, lack of a physical barrier coupled with poor enforcement of the quarantine law resulted in the red bollworm making an incursion into commercial cotton in the district of Chunya in Mbeya region in 1999 for the first time in cotton cultivation history in Tanzania. Apparently, farmers who had been growing cotton adjacent to the quarantine border in that district had, in utter defiance of the quarantine law, expanded cotton farming into deep inside the *de facto* quarantine area. As a result, the government banned cotton production in Mbeya region in a bid to prevent any further spread of the red bollworm in the country. That decision coupled with the continued ban on cotton farming in the southern and south-western parts of Tanzania, has helped keep commercial cotton-growing areas in Tanzania, Kenya, Uganda, Rwanda, Burundi, Ethiopia and the Democratic Republic of Congo free of the red bollworm.

Cotton farming is entirely smallholder-based and completely rain-fed. The crop is grown on holdings of up to 2 ha, although some farmers in parts of Simiyu, Shinyanga and Tabora regions may grow it on larger acreages using draft animals. Sowing of cotton in the WCGA proper and some parts of Iringa and Manyara in the ECGA takes place between 15 November and 15 December. These time limits, which were recommended over 80 years ago, are, to date, mere guidelines because the timing of the onset as well as duration of rains has changed quite drastically owing to climate change. In the ECGA, cotton is often sown between February and March or a bit later. Along the Rufiji river basin, farmers were, in the past, capable of growing two crops in one year; a rain-fed one followed by another grown on the flood plain after water had receded. Cotton production has now stopped in this area.

Average cotton yields in Tanzania are quite low by world standards. Between 1945/46 and 2009/10, average yields for Tanzania remained in the order of 121–156 kg of lint per ha while the global average yield rose from 210 to 738 kg per ha during the same period. During the 2018/19 marketing season, average yield reached 193 kg of lint per ha. Average yields have tended to be low largely due to the continued use of low-quality seed for planting, farmers' non-compliance with best agronomic practices and their total inability to use inorganic fertilizers.

Pest control: Changes in application technology and insecticide use on cotton

Amongst the major constraints facing cotton production in Tanzania, none has been more daunting than the need to suppress insect pests that ravage this crop. Of all the pests frequenting cotton, *Helicoverpa armigera* is by far the most damaging and to date nearly all the insecticides used on cotton tend to be targeted at this pest. In both the WCGA and ECGA, *H. armigera* occurs as a major pest of cotton from February to as late as May, but also feeds on many other wild and cultivated host plants. The dry season (August–October) is spanned usually by diapause pupae (in the WCGA), the moths emerging from which oviposit on a species of *Cleome*, on which the first generation develops, giving rise to adults in

November–December. Larvae of the second generation feed mainly on early-sown maize, and it is usually those of the third and fourth generations that attack cotton. The increasing severity of the attacks of *H.armigera* on cotton has accompanied the replacement of sorghum by maize in recent years. Early sowing of cotton appears to be the best means of withstanding attack by *H. armigera*, as most of the present long season varieties of cotton are able to compensate by later growth even for a very heavy *H. armigera* attack, provided they are sown early enough to utilize the late rains. However, this is difficult to implement because most farmers tend to sow maize prior to cotton in the interests of food security. Populations of *H. armigera* building up on such maize tend to later move onto cotton. In the ECGA, attacks by *H. armigera* on cotton have tended to be heavier than in the WCGA largely due to the practice of relay intercropping of maize with cotton (Reed, 1965; Nyambo, 1988; Kabissa *et al*., 1997; Kabissa, 2014).

Pectinophora gossypiella and *Earias* spp are the other bollworms that attack cotton from time to time. However, owing to their narrow host ranges, their attacks on cotton tend to be largely a function of the extent to which cotton growers comply with some mandatory cultural practices such as the need for area-wide end of season destruction of crop residues coupled with observance of a three-month closed season. These requirements are hardly ever complied with fully, in part due to the prevailing laxity by extension service to enforce them and a general reluctance on the part of farmers to comply. Cotton stalks tend to be used as a source of energy and some farmers have been ratooning their cotton. These practices have contributed to an upsurge in populations of both *P. gossypiella* and *Earias* spp. (Kabissa, 1990). In the case of sucking pests such as *Aphis gossypii*, spider mites and *Dysdercus* spp., their extreme polyphagous nature tends to make their occurrence on cotton relatively unpredictable.

Chemical control of *H. armigera* and sucking pests on cotton in Tanzania began in the early 1950s when DDT and BHC in a dust formulation were used for the first time. The dusts had to be applied by Swaine dusters at dawn in order to minimize drift. As wettable powders replaced the dusts, spraying using Plantector knapsack sprayers began at flowering in order to suppress populations of *H. armigera* that move onto cotton at this time. In the ECGA, spraying began 8 weeks after sowing as opposed to 10 weeks in case of the WCGA owing to its higher pest pressure. In both areas cotton had to be sprayed 8 times, once every week. In 1972, introduction of oil-based insecticides resulted in spraying intervals being extended to 10 and 14 days in the ECGA and WCGA respectively. Adoption of ultra-low volume spraying significantly reduced spray volumes as well as the time needed to spray one hectare of cotton. However, growers had to contend with higher unit costs of ULV insecticides and other issues relating to the cost and variable quality and availability of batteries (Matthews, 1990).

Between 1980 and 1990 synthetic pyrethroids were introduced alongside new sprayers notably the Electrodyne and ULVA+. Given the high potency of pyrethroids and the potential for these sprayers to be operated on less than 8 batteries and spray volumes of 0.5 to 10 l per ha, it became possible for growers to be flexible in their pest control operations (Matthews, 1990). Such developments resulted in annual pesticide use on cotton climbing to 1 million l by 1993. However, by 1999/2000 availability had slumped to a mere 100,000 l in the aftermath of market liberalization. Such dramatic decline in pesticide use contributed to the decline in annual cotton output nationally between 1993 and 2000 as discussed earlier. Amongst the many steps taken to bring about the recovery of cotton production during the first decade of the 21st century, that of switching pesticide use in the WCGA from oil-based ULV spraying to water-based spraying using the ULVA+ was quite critical indeed as it helped to normalize the availability of insecticides which had been wanting for quite some time.

Successful suppression of insect pests on cotton has for a long time depended on the use of host plant resistance as illustrated by the classic control of jassids, and cultural practices as has largely been the case for *P. gossypiella*. To help incorporate biological control on cotton, it became necessary to reduce the scale of chemical control by adopting practices such as threshold spraying and scouting that seek to enable growers to use insecticides on cotton on a need basis.

The use of damage thresholds in determining the need for insecticide sprays against *H. armigera* on cotton was evaluated in 1984 and

1985 in eastern Tanzania. Spraying at thresholds of 10%, 20%, 40% and 80% damaged squares was compared against spraying on a recommended prophylactic regime of six fixed-interval sprays. In both years, spraying on damage thresholds reduced the number of insecticide applications from six to between one and five without significantly affecting control of *H. armigera*, seed cotton yield or overall levels of damage to cotton (Kabissa, 1989). Similar findings were observed in the WCGA (Nyambo, 1989). These results went on to form the basis for scouting-based spraying on cotton in Tanzania.

The use of pesticides in order to manage pests continues to be *ad hoc* and often quite limited for lack of input credit. After liberalization, increased use of unregulated pesticide imports has worsened pest control on cotton. The impact of these factors tends to be further exacerbated by the extreme paucity of extension service in the country. In 2007, only 3379 extension officers were available for the entire country as opposed to a national requirement of 15,082.

Another characteristic of Tanzanian cotton is that total national output tends to fluctuate quite drastically between seasons (see Fig. 9.1).

Most of the fluctuations are due to the rain-fed nature of cotton production and hence its vulnerability to variations in the timing and quantity of rainfall. Global price trends for cotton are also important because, as a rule, planting intentions of smallholders in any year tend to be adjusted upwards or downwards depending on the producer prices paid in the previous season. Cotton prices have other implications such as farmers' decisions to grow cotton or to shift to another crop with a higher per-unit price. Frequent government interventions in the workings of the cotton market have also tended to have a bearing on farmers' appetite to either reduce or increase acreage under cotton. On a final note, the bulk of Tanzania's cotton has tended to be exported annually as lint cotton. Continued export of raw cotton does not reflect well on the health, and hence performance, of the textile and apparel industries. Domestic utilization of lint stood at 27,000 metric tons of lint in 2016 and 23,557 metric tons in 2017. Such decline came in tandem with declining national output starting from the 2012/13 marketing season.

Cotton R & D: The Need for Solutions to New Challenges

As stated earlier, cotton R & D in Tanzania was initiated by the Empire Cotton Growing Corporation and has been reorganized by the launching of a countrywide semi-autonomous body, the Tanzania Agricultural Research Institute (TARI). The change was largely necessitated by the need to enhance Public-Private Partnerships in agricultural R & D in order to increasingly tap

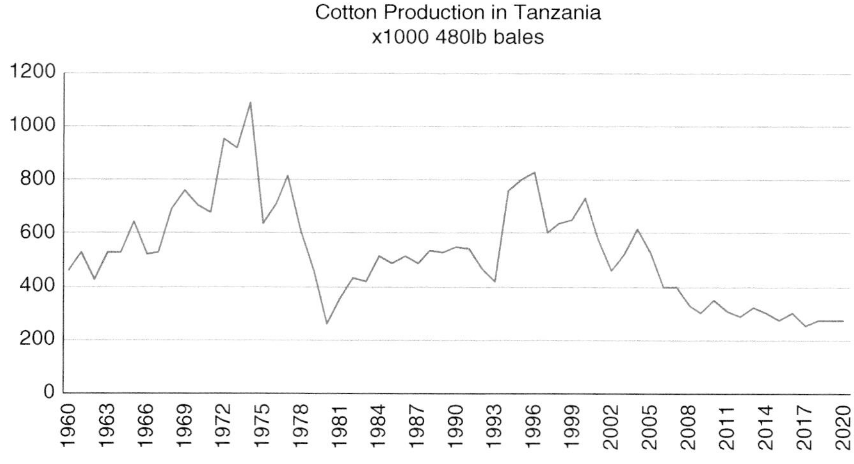

Fig. 9.1. Cotton production in Tanzania.

into the financing and expertise available from both public and private sectors and enhance the capacity of R & D to mobilize financial resources, as well as the autonomy to commercialize some of its outputs and services. Nevertheless, cotton R & D continues to be undertaken along the traditional disciplines of plant breeding, agronomy, soil science, pathology, entomology and fibre technology. Nonetheless, the approach to solving production constraints has become increasingly client-oriented, demand-led and multi-disciplinary in nature.

To date, the overriding R & D agenda is focused on addressing the needs of cotton growing in new areas and how to deal with emerging pest and disease problems in the aftermath of market liberalization. Furthermore, R & D is increasingly playing a more active role in the commercialization of cotton varieties coupled with taking a more proactive role in addressing and articulating new and topical issues relating to climate change, transgenic cotton and robotics, in addition to providing solutions to the more traditional needs of smallholders, ginners and spinners.

On breeding, the challenge is how best to undertake varietal development in a post-liberalization era characterized by the demise of previous seed-multiplication zones and measures then in place to prevent the mixing of varieties and spread of diseases across them. Adoption of a zone-one variety system in 2005, starting with the variety UK91, has continued until 2008 when UKM08 was released. The latter is currently being grown all over the WCGA. The need to revert to the previous subdivision of the WCGA into two or more varietal zones is still valid in view of the WCGA being too diverse in terms of soil conditions, rainfall regimes, patterns of occurrence and distribution of pests and diseases to be considered as one zone. Attempts to resolve this issue must, however, go hand-in-hand with the need by the industry to come up with a functional and acceptable seed system for cotton. The current seed system is being constrained by co-ordination problems owing to the competitive nature of the Tanzanian cotton market coupled with politicization and conflict of interests.

Agronomic research is increasingly being called to address the issue of declining soil fertility both in the WCGA and ECGA. As a result of the continued non-use of inorganic fertilizers on cotton, coupled with the lack of either rotation or fallowing, soil mining has resulted in most soils of the WCGA needing alternative ways of raising fertility while maintaining optimal soil structure. Secondly, following the recent introduction of the use of acid-delinted seed for planting, scientists need to provide recommendations on how best such seeds should be used in order to optimize plant populations per ha and hence yields. Current spacing recommendations, which came into force 80 years ago, may need to be adjusted in the context of facilitating the inclusion of other practices such as mechanical planting, fertilizer use, intercropping or even attaining higher yields through increased plant populations per hectare.

Due to increasing pressure for arable land and the imperative for farmers to produce both food and cash crops, agronomic research needs to come up with answers to an age-old debate on the ban of all forms of intercropping involving cotton in Tanzania. During both the colonial and post-independence periods, the compulsion to have cotton grown as a monoculture was not well received by most farmers. It was perhaps for this reason that the ECGC, in the 1940s, put up a recommendation requiring cotton R & D to find out how cotton could be intercropped with crops with which it competes for land, labour and other resources. To date smallholders continue to find growing cotton as a monoculture incompatible with their mixed-cropping practice.

On plant protection, perhaps the biggest challenge remains how to optimize pesticide use on cotton in the aftermath of the expansion of cotton growing into entirely new areas, the emergence of the fall army worm *Spodoptera frugiperda* as a new cotton pest and the recent elevation to economic status of thrips, jassids and mealybugs. Usage of pesticides on cotton has risen significantly in recent years owing to the need to suppress these pests in addition to the more traditional ones such as *H. armigera*, *P. gossypiella*, *A. gossypii*, *Earias* sp., among others. Most of the insecticides in use are generic pesticides entering the country via both formal and informal channels. Increased use of such pesticide products has tended to trivialize the role of other complimentary pest-management measures such as cultural practices, e.g. mandatory uprooting and destruction of crop residues after harvest coupled with observance of the closed season and promotion of biological control.

If Tanzania is to steer away from a potential pesticide treadmill, adoption of IPM practices is urgently needed. To pave the way for this paradigm, cotton R & D needs to urgently address three major issues. The first one is the need to review current pesticide-use practices on cotton and hence determine how to optimize the use of synthetic pyrethroids and non-pyrethroid insecticides. Emphasis should be on how to deal with both traditional pests as well as new ones. The second issue is the need to address the burgeoning paucity of extension services in general and the plant protection service in particular. One approach of dealing with this would be for cotton R & D to establish a training of trainers programme to address the training needs of the farming community in areas specifically relating to pesticide handling, pesticide spraying, scouting/threshold spraying and IPM.

Thirdly, there is an urgent need for cotton R & D and other affiliated institutions to take up a leading role in the promotion of an informed, unbiased and science-based public dialogue on the topic of biotechnology in general and its roadmap in relation to cotton in Tanzania in particular. To date, the myth on 'fear of the unknown' no longer exists because of field experiences on Bt cotton gained in other countries. The global consensus on transgenic cotton is that where it is specifically deployed to deal with bollworm, use of integrated pest management (IPM) should always be the motto, because nowhere in the world has transgenic cotton proved effective against pests when used unilaterally. There is an impending danger that unless and until the problems facing the current seed systems for cotton and other selected crops are faced, transgenic varieties from Tanzania's neighbours may easily find their way into the country unofficially through illegal imports.

Marketing: The Return of AMCOs

The return of the private sector to the cotton market in 1994 resulted in farmers being paid promptly and given higher prices for their seed cotton than during the era of co-operatives and marketing boards. However, price competition has impacted quite negatively on the co-ordination of the sector and hence the provision of key collective goods such as input and output quality as well as input credit, seasonal inputs and extension services that are needed to make cotton farming more sustainable in the longer term (Tschirley *et al.*, 2009). In addition, due to the wide diversity of cotton stakeholders, and hence vested interests, politics is increasingly coming into play, thus making these issues even harder to resolve.

In an attempt to reform the current marketing model, the government, in 2017, introduced a new marketing set-up that requires all ginners-cum-buyers to procure their seed cotton via designated AMCOs as their statutory buying agents in place of their individual agents. After their abolition in 1976, and subsequent reinstatement in 1984, AMCOs have been operating at a subdued level largely due to their RCUs being unable to actively compete against private buyers. AMCOs warmly welcomed the new initiative by government because it presented a new lifeline for them via the new business opportunities that it presents. Ginners, too, welcomed the opportunity to use AMCOs as their new buying agents because, to them, AMCOs represent a much more secure avenue for channeling their cash than their previous agents who quite often misappropriated or even stole such cash.

In 2018, a *post mortem* made on seed cotton buying using AMCOs revealed that there was a marked increase in the percentage of clean cotton bought relative to previous seasons. This implied that by selling their seed cotton via AMCOs, farmers became obliged to grade their seed cotton prior to selling it, as was the case prior to liberalization. On another note, while most ginners acknowledged to have obtained their money's worth of seed cotton, cash belonging to some ginners could not be fully accounted for. Apparently, some of the bad elements of defunct AMCOs continued into the structures of some of the newly constituted AMCOs. Furthermore, buying seed cotton via AMCOs unfortunately lacked price competition resulting in all seed cotton being bought at the same price. To engender some form of price competition as well as minimize theft and other burgeoning malpractices, the current system involving private buyers buying seed cotton via AMCOs still needed to be reformed.

Ginning: The Need to Extend the By-product Chain

Between 1961 and 1993, the number of operating ginneries rose from 27 to 37 owing to stagnating output. Between 1994 and 2014, the number of ginneries had risen to 84 in the aftermath of market liberalization, partly because of increased cotton output to beyond 100,000 tons of lint on five occasions. Most ginners opted for saw gins owing to their faster operation and lower labour and energy costs relative to the roller-gin technology. To date, ginning equipment is equally split with 50% being roller and the remainder belonging to the roller type as a result of some ginners reverting back to roller gins in view of their ability to produce lint with better fibre length, higher ginning out-turn and being cheaper to operate relative to saw gins, in spite of having higher labour and energy costs.

Fluctuating cotton output meant that only up to 40 ginneries were operational during the ginning season and these faced operational hardships with contaminated and adulterated seed cotton being delivered at ginneries. Until recently, the continued use of outdated, mixed varieties with ginning out-turn values below 40% also negatively impacted on their margins owing to discounts on sales of lint that was contaminated due to within-bale variability. Up until 2010, all ginners were required, under the Tanzania Cotton Industry Act No. 2 of 2001, to reserve up to 14% of the cottonseeds produced for planting of the next season's crop because, until then, the use of recycled 'fuzzy' seed for planting was the rule (Kabissa, 2014). However, such practice has, since 2017/18, stopped, following the adoption of a new variety called UKM08 and institutionalization of a 'one-zone-one-variety' system for seed multiplication.

Because the value of lint obtained from a ton of seed cotton is up to four times the combined value of oil and cake derived from the processing of the seeds, ginners have tended to prioritize the sale of lint in spite of cottonseeds accounting for up to 60% or more of the weight of each kg of seed cotton. Nevertheless, most ginners do, as a general rule, undertake oil milling in addition to ginning because the two processes have a huge cross-subsidization potential in their overall business operations. Tanzania produces an estimated 20 million metric tons of cottonseed oil and 52,000 tons of cottonseed cake per annum from just 32 ginneries utilizing only 17% of their installed capacity. Thus, the production and utilization of cottonseeds, cottonseed oil, cake, husks and linters remain largely under-developed in the country. In 2009, cottonseed oil in Tanzania accounted for only 8% of total national oil consumption, and although cotton ranked second after groundnuts in terms of volumes of cottonseeds produced on an annual basis, its contribution to total national edible oil production remained quite low, due to their oil content being low relative to other oilseeds, stagnating production and volumes of seeds, and the continued use of less-efficient oil-extraction methods (Kabissa, 2016).

Another by-product with considerable economic potential in Tanzania is cotton biomass. To date, cotton biomass, simply the masses of plants remaining after harvest, has either been burned down after harvest or used as firewood in all areas where cotton is produced. So far, there has been no attempt to use such raw material for the production of value-added products such as particle boards, paper, pulp, hardboard and briquettes, as is currently the case in other countries, notably India. Efforts that address the increased utilization of the cottonseed-based by-products value chain, including cotton biomass, have the potential to promote Tanzania's drive towards economic development and poverty alleviation.

Textiles and Apparel: The Quest for Increased Value Addition and Job Creation

Since independence, the government has been promoting industrialization via the textile and apparel sector mainly because of the high potential for value addition within the cotton-to-clothing chain. Furthermore, the cotton-to-clothing chain tends to facilitate the diversification of the economy and, given its vast export potential, has the capability to become a dependable source of foreign exchange earnings. It was perhaps on this understanding that the number of textile and garment mills in the country rose from 8 to 50 between 1968 and 1980, employing some 40,000 people, thus making manufacturing the second largest sector after agriculture.

After SAP, the number of mills declined to 17, by 2019, so the overall contribution of the textile and apparel sector in terms of the total value added to manufacturing in Tanzania fell from 30% to less than 6% between 1980 and 2010, respectively. This was a consequence of cuts in government subsidies, removal of restrictions on imports and depreciation of the Tanzanian shilling, so domestic producers of textiles and apparel became exposed to international competition. The latter intensified further following the advent of trade in second-hand clothing (SHC) as well as an increased availability of cheap imports of new clothing following an increased presence of competitive Asian countries in global garment value chains, all of which had a negative impact on domestic apparel production (Calabrese *et al.*, 2017).

Tanzania seeks to become a mid-income economy by 2025, through industrialization, and intends to promote the development of a viable domestic textile and apparel industry in order to maximize the potential for value addition within the cotton-to-clothing chain, facilitate diversification of the economy, and utilize its vast export potential as a dependable source of foreign exchange earnings. To date, though, there are only ten operational textile and apparel mills in the country that produce garments and textiles for both domestic and export markets. Other textile and apparel mills have closed down with little or no hope of reopening.

In spite of Tanzania's vast comparative advantages, such as having plentiful supplies of raw cotton produced locally; eligibility for preferential market access to the USA, the EU and China; abundant and cheap labour; political stability; and geographical location, the country has found it very difficult to further expand its textile and apparel industry and continues to export most of its raw cotton, up to 70% annually.

The continued export of cotton lint implies that Tanzania misses out on both job creation and value-addition opportunities, as well as not getting its value for money for the lint exports due to low prices paid for its cotton and discounts on lower-quality lint. Surveys conducted by the Textiles Development Unit (TDU) between 2015 and 2018 identified some of the key challenges

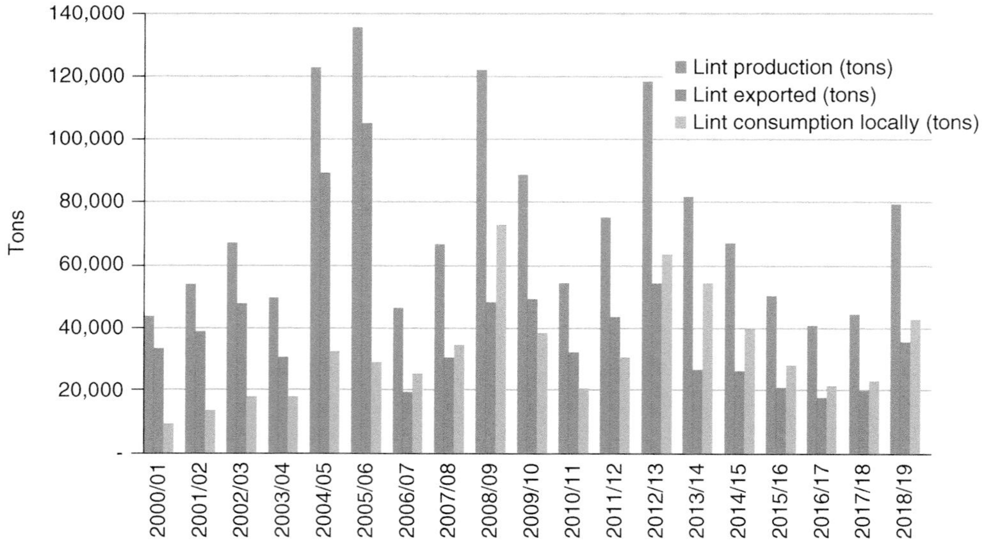

Fig. 9.2. Trends on lint production, exports and consumption for the period 2000/01–2017/18). (From Tanzania Cotton Board.)

facing the performance and growth of the textile and apparel industry in the country. Most local mills have had to actively compete with overseas buyers of Tanzanian cotton for their share, which emphasizes the need for a stable supply of good-quality raw cotton for the domestic industry.

Business registration is cumbersome owing to high transaction costs, arising from regulatory requirements in licensing and compliance coupled with a multiplicity of taxes, fees, charges and levies. Furthermore, time-consuming and procedural bottlenecks at the port of Dar es Salaam and other custom deficiencies tend to allow the market to be flooded with cheap imports and smuggled second-hand clothing (SHC), which tend to undercut the competitiveness of locally produced goods due to rampant under-declaration and undervaluation at ports of entry.

Many other factors make Tanzania uncompetitive for foreign direct investment (FDI) in garment assembly for exports, but there is now a Cotton-to-Cloth (C2C) strategy, which seeks to address most of the bottlenecks to raise the profitability of cotton production and boost competitiveness of Tanzania's textile and clothing firms through enhanced productivity and product diversification.

Future Prospects of Cotton in Tanzania

Cotton's potential to contribute to economic development and poverty alleviation in Tanzania has remained largely untapped. The single-channel marketing system had to be replaced in the early 1990s by a competitive private sector-led market because it could not deliver on both production and productivity. In spite of numerous hiccups, the competitive market structure resulted in increased cotton output on several occasions during the first and second decades of the 21st century. Increases in output were, however, not sustained over time and did not result in a vibrant cotton-to-cloth value chain. If the cotton industry is to sustainably support economic development via the cotton-to-cloth value chain, the government needs to resolve co-ordination problems that currently make the market unable to provide inputs, input credit and extension services, as well as support the development of a robust seed system. In addition, the government must address policy and institutional and related issues that currently make the cotton-to-clothing value chain in the country uncompetitive and thus unable to contribute significantly to the country's economy.

Ethiopia

Joe Kabissa

Introduction

In Ethiopia, as in most African states, agriculture continues to be the most dominant sector in the economy of the country, supporting the livelihoods of well over 80% of the rural population and accounting for 42.9% of the Gross Domestic Product. In view of the strategic importance of agriculture, the government of Ethiopia recently embarked on the so-called Growth and Transformation Plan – 2025, which seeks to move the country to a middle-income status by 2025 via agricultural transformation. It may be recalled that Ethiopia, unlike other African states, was never colonized as such but only briefly occupied by the Italians between 1935 and 1941. Thus it never really benefitted from the colonial influence on commercialization and modernization of its agriculture whereby agricultural production was geared mostly at meeting the raw material needs of the colonial administration.

Anecdotal evidence has it that the practice and culture of growing, processing and utilizing cotton for production of cotton-based garments using handlooms dates far back in history, long before the commercialization of cotton began. To date, Ethiopia is recognized as a global centre of origin for some key crops such as coffee and probably the cotton species *Gossypium herbaceum* as well.

Despite Ethiopia's long cotton history and being endowed with an estimated 2.6–3.2 million ha of land that is suitable for cotton farming, only about 3–7% of such area is currently under cotton cultivation. Important cotton-growing areas include the Awash Valley, Humera, Metema, Gode, Arbaminch, Sille, Woyito, Omorate, Gambella and Beles. Aerial spraying of cotton was used in the Awash Valley. Cotton farming is the undertaking of both large-scale and smallholder farmers in the proportions 67% (45% is by private large farmers and 22% by large state-owned farms) and 33%, respectively. Smallholders have average acreages of between 0.5 and 3 ha.

In the major high-potential areas, cotton tends to be grown as a sole crop under both irrigated and rain-fed conditions. One of the large farms was at Tendaho. It was managed by a UK company for many years until it was nationalized and was where aircraft pilots were practising spraying the crop. Key issues were the need to restrict the period of sowing to have a distinct closed season and improve crop monitoring for better timing of sprays. One of the problems resulting in range of sowing dates was the availability of water for irrigation at the key time. Rain-fed crops are often grown as part of a farming system incorporating sesame, sorghum and maize. Cotton can be sown either between mid-April and late June or between June and August, depending on whether it is irrigated. Between the 2010/11 and 2019/20 seasons, the area under cotton fluctuated between 60,000 and 100,000 ha, 35,000–54,000 ha of which was irrigated, with the remainder under rain-fed conditions.

There are an estimated 500,000 households that depend on cotton for their livelihoods, 2000 of which are entirely involved in the production of organic cotton in the Arbaminch area. Organic cotton production is promoted by a local NGO called Pesticide Action Nexus (PAN) Ethiopia in collaboration with PAN-UK (Tadesse *et al.*, 2017). There is considerable potential for increased organic cotton production in Ethiopia, partly because many farmers are unable to afford the use of agrochemicals. The seed used for planting is derived from one or more of the four main varieties that are currently in use, namely: Mid-Awash, Bazen, Hiwot and Semen OMO. Mensah *et al.* (2012) developed a supplementary food spray product (Benin Food Spray) from local ingredients and used it to manage pests and beneficial insects on organic cotton crops in Benin, resulting in higher cotton yields and profitability. Similarly, applying the spray to organically grown cotton crops in Ethiopia conserved the populations of beneficial insects naturally, reducing the number of pests, resulting in higher cotton yields and higher gross returns to the farmers (Amera *et al.*, 2017).

Recently, the Ethiopian government approved the commercialization of Bt cotton with effect from June 2018. Two varieties called JKCH 1050 and JKC 1947 are currently being promoted. During the 2019/20 planting season, the country expected to produce a crop of between 57,000 and 64,000 tons of lint.

Cotton R & D

Cotton R & D was initiated by Italians at the turn of the 20th century but their attempts to introduce the extra-long-staple cotton called *Gossypium barbadense* failed, so *Gossypium hirsutum* is grown. Large-scale production of cotton was found feasible and by 1964 a cotton research station was established at the Werer Agricultural Research Centre, focusing on irrigated cotton. In 2016, the Asosa Agricultural Research Centre was charged with the responsibility of addressing the R & D needs of rain-fed cotton.

The crucial task is providing information and advice to enable large and smallholder farmers to increase production and boost productivity. To date, average yields for rain-fed and irrigated cotton are estimated to be in the order of 1.5 and 2.5 tons of seed cotton per ha, respectively. For both smallholder and large-scale farms, weeds and insect pests are the major production constraints. Yield losses can be enormous due to limited use of insecticides. Herbicide use is uncommon. The recent introduction of transgenic cotton may help to deal with both problems, although the recently introduced varieties do not yet have stacked genes. The two current cotton R & D centres tend to adopt a multidisciplinary approach to tackling production constraints facing irrigated and rain-fed cotton but do not as yet have their work replicated in all six major agroecological zones where cotton farming takes place.

Challenges Facing Cotton Production

In spite of the huge potential for increased cotton that exists, Ethiopia's capacity to increase annual output is being constrained by several key factors. Owing to the lack of mechanized harvesting, large-scale farming is critically dependent on human labour whose availability tends to be quite problematic. During the harvest period, for example, up to 3000 labourers may be needed per day. However, due to both ethnic and tribal conflicts, especially in the regions of Amhara, Benshagul Gumuz, Gambela and Afar, this is a major problem, so harvesting is quite often delayed. For security reasons, movement of people tends to be impeded by road blockages and checks. In addition, both large and smallholders tend to face limited access to credit, improved seeds and the requisite necessary farm machinery and implements.

Cotton farming also tends to face stiff price competition from crops such as maize, sesame, sunflower and mung beans grown in the same farming system. In the Awash Valley, where cotton used to be the most important crop, accounting for up to 64% of the country's total cotton output, cotton farming is being replaced by plantations of both sugarcane and sesame. Institutionally, cotton has, until recently, been lacking the requisite organizational structure for overseeing and promoting the overall development of the cotton industry.

The Ethiopian Lint-to-textile and Apparel Value Chain

In Ethiopia, cotton is the raw material for the booming textile and garment industry as well as traditional local handicraft and cottage industries. However, because the country's annual cotton output of 64,000 tons of lint is only 58% of the total annual requirement of 111,081 tons, the existing shortfall is being met by imports of lint. Now that the textile and apparel industry is being accorded top priority in the Ethiopian Industrial Development Strategy, the need to boost cotton output at the local level is being emphasized. The value chain largely comprises 21 ginneries (18 saw gins and 3 roller gins) and 122 textile and garment factories, 3 spinning

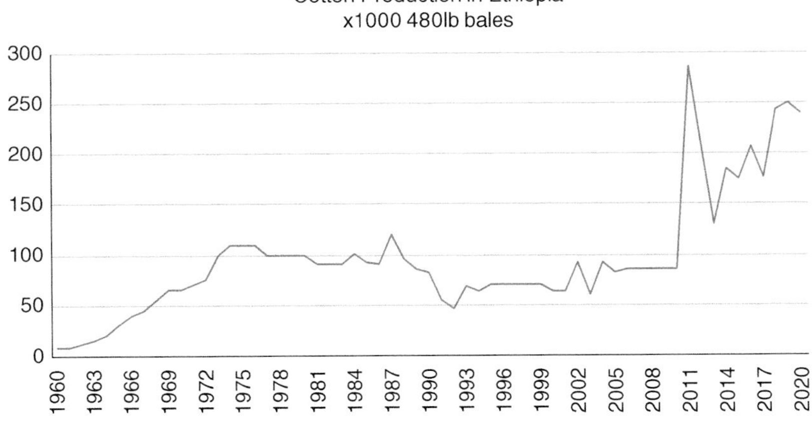

Fig. 9.3. Cotton production in Ethiopia.

mills, 18 weaving and knitting factories, 13 integrated mills, 60 garment factories and 7 traditional handloom companies, and has been performing well below the installed capacity.

The ginning capacity is quite low owing to the age of some ginneries and, perhaps more importantly, the inadequate supply of raw cotton. Between 2010 and 2017, ginners in Ethiopia were banned from exporting their lint to help satisfy the existing demand for lint cotton by the domestic textile and apparel industry.

Future Outlook of Cotton in Ethiopia

Under its 2015–2020 economic plan, the Ethiopian government is investing quite considerably in the development of the textile and apparel sector. Under the Growth and Transformation Plan (GTP II), the government intends to create an estimated 140,000 new jobs as well as raise export revenues to one billion US$ by 2020. Looking well beyond GTP II, the Ethiopian government envisions making the country the textile and apparel manufacturing hub of Africa with annual exports worth 30 billion US$ by 2025, by investing quite substantially in the development of industrial parks. In these parks are 21 textile and apparel manufacturers that capitalize on benefits from inexpensive labour and electricity, among other incentives.

The Ethiopian government, under its New Cotton Development Strategy (NCDS), seeks to make the country one of the world's top cotton-producing countries with an annual production of 1.1 million metric tons of lint. An Ethiopian Cotton Development Authority (ECDA) is planned to oversee and implement plans to make the cotton sector more competitive (ATA, 2012), involving three major recent developments: the introduction of transgenic cotton cultivation; an ongoing effort to further advance contract farming arrangements and help farmers obtain relatively better access to seasonal inputs as well as extension services and assured market outlets for their seed cotton and lint; and the ongoing construction of the Ghibe-3 dam on the Omo river, which will help expand cotton cultivation under irrigation.

Uganda

Pius Elobu

Introduction

Cotton production was introduced in Uganda by the British in 1903, initially in the central region, with production later spreading to the rest of the country. While production was carried out in Uganda by small-scale farmers at the start, ginning was done in Kenya and the lint was then exported to Liverpool to service the British textile mills. The policy framework governing cotton production and marketing was embedded in the revised Cotton Act (1964) and the Lint Marketing Board (LMB) Act (1959, amended 1976). The Board monopolized cotton lint and cottonseed trade at domestic and international levels, while ginning and marketing functions were assigned to co-operative unions.

The country then witnessed a continuous increase in the production of cotton until the

Address for correspondence: piuselobu@gmail.com

1970s. The government required farmers to allocate at least 0.5 ha to cotton production (Baffes, 2009a; Lugojja, 2017). Other factors that contributed to the success include the introduction of ginneries, textiles mills and other processes across the country, both by government and the private sector. Heavy government involvement in the sector also contributed significantly to its post-independence success, largely in the form of protecting farmers from price volatility. Such efforts included the establishment of the Cotton Control Board (CCB) in the 1930s and the passage of the Cotton Zone Ordinance, which established 14 production zones, each allotted a ginnery (Lugojja, 2017). However, after the 1970s, political instability and poor policies led to the sector's demise. Attempts to revive the sector during the Economic Recovery Programme of the World Bank in the 1990s set the pace for policy reforms. These reforms, combined with a lending operation and the high cotton prices of the 1990s, led to a revival of the sector (Baffes, 2009b).

In 1994, the Ugandan government liberalized the ginning and marketing of cotton by passing the Cotton Development Act, and creating the CDO, a semi-autonomous agency under the Ministry of Agriculture, Animal Industry and Fisheries. Cotton ginning and marketing was opened to private sector participation, while the CDO was responsible for monitoring, promotion, processing, marketing and regulation of the cotton sub-sector. Its mandate includes: to enhance the quality of lint and cotton products, both for export and domestic consumption; to promote the distribution of high-quality planting seed; and generally to facilitate the development of the cotton industry. Only 5–10% of production is used domestically (Masiga and Ruhweza, 2007). Samples taken from each bale are analysed either by manual classing or high-volume instrument (HVI) classing by CDO.

The National Agricultural Research Organisation (NARO) is responsible for research to enhance the development of production technology, through the National Semi-Arid Resources Research Institute (NaSARRI). Research on production technology is aimed at developing new cotton varieties, yields, fibre quality, initial seed multiplication, improving existing varieties, controlling pests and diseases, and improving current farming practices.

The Uganda Investment Authority (UIA) has identified a number of investment opportunities in the cotton sector. The key areas for investment include: cotton ginning, cottonseed oil, animal feed, absorbent cotton wool and cotton yarn. According to UIA, investments in the cotton sector can be located near the cotton-growing areas and where large tracts of land can also be leased to investors for them to grow their own seed cotton. Moreover, UIA can also allocate the investor industrial land to construct new facilities in planned industrial parks in the North and West Nile regions. In the recent past, Uganda has developed two policies that are important for developing cotton products and by-products in Uganda. First, the Buy Uganda Build Uganda (BUBU) policy was initiated in 2014, based on existing government policies to support and encourage the consumption of locally produced goods and services. The BUBU policy vision is to develop a vibrant, dynamic and competitive private sector that transforms local products through the value chain to meet the required standards. This is to be realized by its mission statement, which is to support the production, purchase, supply, and consumption of local goods and services (Government of Uganda, 2014). Secondly, the National Textile Policy of 2009 was designed to enhance the performance of agro-based industries in the country, so as to increase value addition on locally produced raw materials and export of manufactured goods. The policy vision is to create a strong and vibrant textile and clothing industry, with sustainable capacity utilization and enhanced investment, throughout the cotton-to-clothing value chain.

Overall Importance in Uganda's Agricultural Economy

Cotton is grown in rotation with other crops in over 30 of the 56 administrative districts and generates disposable cash income to farmers in rural settings. The deep-rooted crop improves yields of subsequent crops, such as finger millet, sorghum, sesame, groundnut and cowpea.

The cotton sub-sector employs an estimated 2.5 million people, directly and indirectly, as farmers and farm labourers, seed cotton and cottonseed buyers, transporters, ginnery workers, cotton exporters, textile and garment manufacturers, oil millers, etc. In 2007, cotton was ranked as the third most important export crop after coffee and tea (Cotton Development Organization, 2007).

According to UNCTAD, Uganda still has several impediments to overcome for further development of the cotton industry:

- high energy costs;
- high transport costs: Uganda's landlocked position significantly increases trade costs;
- outdated machinery and equipment: ginneries, textile mills and oil mills throughout the country use outdated machinery, constraining their ability to produce quality goods at competitive prices;
- high cost of credit: the high cost of capital limits investments in upgrading existing machinery or purchasing new equipment;
- shortage of cottonseed: the low production of cotton has resulted in low utilization of the installed capacity at all the cottonseed-based processing firms;
- limited capacity to stock lint: according to the CDO, domestically produced lint is only available on the Ugandan market from January to May each year. Outside of this period, ginners are unwilling to hold lint bales without payment;
- shortage of specialized technicians results in inefficiency in operations and a high dependence on expatriate technicians, adding to the cost of production.

Cotton's Impact on Food Security

Food security exists when all people, at all times, have physical and economic access to sufficient, safe and nutritious food to meet their dietary needs and food preferences for an active and healthy life at the individual, household, national, regional and global levels (World Food Summit, 1996). Cotton production contributes to food security in several ways. At the national level, export revenue makes it possible for a country to access food through imports. When cotton export revenue is the major source of foreign exchange, its contribution to food security is obviously of primary importance. Declining cotton prices and earnings resulted in a reduction of merchandise imports. At the rural household level, the contribution of cotton production to food security is mainly through income. When households specialize in cotton production, there is a direct link between cotton production and their ability to buy basic foodstuffs and goods.

Cotton in Uganda is mostly grown on small family farms with an average area of 0.5 ha. Nevertheless, it is an important cash crop for the diversified small farms as income is needed by rural households to acquire healthcare, food and many other services. According to the survey, on average, households used about 35% of their total cash income to buy food, 10% to buy clothing and 40% for many other needs such as medical care, communications and education. The export revenue from textiles and clothing and employment opportunities in the cotton ginning, spinning, weaving, processing, marketing and transporting all have very important income and food security implications for the country and households living both in rural and urban areas.

In Uganda, cotton is the opening crop for most resting fields. In most of the cotton-growing areas of eastern and northern Uganda, there is a direct link between cotton acreages and those of finger millet. Finger millet follows cotton in the rotation and then cowpea. Some farmers actually grow cotton in order to have good seedbeds for finger millet in the subsequent season. Among many ethnic groups within these areas, millet is a very important food crop because of its nutritional status and its ability to store for long periods. Other important food crops that follow cotton in the rotation are groundnut and sesame, while beans and maize are often intercropped with cotton by many farmers. Large acreages of cotton therefore mean more acreage of finger millet, cowpea, beans, maize and groundnut for the household, either at the same time or, in most cases, in the subsequent seasons.

Location-specific Advantage of Cotton in Uganda

The total water requirement for cotton is 900–1200 mm, 20% of which is needed from establishment to flowering, 40% during flowering, 30% from flowering to boll bursting and 10%

from boll bursting to maturity (Ikisan.com, 2000). According to Uganda's National Water Development Report 2005, the average annual rainfall for Uganda varies from 900 mm in the north-eastern semi-arid areas of Kotido to 2000 mm on Sese Islands in Lake Victoria.

Uganda grows Bukalasa Pedigree Albar (BPA) cotton varieties. The Allen Black Arm Resistance (Albar) genes were fully incorporated into them 50 years ago and have given the varieties ample resistance to the bacterial blight disease, which is normally a big threat to most cotton industries worldwide. These varieties are extensively adapted to the entire country including western Uganda, West Nile region, northern and eastern regions including the Busoga cotton belt. These areas constitute over 75% of Uganda's arable land. Other vast areas like Karamoja have the potential for cotton production once irrigation is assured and this would increase the acreage for cotton. The crop therefore is not drastically limited by environment as compared to other crops, such as apples, potatoes and coffee.

Market Potential of Cotton in Uganda

All cotton farmers sell their cotton to local buyers spread all over the country. Lugojja (2017) outlines the different cotton by-products in Uganda. The ginners process the cotton (ginning) and export much of the lint to international markets such as Japan, USA, Germany, etc., where the yarn is used for weaving into pure threads or for blending, for knitting and for making threads. Some of the lint is consumed locally for making blankets, mats, carpets, writing papers, toilet tissues, X-ray films, photographic films, upholstery padding, cold/heat insulation, wool and gauze for medical use. The seed is consumed locally. Part of it is bought by Cotton Development Organization (CDO) for further treatment and distribution for farmers, while the other part is sold to the local mills for processing into cooking oil, margarine, soap and animal feed cake, while the hulls can be used for fuel and mushroom growing. Some of these potentials of cotton are not yet fully explored, but they imply cotton has a large potential market locally, regionally and internationally. Cotton's market potential is for a wide array of people ranging from farmers and traders to manufacturers.

Nathan-MSI Group (2001) reviewed the many positive assets that define Uganda's textile and clothing sector. The country's agricultural sector produces raw material (cotton lint) of outstanding quality, medium-to-long-staple fibres, with good colour, low micronaire values, and high-quality ginning. Private ginneries process some 100,000 bales of lint each year (compared with a peak year of 400,000 bales), over 90% of which is immediately sold to international markets. Ginneries have existed in Uganda for a long time and all are now in private hands. Most integrated mills were begun just after independence in the mid-1960s. Despite the country's turbulent history and the ensuing effects on its industrial operators, an installed capital equipment base still exists in Uganda. There is ample supply of ready-to-work, inexpensive labourers. Many of the managers of the country's larger textile and garment firms display sound managerial capacity and reasonable technical and marketing skills. Often, this skilled manpower has already been abroad for professional textiles training or has worked overseas. Some of these are Asian-Ugandans who were exiled in the 1970s and have since returned to take up where they left off; others are Africans who have been overseas or in neighbouring Kenya and gained valuable training there.

Performance of the Cotton Sector

Right from independence, cotton production in Uganda has fluctuated so much as shown in Fig. 9.4. The fluctuations relate closely to the political scenarios the country has passed through.

Most areas of Uganda have the potential to produce cotton. Currently, the main production areas include the north, east and south-east of Lake Kyoga, and in the Kasese area in the west. Different varieties are produced in the country, which yield finer fibre of medium to long staple length, with excellent characteristics, being silky and well suited for spinning. The total area under cotton cultivation in Uganda has fluctuated over the years due to various factors, including weather conditions, price expectations and the provision of farm inputs (Ahmed and Ojangole, 2012). For instance, from 2008 to 2009, total area under cultivation reduced

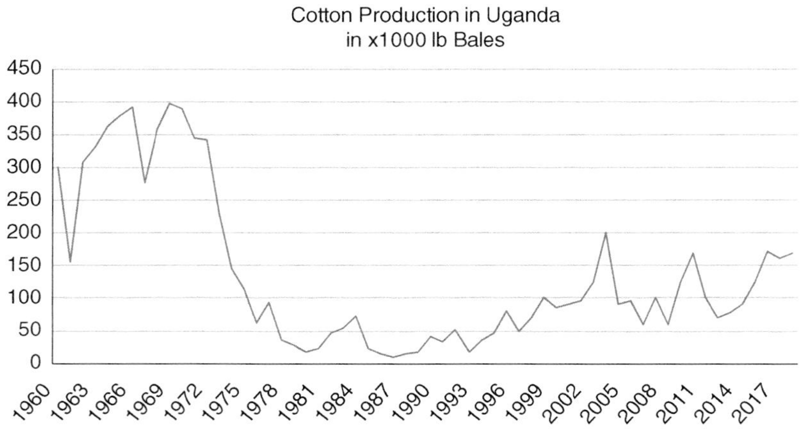

Fig. 9.4. Cotton production in Uganda.

from 101,215 ha to 68,000 ha, before increasing to 80,000 ha in 2010 (MAAIF, 2012). The area has since increased to 109,312 ha by 2017 according to preliminary estimates by the CDO. It is also important to note that cotton production has not reached its historic peak of the late 1960s. In fact, production is still far below the estimated potential of 1,000,000 bales at full employment level.

Changing weather patterns due to climate change and over-reliance on rain-fed production caused yields to vary between 700 and 850 kg/ha between 2013 and 2017. The average yield is far below that of the main cotton-producing countries, such as China with 1270 kg/ha and Israel with 1700 kg/ha. This is partly due the unpredictable rainfall patterns, but also due to smallholder producers lacking appropriate farming skills. Climatic changes have also adversely affected the timing and duration of the planting, leading to poor harvests. Low usage of inputs like pesticides and fertilizers, coupled with declining soil fertility, are other factors affecting productivity. Most of the production inputs are imported and are relatively costly to an average farmer (Lugojja, 2017). Efforts by the government to improve productivity have yielded some positive results, but having new varieties to help farmers adapt to changing needs takes time. Other factors include constrained extension services and limited credit for purchasing inputs.

These farmers are rarely dedicated, professional cotton growers and allocate their land and labour among, for example, other cash crops, subsistence food crops, as well as other non-farming activities. The allocation of smallholder resources to cotton thus depends on the comparative returns from cotton compared with the returns from competing crops. Indeed, a profitability analysis carried out by Baffes (2009a) revealed that cotton has low profitability when compared with food crops like beans, maize, bananas (matooke) and sorghum. During the period 1989/90–2003/04, nominal cotton prices received by farmers increased by an annual average of 3.1%, while the corresponding price increases of the four food crops were much higher: 5.3% for beans, 6.1% for maize, 5.3% for matooke and 7.8% for sorghum. Baffes (2009a) noted that this was because food commodities respond to domestic demand conditions, whereas cotton produced in Uganda is mostly exported to Bangladesh, China, Switzerland, United Arab Emirates, Singapore, the UK and Hong Kong (China).

Cotton has the potential to improve the welfare of about 250,000 low-income farming households (Gordon and Goodland, 2000; Baffes, 2009a). The CDO has encouraged the formation of associations to support the development of cotton by-products in Uganda. The three key associations are: the Uganda Ginners and Cotton Exporters Association (UGCEA), formed in 1997; Uganda Oilseed Producers and

Processors (UOSPA), formed in 2007; and Textile Manufacturers Association of Uganda (TEMAU). The UGCEA is an umbrella body that brings together all the cotton ginners and exporters in Uganda. The Association seeks to: promote, protect and co-ordinate the interest of ginners, farmers and exporters in Uganda; initiate discussions and exchange of information amongst members on cotton issues; advise government on key policies affecting the cotton sub-sector; provide a united forum for people interested in cotton ginning and export; and, generally, the improvement of the cotton industry.

According to the Government of Uganda, UGCEA adopted the Cotton Production Support Programme (CPSP) and initiated a Cotton Development Fund (CDF) to finance the CPSP. Through these interventions, the association works with CDO to provide production inputs such as planting seed, pesticides, spray pumps, fertilizers and herbicides to mobilize and sensitize farmers, as well as provide them with extension services. UOSPA, on the other hand, is an umbrella organization of cottonseed processors. They procure cottonseed from ginneries for milling into cottonseed cake and oil. Cottonseed oil is used primarily as cooking oil by bakeries and confectioners, while the cake is used in compounding livestock feeds (Lugojja, 2017). With all the cotton being harvested by hand, the adoption of strict quality control and ginnery monitoring, and the refurbishment of the CDO cotton-classing laboratory, the quality of lint has been greatly improved with better prices for premium-grade Ugandan cotton. Under the Cotton Subsector Development Program (CSDP), a Ginning Training School was established at Busitema National College of Agricultural Mechanization in 1998. The college is now a university and offers a Diploma in Ginning Engineering and a degree in Textile Engineering. There are 40 ginneries, all equipped with roller gins with only two having saw gins.

Cotton Varieties

Before 1990, Uganda used to grow two cotton types, namely Bukalasa Pedigree Albar (BPA), with longer fibres, and Serere Albar Type Uganda (SATU), with shorter and finer fibres. The cotton types were earlier selected by Allen in Uganda during the 1960s for their resistance to bacterial blight (*Xanthomonas malvacearum*). The acronym 'Albar' in the names of the two cotton types therefore stands for Allen black arm resistance. BPA was grown in the cotton areas south of the River Nile (mainly central and western Uganda), while SATU was grown in the northern and eastern parts of the country (north of the Nile).

With increased worldwide demand for longer fibres, NaSARRI scientists, funded by the World Bank Cotton Sub-sector Development Programme (CSDP), rigorously re-evaluated the two cotton types under multi-location trials in the 1990s. These trials led to the adoption of only one cotton type (BPA) across the whole country, with breeding concentrated on BPA that left SATU at maintenance level (Serunjogi *et al.*, 2000). A number of varieties were subsequently released to farmers (Table 9.1). The naming of cotton varieties follows the year in which the foundation seed of that lot was grown

Table 9.1. Desirable attributes of past and currently released cotton varieties in Uganda (Amoding *et al.*, 2019).

Variety	Release Year	Ginning out-turn (%)	Days to maturity	Boll size (g)	Seed cotton yield (kg/ha)	Fibre length (mm)	Fibre strength (g/tex)	Micronnaire	Lint index
BPA 95	1995	40.0	140–165	5.1	1440	28.6	29.5	3.9	7.0
BPA 97	1997	39.4	135–155	4.6	1008	28.3	28.6	3.9	6.7
BPA 99	1999	38.6	135–150	5.1	1927	29.1	28.7	4.2	6.7
BPA 2000	2000	39.2	130–150	5.2	1488	28.9	26.7	4.3	7.0
BPA 2002	2002	38.0	135–155	6.5	1940	27.0	28.0	3.7	6.7
BPA 2015 A	2015	41.0	120–150	6.5	2400	30.0	29.0	3.9	7.8
BPA 2015 B	2015	40.0	100–135	5.6	2000	27.0	29.0	3.4	8.2

in NaSARRI's main farm for release to farmers, e.g. BPA 90 was grown at the main NaSARRI farm and released in 1990. Later varieties from Zimbabwe were introduced to Uganda and some of them have been adopted as commercial varieties. Currently, Uganda grows BPA 2015 A, BPA 2015 B, both of Ugandan origin, QM 301 and SZ 931Y, both from Zimbabwe.

Agronomy of Cotton

Agronomic considerations for cotton growing include proper planting period, plant spacing, fertilizer use, intercropping with food crops, management of vegetative growth and harvesting.

Cotton is sown from May to July in the eastern and northern regions, while for farmers in the west, sowing is from June to August. The bimodal rainfall generally starts in March/April, reduces in July to give a dry spell before starting again in August until November/December. Most farmers pick and sell their cotton between December and March, which is generally a dry period during which all the ginneries are active.

The recommended plant spacing is 75 cm between rows and 30 cm within rows (75 x 30) for the east and north, and 90 x 30 cm for the west. However, farmers in both zones often do not follow these recommendations and prefer to plant their cotton at wider spacing, which may be why farmers' yields are often lower that what is obtained at research institutes.

Over-cultivation of soils has led to drastic decline in productivity, warranting the need to apply organic and inorganic fertilizers, but not many Ugandan farmers use fertilizers on their cotton. They have been advised to do soil testing to guide them in fertilizer application. However, many farmers, who are able to use fertilizers, apply 250–300 kg/ha of NPK at second ploughing or at planting and then top-dress with 100 kg/ha of urea at the start of flowering. Organic materials of animal origin like kraal, goat, sheep and poultry manures are applied at rates of 1–3 t/ha while those of plant origin like *Mucuna pruriens* (Elobu *et al.*, 2014), *Tithonia diversifolia* (Elobu *et al.*, 2016; crop production – ICAC.ORG) etc. may be used. *Mucuna* is first planted in a field, allowed to grow for at least three months then ploughed down for cotton to be planted. *Tithonia diversifolia* grows wild in many parts of Uganda and the leaves are harvested and applied under the young cotton crop at a rate of 1–5 t/ha. The limitation with use of organic materials is their bulkiness, making transportation to the fields costly.

Weeding cotton crops by hand using a hand hoe starts 30–45 days from sowing, and is usually repeated 3–4 times before the crop is harvested (Elobu *et al.*, 1998). Some farmers who have oxen have used animal-drawn weeders, but lack of weeders has limited this method for weed control. Only large-scale farmers apply herbicides for control of weeds in cotton.

Pest Control

Cotton in Uganda is attacked by different pests at different growth stages, from emergence to picking, as shown in Table 9.2.

Various efforts to address these pests date back to the establishment of a cotton research station at Namulonge, Uganda, in 1947 and operated by the Empire Cotton Growing Corporation. Following the success of DDT against mosquitoes, the vector of malaria, Pearson started a trial in 1950 using DDT sprayed weekly to protect cotton from attack of *Lygus vosseleri* bugs damaging leaf and flower buds. A simple

Table 9.2. Insect pests occurring at different cotton growth stages in Uganda.

Stage of plant growth	Common pests	Occasional pests
Emergence to first square of cotton	thrips, aphids, mites, jassids	cutworm, plant bugs
First flower bud to first bloom	plant bugs, stainers, lygus bugs, whitefly	aphids, spider mites, bollworm, tobacco budworm, fall armyworm
After first bloom	bollworm, tobacco budworm, plant bugs, stink bugs, stainers	aphids, loopers, fall armyworm, spider mites, whitefly

syringe sprayer (Plantector) was used. A 25% increase in yield was obtained, with the number of lygus bugs decreasing after each spray. This decrease was also noted in the untreated plots, probably due to their flight activity ranging between the individual plots. Later, farmers used a knapsack sprayer with lance to apply insecticides.

In countries using a syringe sprayer or a knapsack sprayer with a lance, yields were not increased so much as with a tailboom or low-volume applicator, referred to in Chapter 8. The operation of the syringe sprayer was arduous and distribution of spray was not consistent. Early recommendations often required sprays at two-weekly intervals, despite the rapid increase in plant height, with unprotected foliage and the extent of damage that occurred when larvae could get established in buds or bolls within a few days. Using a simple lance was shown to increase the amount of spray collected on the operator, especially on a fully grown crop as walking between the rows resulted in the operator's body touching sprayed foliage.

Farmers were supplied with packs of DDT to spray four times against *Lygus* and early *Helicoverpa armigera* (bollworm) using 1.1 kg of active ingredient per ha in 90 litres/ha of water, starting five weeks after germination and repeated at two-week intervals. About 400,000 acre-packs were distributed annually, but this covered less than a third of Uganda's cotton area (Reed, 1976).

Later trials (Davies, 1976) compared the effects of applying four, eight and 12 sprays of various insecticides on control of *Helicoverpa armigera* Hub. and *Earias biplaga* Wlk., and the consequent effect on yield and quality of seed cotton produced. Yields of unsprayed cotton ranged from 300–900 kg/ha, while four sprays of DDT gave 1000–1200 kg/ha. Increasing the number of sprays to 12, using three insecticides (DDT, endosulfan and carbaryl) gave increments of 300–600 kg/ha over the standard four-spray treatment. Eight sprays gave smaller but significant yield increases, and endosulfan, dicrotophos and carbaryl were particularly promising in such regimes. A mixed DDT/phenthoate spray was not superior to DDT used alone. Spraying reduced the percentage of stained cotton.

The false codling moth *Cryptophlebia leucotreta* (previously named *Argyroploce leucotreta*), a pest of citrus, was reported to be causing damage on cotton by Pearson (1958), but insecticide sprays had little impact on this pest (Reed, 1976), possibly because the larvae feed inside the fruit.

IPM farmers were expected to spray, initially, soapy water applied for aphid control, followed by 3–4 applications of insecticides based on scouting 50 plants every two weeks, although farmers could sample 25 plants/week. It was expected that the ginneries would have sufficient dimethoate for two applications and sufficient pyrethroid for a further two applications. A sprayer with a rotary nozzle (Micron ULVA+ sprayers) was distributed, although it was not suitable for applying soapy water.

Recent research recommended use of predators and parasitoids as natural enemies in the fight against cotton pests. They include black ants (*Lepisiota* spp) or *Ngininginі* in the local Lango language (Ogwal *et al.*, 2003), ladybird beetles (Coccinellids), spiders, rove beetles, syphidae larvae, lacewing, pirate bugs, wasps/egg/pupae and larval parasitoids (Amoding *et al.*, 2019). These can be reared and released to farmers' fields for the control of cotton bollworm, cotton stainers, lygus bugs and spider mites.

New and effective synthetic and biopesticides for the control of bollworm, lygus bugs, stainers and spider mites have also been identified on the basis of a preliminary field experiment and recommended for control of cotton pests in Uganda (Gayi *et al.*, 2016). The synthetic pesticides recommended for use in cotton include: thiodicarb, methomyl, profenofos + cypermethrin, thiamethoxam + lambda-cyhalothrin, lambda cyhalothrin, profenophos, bifenthrin lambda cyhalothrin + profenophos and acetamiprid + cypermethrin. Biopesticides included azadirachtin (neem oil), emamectin benzoate (an actinomycete) + abamectin (a bacterium). They both control cotton pests and are associated with significantly higher survival values of natural enemies (ladybirds, ants and syrphid larvae) in the cotton fields. The addition of fertilizer on sprayed cotton fields increased yields (Hillocks, 2005).

Disease Control

A number of diseases affect cotton in Uganda as summarized in the cotton production guide for farmers (Amoding *et al.*, 2019) and they include cotton bacterial blight (*Xanthomonas axonopodis* pv *malvacearum*), fusarium wilt (*Fusarium oxysporum* forma *specialis vasinfectum*), verticillium wilt (*Verticillium dahliae*) and alternaria leafspot (*Alternaria macrospora*). Control measures of bacterial blight disease include breeding for resistance, crop rotation with crops that are not infected by the bacterium, e.g. cereals like finger millet, use of disease-free seeds and resistant cultivars, supplied by CDO, deep ploughing to bury plant residues and followed by fallowing the area, treating seeds with Bronopal at rate of 200 kg of delinted seed/1 kg of chemical or 150 kg of fuzzy seed/1 kg of chemical, or treating seed with fungicides plus thiamethoxam at a rate of 300 ml of the chemical to 100 kg of delinted seeds.

Fusarium wilt disease is managed by use of the resistant cotton varieties available, proper control of weeds during and between crops because weeds act as alternate hosts to the pathogen, avoiding of mechanical inter-row cultivations if possible since any root damage during the process may become points of infection by the fungus, practising proper crop rotation, and dressing cottonseed with fungicides such as fludioxonil and mefenoxam. No single method is highly effective in controlling verticillium wilt disease, but an integrated management system is necessary to minimize losses from the disease. Start with selection of a variety with good resistance to wilt and good adaptation to the geographical location. This should be followed with cultural practices such as rotation with cereals (finger millet, sorghum, maize) and legumes (beans, soybeans) for two seasons to reduce the soil fungus levels. Potassium fertilizers can also be applied as P-deficiency increases the severity of wilt. Combining organic amendments with inorganic fertilizers decreases wilt compared with the inorganic fertilizers alone. Ensure that seed is dressed with fungicides, as this protects the seed from early infection and minimizes yield losses. No control measure is currently directed towards alternaria leaf spot disease, as it is managed with other measures that are applied in the management of bacterial blight and wilt diseases. Such measures include selection of resistant varieties, cultural management like destroying infested crop residue and dressing delinted cottonseed with fungicides.

Kenya

Anthony Muriithi

Introduction

Some pioneer settlers thought cotton might succeed in East Africa, which coincided with the development of the Kenya/Uganda Railway, and in 1906, the British East Africa Corporation Ltd was established to spread the work of the British Cotton Growing Association's aim of establishing cotton growing with cotton ginneries and plantations. Capital of £500,000 was raised, a considerable sum in those days, which encouraged the Colonial Office to provide 30,000 acres for government farms and an annual grant for experimentation and education. Ginneries were built at Malindi, Kilindini and Kisumu and experimental farms

Address for correspondence: gikandimuriithi@yahoo.com

were established at Malindi, Voi, Mombasa, Kibos and Kisumu. Due to unfavourable weather, the plantations in Kenya were not very successful, whereas in Uganda, on good soil, cotton growing spread rapidly.

Later, some cotton growing began to thrive in areas near Lake Victoria, and in 1955, the Cotton Lint and Seed Marketing Board (CL and SMB) was established. Production reached a peak in 1985, when about 100,000 bales annually were produced compared with 400,000 bales in Uganda and 700,000 in Tanzania. Then as cotton farmers switched to maize and other crops, Kenya produced only 25,000 bales by 2014. It was suggested that the industry's poor performance was due to weak farmer organizations, high costs of production, inadequate quality inputs and over-reliance on rain-fed production. During the cotton boom in the late 1970s and 1980s, the government had supported the industry, when the 1955 Cotton Lint and Seed Marketing Act was repealed and replaced with the Cotton Act Cap. 335 No 3 of 1988, and later revised in 1990 creating the Cotton Board of Kenya, which had an organized marketing system that paid farmers promptly. The board invested heavily in factories, such as Raymonds and Rift Valley Textiles (Rivatex) in Eldoret, as well as Kicomi. But the Cotton Board became ineffective, when the textile market was liberalized in 1991. Second-hand clothes were imported, providing a serious challenge to local cloths. Gradually firms closed down and cotton ginneries lay idle. Previously one of Kenya's main exports, cotton seemed to be in its dying days. Raw textiles were, instead, purchased from Taiwan and Singapore. In 2006, the Cotton (Amendment) Act was enacted creating the Cotton Development Authority (CODA) whose mandate is to regulate, coordinate and promote the cotton industry. In 2014, the Cotton Development Authority was merged with Kenya Sisal Board to form the Fibre Crops Directorate under the Agriculture and Food Authority (AFA). Kenya's national economy has continued to rely on the agricultural sector, as it contributes 24% of Gross Domestic Product (GDP) and 65% of the export earnings. In addition, the sector provides the livelihood of over 80% of the Kenyan population and their food security. The strengthening of the agricultural sector is essential to achieve economic recovery and growth.

Eighty per cent of Kenya's land area is arid or semi-arid with limited economic opportunities for the communities in these areas. Cotton is now considered under the Country Vision 2030 and in the Agricultural Sector Development Strategy (ASDS) as one of the most important industries to implement the long term Arid and Semi-Arid Lands (ASAL) development initiatives and industrialization strategy. The aim is for agriculture to deliver 10 per cent annual economic growth envisaged under the economic pillar of the Vision 2030 strategy. Hence the government is implementing initiatives aimed at reviving the Cotton Industry, comprising the cotton growers, ginning, spinning and textile sectors, with the government providing targeted support to the smallholder farmers, by providing planting seeds, an advisory service through public funded extension service and research. It is also supporting rehabilitation of irrigation schemes to reinstate irrigated cotton production in a crop rotation with other food crops. The government has also developed a reliable method for testing cottons through an instrument-based classing system and assistance to the textile industry and export of cotton.

In 2006, the Government set up the Cotton Amendment Bill to provide the legal framework with which to support re-organization of the cotton sector. The initial impact was an increase in production, but this was achieved by an increase in the number of producers (hectares under cultivation) rather than any substantial increase in productivity. Average yields remain at 400–600 kg/ha of seed cotton.

The country has an estimated potential of 350,000 hectares suitable for rain-fed crop production and 35,000 hectares of irrigated cotton. This combined potential can produce an estimated 200,000 MT of seed cotton. However, this potential has not been achieved due to inefficiencies and inadequate integration from production to textile manufacturing and consumption. In particular, farmers lack affordable credit to purchase inputs, good quality planting seed, as well as lack of information on appropriate use of pesticides, fertilizer and manure.

Seed planted has not been certified, but the plan is that all farmers will be able to receive certified seed by 2020, and an inspection system will ensure that ginners do not make uncertified seed available to farmers. There are over

150 cotton lines available in the National Gene Bank of Kenya, managed by Kenya Agricultural and Livestock Research Organization (KALRO). Two varieties were developed and evaluated in the 1980s. They were released and are commercialized: KSA 81M and HART 89M. Other varieties are undergoing National Performance Trials with Kenya Plant Health Inspectorate Services. KSA 81M, currently grown in Nyanza, Western Kenya and parts of Rift Valley has a potential seed cotton yield of 2500kg/ha, under irrigation with a lint percentage of 35–37 and is fairly resistant to verticillium wilt and jassids. Hart 89M is grown in Central, Eastern Kenya and Coast province. It has a potential seed yield of 3000kg/ha under irrigation with a lint percentage of 36–40. It is resistant to bacterial wilt and jassids.

Production of basic and foundation seed of these two commercial varieties is carried out every season at Kibos and Mwea research stations to maintain genetic purity of seed continuously on a year to year basis. The potential yields compare with the current production of only 572kg/ha.

Continuous evaluation of new varieties was revived from the 2006–07 cotton season. Deltapine lines/varieties (34) from Monsanto have been screened, out of which 3 varieties were submitted for National Performance Trials (NPT). One of the varieties 06K486 is at advanced stage of approval for commercial multiplication alongside KSA 81M and Hart 89M. Other varieties under analysis are Scala v – 1 and L142.9 in addition to some early maturing varieties from Pakistan. H211 Hybrid variety from Israel has also been released for planting under irrigation and is currently produced in Hola, Bura, Perkera and Galana Kulalu Irrigation Schemes.

Crop Protection

Cotton farmers are advised to use Integrated pest management. Scouting and identifying pests is required to adopt a combination of different strategies including chemical, biological and cultural control methods. A booklet was published to illustrate the various insect pests, as well as indicate the predators, parasitoids and pathogens that affect the pests (van den Berg and Cock, 2000). Midega *et al.* (2012) reported that in western Kenya, most farmers could only identify the bollworm (*Helicoverpa armigera*) and cotton stainers (*Dysdercus spp.*), known as 'Nyanginja' and Oero yyore', respectively. Aphids, red spider mite and a few other insect pests can occur.

Farmer field school training has been advocated, but application of insecticides is the most commonly used method to control the different insect pests. Unfortunately, most farmers did not apply them effectively due to high costs and shortage of appropriate insecticides, lack of sprayers and limited knowledge of all pest management. It was pointed out that farmers generally did not receive any extension services on pest management based on the practice of mixed cropping. Choice of insecticide has largely depended on price and availability, although a large number of pesticides are registered for use.

Just over half the farmers surveyed applied a pyrethroid, 27% used another type while 17% did not know the names. The majority only used one insecticide, and when asked about the frequency of sprays, less than 10% followed a schedule. Part of this was due to fewer than 5% of farmers owning a sprayer, while others relied on a single sprayer shared by about 60 farmers. None had any protective clothing or had received any training.

Mambiri (1987) reported on trials using different ways of applying insecticides. The yield of seed cotton was increased by 534 kg/ha using an Electrodyn sprayer, but the following year there was no significant difference between the sprayers tested; however, in 1985, using the ULVA sprayer gave the highest increase in yield of 592 kg/ha compared to the control. The Electrodyn sprayer was considered to be the best judging by yields on farmers' plots, but the equipment was withdrawn from the market. The most commonly used sprayer was a lever-operated knapsack sprayer (Fig. 9.5).

Currently, about 30,000 bales are produced annually, but the domestic market demand is 140,000 bales with a potential to grow to 260,000 bales. The country is therefore only utilizing 31% of the current ginning capacity of 140,000 bales. Similarly, out of 400,000 ha of land potentially suited for cotton, only about 45,000 ha are utilized.

In many countries, genetically modified cotton varieties have been grown, but in 2006, the African Union adopted a resolution stating

Fig. 9.5. Spraying cotton with single nozzle. (Photo: Pius Elobu)

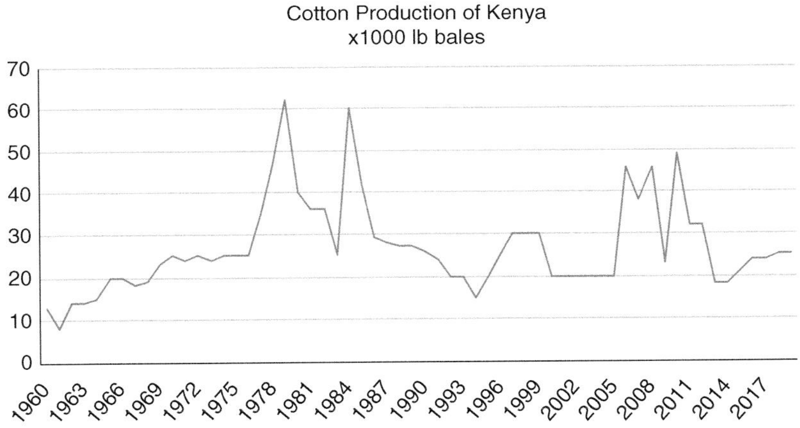

Fig. 9.6. Variation in cotton production in Kenya.

that genetically modified organisms (GMOs) were not welcome on the continent. Soon it was realized that GMOs have the potential to redefine agriculture. Subsequently, Kenya enacted the Biosafety Act in 2009 and established a National Biosafety Authority (NBA) in 2010. Since its establishment, NBA has reviewed over 28 contained-use applications, approved 14 confined field trials with environmental release applications for Bt cotton and Bt maize. In April 2020, the distribution of free cottonseeds to farmers, including a genetically modified variety, was started in the latest push to revive the cotton industry in the country. According to the Ministry of Agriculture, 16 metric tonnes of hybrid seeds and one metric tonne of BT cottonseeds will be distributed across 24 counties. With control of *Helicoverpa* bollworm, the single most destructive cotton pest, causing up to 100% loss, this consignment, it is anticipated, will produce over 2000 kg of seed cotton per ha against the current average yield of 572 kg per ha. It is also expected that Kenyan farmers will enjoy a four-fold reduction in production costs.

Cotton Marketing

The cotton market has been fully liberalized and marketing is now wholly in the hands of the private sector. Under the arrangement, Fibre Directorate as a regulatory body is charged with co-ordinating activities to promote transparency in marketing operations. However, cotton marketing is facing difficulties, emanating from weak cotton co-operative unions/societies or organized farmer groups, resulting in poor bargaining power of farmers and no economies of scale. Farmers are being encouraged to form and strengthen farmer organizations to undertake marketing activities on behalf of members.

To hasten recovery of the co-operative societies and farmer organizations, the government waived the non-performing loans owed to the Cooperative Bank of Kenya by the co-operatives totalling Ksh 237 million (3 million US$) in 2007/08 thinking that they would be revived and play their rightful role in the revitalization of the industry. The dormant co-operative societies needed to be revived since most of the infrastructure is still intact or requires minimal rehabilitation. The Fibre Crops Directorate is assisting the groups to develop and implement business plans.

Since 2013, decisions on seed-cotton pricing are taken in conjunction with the stakeholders comprising the Kenya Cotton Growers Association, Kenya Cotton Ginners Association and the textile manufacturers, using the world cotton prices as a benchmark price for lint, who then assign figures for freight, prevailing exchange rates, ginning costs (transport, ginning), ginners' profit margin and the ginning out-turn to work out a minimum seed cotton price that the farmer should receive in the current season. Market information is relayed to farmers to promote fair trade and competitive marketing for seed cotton. All seed-cotton buyers are now registered by Fibre Directorate to improve transparency in seed-cotton purchases. Three operational ginneries are located in the eastern area and one in the Rift Valley, but there are several that have closed.

Cotton Testing and Classification

The Authority has established a high-speed cotton-testing instrument with a capacity to handle 700 bales per day to assure spinners and traders of actual lint quality. This will improve competitiveness in processing of lint by optimization of manufacturing process. The system has paved the way for branding of Kenyan cottons in line with the global trends.

References

Ahmed, M. and Ojangole S. (2012) *Analysis of Incentives and Disincentives for Cotton in Uganda*. Technical Notes series. MAFAP, FAO, Rome.

Akiyama, T., Baffes, J., Larson, L. and Varangis, P. (eds) (2001) *Commodity Market Reforms: Lessons of Two Decades.* World Bank.

Amera, T., Mensah, R.K. and Belay, A. (2017) Integrated pest management in a cotton-growing area in the Southern rift valley region of Ethiopia: development and application of a supplementary food spray product to manage pests and beneficial insects. *International Journal of Pest Management* 63, 185–204.

Amoding, G., Elobu, P., Takan, J.P., Gayi, D. and Orawu, M. (2019) *Increasing Productivity and Quality of Cotton for Improved Livelihoods of Farmers: A Production Guide for Cotton Farmers in Uganda.*

ATA (Ethiopian Agricultural Transformation Agency) (2012) *Program Update*, vol. 1, no. 2. July. Ethiopian Agricultural Transformation Agency, Addis Ababa.

Baffes, J. (2009a) The full potential of cotton. *Development Policy Review* 27, 67–85.

Baffes, J. (2009b) *The Gender Dimension of Uganda's Cotton Sector*. World Bank.

Bates, R.H. (2015) *Markets and States in Tropical Africa: The Political Basis of Agricultural Policies.* University of California Press, Oakland, California.

Beckert, S. (2015) *Empire of Cotton: A Global History*. Vintage Books.

Calabrese, L., Balchiu, N. and Mendez-Parra, M. (2017) The phase-out of second-hand clothing imports: what impact for Tanzania? MPRA Paper No. 82175. Available at: https://mpra.ub.uni-muenchen.de/82175 (accessed 5 August 2021).

Cotton Development Organization (2007) Cotton production and farmers' earnings by district (2002–2005). CDO.

Cotton, Textile and Apparel Sector Investment Profile-Ethiopia. http://financedocbox.com/Tax_Planning/71657714-Cotton-textile-and-apparel-sector-investment-profile-ethiopia.html (accessed 5 August 2021).

Davies, J.C. (1976) Trials of spraying and cultural practices on cotton in Uganda: II. Use of extended protection. *Experimental Agriculture* 12, 163–176.

Elobu, P., Orwangga, J.F., Ocan, J.R., Okwii, A. and Alepo, B.H. (1998) Weeding initiation time and frequency: their effect on performance and the economics of weed control in cotton. *Proceedings of Second World Research Conference* Athens, 6–12 September 1998, pp. 392–395.

Elobu, P., Nalunga, J., Musunguzi, P., Ocan, J.R. and Olinga, J. (2014) Use of Mucuna pruriens to improve soil fertility for cotton production in Uganda. *NARO Conference*, August 2014, Entebbe, Uganda.

Elobu, P., Ocan, J.R. Olinga, J. and Ogabe, P. (2016) Use of *Tithonia diversifolia* to improve cotton productivity under marginal soils in Uganda. Paper presented at the 6th World Cotton Research Conference in Goiana, Brazil, 2–6 May 2016. Available at: http://staging.icac.org/meetings/wcrc/wcrc5/Pdf_File/154.pdf (accessed 5 August 2021).

Gayi, D., Ocen, D., Lubadde, G. and Serunjogi, L. (2016) Efficacy of bio and synthetic pesticides against the American bollworm and their natural enemies on cotton. *Journal of Agricultural Sciences* 17, 67–81.

Gordon, A. and Goodland, A. (2000). *Production credit for African small-holders: conditions for private provision*, January.

Government of Uganda (2014) *Harnessing Uganda's Demographic Dividend for Socio-economic Transformation. The State of Uganda Population Report.*

Hillocks, R. (2005) *Promotion of IPM for smallholder cotton in Uganda (R8197 [ZA0516]).* Final Technical Report. Natural Resources Institute.

Ikisan.com (2000) http://www.ikisan.com/links/ap_cottonWater%20Management.shtml

Kabissa, J.C.B. (1989) Evaluation of damage thresholds for insecticidal control of *Helicoverpa armigera* (Hubner) (Lepidoptera: Noctuidae) on cotton in eastern Tanzania. *Bulletin of Entomological Research* 79(1), 95–98.

Kabissa, J.C.B. (1990) Seasonal occurrence and damage by *Pectinophora gossypiella* (Saunders) (Lepidoptera: Gelechiidae) to cotton in eastern Tanzania. *Tropical Pest Management* 36, 356–358.

Kabissa, J.C.B. (2014) *Cotton in Tanzania: Breaking the Jinx*. Tanzania Educational Publishers Ltd.

Kabissa, J.C.B. (2016) *Cotton and Its By-Products in The United Republic of Tanzania.* United Nations conference on Trade and Development (UNCTAD).

Kabissa, J.C.B. and Nyambo, B.T. (1989) The red bollworm, *Diparopsis castanea* Hamps (Lepidoptera: Noctuidae) and cotton production in Tanzania. *Tropical Pest Management* 35, 190–192.

Kabissa, J.C.B., Temu, E., Ng'homa, M. and Mrosso, F. (1997) Control of cotton pests in Tanzania: Progress and prospects. In: Adipala, E., Tendwa, J.S. and Ogenga-Latigo M.W. (eds) *African Crop Science Conference Proceedings*, Pretoria, 13 – 17 January 1997, pp. 1159–1160.

Lugojja, F. (2017) *Cotton and Its By-Products in Uganda.* Background paper for the United Nations conference on Trade and Development (UNCTAD).

MAAIF (2012) *Joint Agricultural Sector Annual Review (JASAR).* Workshop Report, November.

Mambiri, A.M. (1987) Evaluation of some crop sprayers in the application of insecticides on cotton in Kenya. *Tropical Pest Management* 33.

Masiga, M. and Ruhweza, A. (2017) *Commodity Revenue Management: Coffee and Cotton in Uganda.* International Institute for Sustainable Development.

Matthews, G.A. (1990) Changes in application techniques used by small-scale cotton farmers in Africa. *Tropical Pest Management* 36, 166–172.

Mensah, R.K, Vodouhe, D.S., Sanfillippo, D., Assogba, G. and Monday, P. (2012) Increasing organic cotton production in Benin West Africa with a supplementary food spray product to manage pests and beneficial insects. *International Journal of Pest Management* 58, 53–64.

Midega, C.A.O., Nyang'au, I.M., Pittchar, J., Birkett, M.A., Pickett, J.A., Borges, M. and Khan, Z.R. (2012) Farmers' perceptions of cotton pests and their management in western Kenya. *Crop Protection* 42, 193–201.

Nathan-MSI Group (2001) *Agoa Textile and Garments: What Future for Uganda's Export?*

Nyambo, B.T. (1988) Significance of host plant phenology in the dynamics and pest incidence of the cotton bollworm, *Heliothis armigera* (Lepidoptera: Noctuidae) in western Tanzania. *Crop Protection* 7, 161–167.

Nyambo, B.T. (1989) Use of scouting in the control of *Heliothis armigera* in the Western Cotton Growing Area of Tanzania. *Crop Protection* 8, 310–317.

Ogwal, S., Epieru, G., Bwarogeza, M. and Acom, V. (2003) Effects of cotton inter-cropping on establishment and biological control efficacy of *Lepisiota* spp. predator and on major insect pests. *Uganda Journal of Agricultural Sciences* 8, 67–74.
Pearson, E.O. (1958) *The Insect Pests of Cotton in Tropical Africa*. Empire Cotton Growing Corporation, London.
Reed, W. (1965) *Heliothis armigera* (Hb.) (Noctuidae) in Western Tanganyika. II.—Ecology and natural and chemical control. *Bulletin of Entomological Research* 56(1), 127–140.
Reed, W. (1976) Entomology. In: Arnold, M.H. (ed.) *Agricultural Research for Development.* Cambridge University Press, Cambridge, UK.
Serunjogi, L.K., Mukasa, S.B., Odeke, W. and Ochola, G. (2000) Changes in breeding strategy for needs in a liberalized cotton industry in Uganda. In: Gillham, F.M. (ed.) *Proceedings of World Cotton Research Conference 2: New Frontiers in Cotton Research* 1, 177–178.
Tadesse, A., Mensah, R.K. and Belay, A. (2017) Integrated pest management in a cotton-growing area in the Southern Rift valley region of Ethiopia: development and application of a supplementary food spray product to manage pests and beneficial insects. *International Journal of Pest Management* 63, 185–204.
Tschirley, D., Poulton, C. and Labaste, P. (eds) (2009) *Organization and Performance of Cotton Sectors in Africa: Learning from Reform Experience*. World Bank.
Van den Berg, H. and Cock, M.W. (2000) *African Bollworm and Its Natural Enemies in Kenya*, 2nd edn. CAB International, Nairobi.
World Food Summit (1996) *Report of the World Food Summit*, 13–17 November, Rome.

10 Cotton Growing in West Africa

Germain Ochou Ochou*, S.W. Avicor and G.A. Matthews

When cotton was first cultivated in West Africa is not known, but nomads crossing the Sahara from the Middle East through Egypt no doubt introduced cotton over a millennium ago, and by the 10th century cotton was being used more widely (Kriger, 2005). Some of the earliest cotton was probably in the area of the Senegal and Gambia rivers, going back well over a millennium. *G. arboretum* and *G. herbaceum* cotton plants were grown on a small scale with many houses having their own 'cotton tree', indicating that it was being grown as a perennial (Kriger, 2005). Schwartz (1996) reported that yields were low during this period, rarely reaching 150 kg/ha seed cotton. The fibre was spun and woven into fabric to make clothes for local use. Some of the fibre was also used to barter with Saharan traders. When European mariners started to explore the coast, there were local fabrics to export, including quaqua and other cloths woven on treadle looms inland and brought to the coast by traders, which continued during the slave trade era.

However, when the need for labour in the USA was apparent, ships from Europe would trade goods such as cloth, guns, gunpowder and brandy worth about £3 for a slave captured by local tribal chiefs. These slaves taken to America *c.*1700 were then sold for £20, so despite the risks, a good profit was made. Indian cotton textiles of various types were important for European merchants to purchase slaves along coastal areas of West Africa (Kazuo, 2016). Dark-blue cotton textiles produced in Pondicherry, called *guinées*, were also important in the trade in gum Arabic.

The rapid expansion of cotton grown in the USA resulted in most European countries relying on shipping cotton from there to supply their own developing textile industries. However, when the civil war in America reduced supplies, the European colonists decided to try to develop cotton growing in Africa, as the cost of transport would be less than buying from India. However, the war in America did not last very long, so interest in cotton growing in Africa declined. It was primarily after the First World War that a more serious effort was made by the British and French to get more cotton produced in Africa. Subsequently, there was a will to promote West African cotton, but was there a way? A number of projects were promoted from the UK, but with limited funding and knowledge, these failed (Ratcliffe, 1982).

Yields of cotton remained low, mainly due to insect pests, until the arrival of insecticides, notably DDT, when efforts were made to get farmers to use this new technology to increase yields. Adoption was very poor as the rain-fed

*Corresponding author: ochougermain@yahoo.fr

 Pest Management in Cotton: A Global Perspective
(eds G.A. Matthews and T. Miller)
DOI: 10.1079/9781800620216.0010

crop did not have an easily accessible supply of water for spraying with a knapsack sprayer. The arrival of ultra-low-volume spray, formulated in an oil in the francophone countries in 1975, started a period of increasing yields (Figs 10.9 and 10.19). The higher cost of UL formulations ultimately led to a reintroduction of using water-based formulations in 1995, but the spinning disc sprayer, now improved and requiring less power and with an advanced design of the rotary atomizer, was retained to apply a very-low-volume spray using ten litres of water per hectare (Fig. 10.3). An electrostatic ULV sprayer (Electrodyn) was used briefly in Nigeria (Fig. 10.21) and several other countries, but was discontinued by the manufacturer.

Information about the cotton-growing countries is provided in the following sections.

Cotton Growing in Mali

Mali was once part of three West African empires which controlled trans-Saharan trade in gold, salt, slaves and other precious commodities. As elsewhere in West Africa, before the introduction of *G. hirsutum*, cotton found locally was mainly *G. arboreum* and, maybe, *G. herbaceum* grown by villagers. Cotton was a perennial and some grew into a tree. In the early 1950s, the French parastatal Compagnie Française pour le Développement des Fibres Textiles (CFDT) extended its improved production, processing and marketing operations in Mali from Central Africa, where Belgium was the initiator of the cotton-zone system in Zaire (Democratic Republic of Congo), endowing monopoly right to a cotton company in a specific zone. However, it was not until political independence was gained in 1960 that the Keita regime encouraged a continuation of CFDT investments through an agreement that maintained the CFDT monopoly control over cotton production and processing. Later, in 1974, the Malian government and the CFDT agreed to create the Compagnie Malienne pour le Développement du Textile (CMDT), with 60% Malian government and 40% CFDT capital. This joint Malian-CFDT decision enabled the Malian government to attract foreign public capital (initially from France and later from the World Bank) for important rural development and infrastructure activities (e.g. roads, literacy etc.) throughout the cotton zone. A village extension agent with the CMDT helped solve the issue arising from villagers' withdrawal from cotton production as a protest against being cheated by dishonest agents at cotton marketing. The solution suggested was to let villagers to market by themselves after proper training. What had been accepted for trial led to a rapid transfer of responsibility for cotton grading and weighing, equipment, supply orders and credit management to designated village groups, which were later established as Village Associations (Associations Villageoises, AVs). In collaboration with

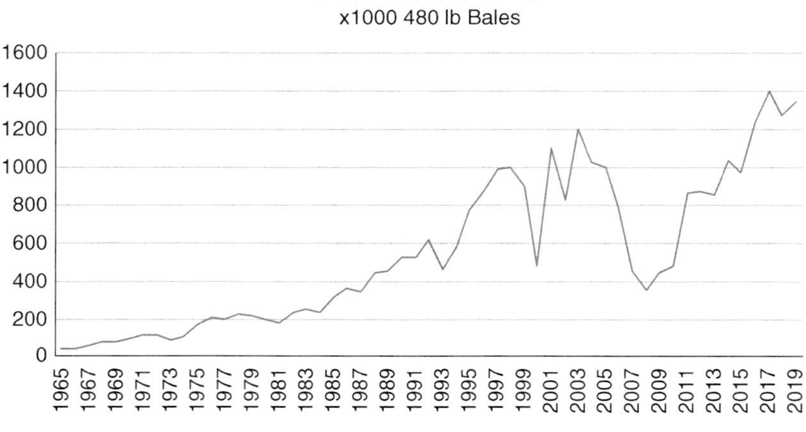

Fig. 10.1. Changes in production in Mali.

the government, CMDT also secured World Bank financing to support the development of management skills within the AVs. By 1982, about 75% of total production was marketed by AVs. In the Bambara language, a programme was designed to ensure that farmers had the literacy and numeracy skills to fulfil credit and marketing tasks and to prepare account books (Bingen, 1998). In fact, the decision of transferring the seed cotton marketing to villages implied training them in book–keeping, and therefore conducting literacy programmes in the local language as villagers had seldom been to school.

As in Cote D'Ivoire, the production of cotton began to increase when the recommendation to apply insecticides using a ULV sprayer was introduced in 1975. The decision enabled farmers to protect their cotton without the need to take large quantities of water to the field for spraying. Used to the battery-powered sprayer, the change to very-low-volume spraying to reduce the cost of the insecticides was also accepted by 1990.

The Malian government kept actively backstopping cotton production in the mid-1990s in spite of reduction of international support in the absence of alternative cash crops in the cotton-producing areas. The resulting production increase made Mali one of the largest cotton producers on the African continent. The Better Cotton Initiative (BCI) was engaged by 2010 in view of a more sustainable way of production, but it did not get the market premium it expected. In the 2017/18 cotton season, 3879 licensed BCI farmers in Mali produced 6000 metric tonnes of 'better cotton', a small share of the total production on 14,000 hectares. The crop is sown from June to July and harvested from October to January.

The dip in production in 2008 was due to the persisting low world price of cotton, starting in 2001. CMDT had to pay a high price to farmers and yet get a low price for the cotton it exported, which led to a deficit, which created other problems including late ordering of inputs due to the lack of funds resulting in farmers getting chemicals later and complaining.

Mali, along with Benin, Burkina Faso and Chad, had reacted to the unfavourable world price by protesting against the subsidies implemented by some major countries or regions (namely the USA, China and the EU). It gave rise to the 'C4 initiative', submitted by the mentioned four countries in the framework of the WTO ministerial meeting in Cancun (Mexico) in September 2003. The mentioned initiative is the first item feeding the 'cotton dossier' at WTO and which remains unsolved till now. This was considered as a key event in international trade relations (Heinisch, 2006).

The CMDT has a monopoly on cotton processing and marketing in Mali, which has enabled it to maintain and increase cotton export earnings to serve the government's interests and to be able to purchase agricultural supplies, especially fertilizer, on concessional terms, which helped all companies get fairer acquisition prices for inputs. This structure has increased cotton production, with over 100,000 households, largely in the southern part of the country, successfully cultivating cotton under rain-fed conditions. From the mid-1960s, the ginning firms also provided farmers with inputs on credit, extension, transport and other services. Because of the concept of integrated rural development and the perception of the role of the public sector in promoting development, national cotton companies in francophone Africa were entrusted to provide services beyond cotton production, *sensus stricto*, on behalf of the governments, with or without specific funding from the governments. It was when the wind of economic liberalization started blowing in the first years of the 1990s that cotton companies were ordered to stick strictly to the cotton business. Thus cotton production is seen more as a means to achieve other outputs, whether it is other economic activities, the organization of villages to take charge of their own development, opposition to cotton production without considering the long-term consequences on the environment or the maintenance of land capital (Fok, 1993). A major decrease in production from 2005–2008, from a peak of over 600,000 tonnes in 2003/04 to less than a third of this amount in 2008/09, was attributed to factors such as low international prices and unfavourable rainfall patterns, but also to numerous types of institutional malfunctioning, such as decreased and less effective extension systems, and corrupt and disorderly input-procurement methods.

Smallholder farmers have been encouraged to diversify their agricultural production system in order to achieve food self-sufficiency and enhance family income. In addition, having a cash crop and engaging in food-crop production and livestock rearing contribute to the reduction of extreme poverty, malnutrition and food insecurity. Four diversification strategies: cotton and maize; cotton maize and millet; cotton maize, millet and sorghum; and food-crop production have been used by farmers in southern Mali. Those with a larger family were more likely to diversify compared to farmers only engaging in food-crop production. Similarly, farmers owning oxen were more likely to diversify (Dembele *et al.*, 2018). To prevent farmers or their families spending too much time getting water, a drilling programme funded by the World Bank was launched in Mali, which resulted in the first Chinese companies arriving in the country to set up the drills.

Cotton Growing in Cote D'Ivoire

Cotton is grown originally in the savannah region in the northern and central part of the country. At present, the estimated cultivated area averages more than 300,000 ha and involves more than 100,000 individual farmers.

Cotton is a relatively new crop in Côte d'Ivoire. The first trace of cotton growing in Côte d'Ivoire, as described by Hau (1988), dates back to the 18th century with the variety *Gossypium barbadense*. It remained, for a long time, a secondary crop, just suitable for association with other food crops. Availability of insecticides (DDT) in 1960 enabled the cultivation of a more productive variety, *Gossypium hirsutum*. Progressively, cotton became the principal crop with more productive and modern agricultural practices (improved genetic materials, cultural and protection practices). In 1974, the Ivorian government decided to initiate incentives and an effective policy through CIDT (Compagnie Ivoirienne pour le Développement des Textiles), so that cotton is the only major cash crop for farmers in the northern region. The crop, therefore, has an important socioeconomic role, but in the national economy it represents a relatively small part (about 3–7%) among the export crops (cocoa, coffee, cashew and palm tree).

At the beginning, the cotton production system was managed by CIDT, a semi-private society belonging 75% to the government and 25% to the French company CFDT. CIDT was founded in 1974 as a replacement of the French CFDT, and depends on the Ministry of Agriculture; its principal sociopolitical objective is to promote the cotton crop and the associated food crops such as rice, maize and peanuts.

With respect to the issue stated above, CIDT's activities deal with the following: the extension service, the technical assistance to farmers, the broadcasting of cultural techniques and delivery of new technologies; the training and advisory role in crop-management systems; the modernization (intensification) of farms through mechanization; the implementation and assessment of research-development actions; the rural promotion (involvement in organization of farmers' structures; the distribution of inputs to farmers; the production and distribution of cottonseed, maize and rice; the distribution of inputs such as fertilizers, insecticides, herbicides, etc.; the distribution of agricultural materials (tractors, cattle); the collection of the cotton production and its transfer to the ginning industry; and the local and export marketing of fibre and cottonseed.

As for CMDT in Mali, the CIDT used to have a monopoly on cotton processing and marketing in Côte d'Ivoire. Funded by 1974, this structure has helped increase cotton production, with over 100,000 households, largely in the northern part of the country, successfully cultivating cotton under rain-fed conditions. With economic liberalization in the 1990s, the cotton sector was privatized in 1998. The number of private cotton societies fluctuated according to year. At present, six ginning companies (CIDT, COIC, SECO, SICOSA, Ivoire Coton and Gjobal Cotton) are sharing all the cotton-producing zones.

A major decrease in cotton production from a peak of over 400,000 tonnes in 2002/03 to less than a third of this amount in 2008/09 was attributed to factors such as low international prices and unfavourable rainfall patterns, but also to numerous types of institutional malfunctioning, such as decreased and less effective extension systems due to socio-military crisis from 2002 to 2011.

The French colonial government pursued a policy of cotton export promotion through coercion, but when this strategy failed, cotton exports fell dramatically. During the colonial period, prices received by local producers were more attractive than the prices offered by the colonial textile industry. Much later, a dramatic

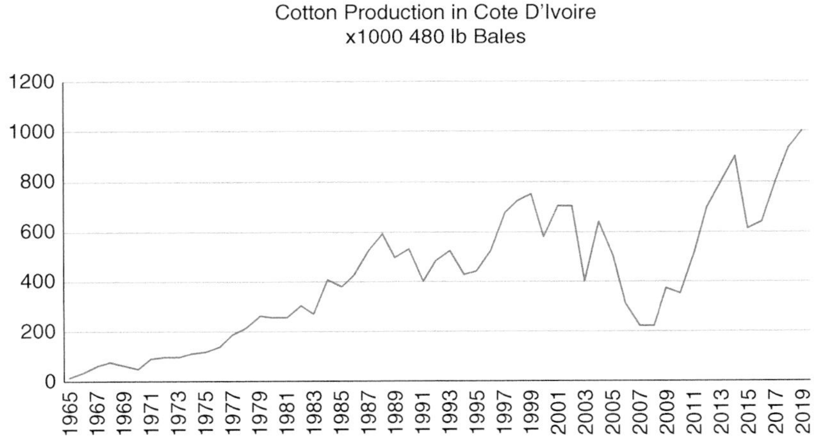

Fig. 10.2. Cotton production in Cote D'Ivoire.

expansion of cotton production was achieved by 'intensification', using more labour, and later by 'extensification', i.e. an increase in the area cultivated (Bassett, 2001).

Production remained low until 1975, when the technique of ultra-low-volume spraying was introduced, applying two litres of oil-based formulation/ha (Cauquil, 1987). This was a major change from the much heavier knapsack sprayer that required considerably larger volumes of water (up to 300 litres/ha). According to the French Democratic Confederation of Labour (CFDT), the diffusion of the ULV technique 'revolutionized' cotton growing (Bassett, 2001).

In 1995, as the agrochemical industry decided to cease supplying the oil-based UL formulation, the farmers had to revert to formulations mixed with water, but by retaining the rotary atomizer sprayers, the volume of water required was 10 litres/ha. In addition, the government has stopped subsidizing cotton insecticides since that year.

Cotton production was seriously affected by the first Ivorian Civil War from 2002–2004, with the country split between the rebel north and the government-held south. Through United Nations intervention, the country calmed down and a peace agreement was signed in 2007. Unfortunately, fighting resumed in 2011 due to delays in elections and disagreement about the election results.

Since 2009, the total land under cotton has more than doubled, from 185,000 to 400,000 ha.

Fig. 10.3. Very-low-volume (VLV) spraying. (Photo: Graham Matthews)

This growth has been stimulated by strong public-private engagement to create the right policy framework and a conducive operating environment (Ajayi *et al.*, 2009).

The International Société d'Exploitation Cotonnière Olam (SECO), a subsidiary of the global agribusiness Olam, introduced a range of new ideas and commitments, such as faster payments and higher-quality cottonseeds, which helped to bring a new stability to village life and significant new benefits to the cotton industry. SECO has benefitted as in 2017/18 the government allocated exclusive zones to each ginner and facilitated long-term investments encouraging modernization of farming practices and improvements in total crop care, leading to improved yields. This more intensive engagement with farmers provided an ideal platform for innovative IT applications to engage farmers and eventually link them to digital payment options.

In Côte d'Ivoire, cotton is attacked by a wide variety of pests including the cotton caterpillar, *Helicoverpa armigera*, the aphid, *Aphid gossypii*, the whitefly, *Bemisia tabaci*, the mite, *Polyphagotarsonemus latus* and the jassid, *Jacobiella facialis*. To protect cotton cultivation from this broad pest complex, farmers have used chemical control.

From 1977, as a replacement for organochlorines, pyrethroids were widely used because of their low-dose efficacy on several pests and their low toxicity to mammals. The susceptibility of *Helicoverpa armigera* to pyrethroids was investigated in West Africa by means of laboratory bioassays from 1985, when pyrethroids were introduced. This revealed that there was a trend for the pest to become more tolerant to pyrethroids, and during the 1996 growing season, farmers using calendar-based spraying programmes reported control failures in various countries. The strong efficacy of cypermethrin on small larvae was confirmed in experimental plots, but the effect decreased quickly with successive instars (Martin *et al.*, 2000). In the early 1990s, the loss of sensitivity to pyrethroids was reported in *H. armigera* in laboratories in Côte d'Ivoire and Burkina Faso. In the field, large infestations of caterpillars of this pest are observed locally, becoming frequent from year to year, earlier and difficult to control. In 1996, *H. armigera*'s resistance to pyrethroids was confirmed in the laboratory, which helped explain the very high larval infestations observed in peasant environments and the numerous cases of treatment failure reported in Benin, Burkina Faso, Côte d'Ivoire and Mali.

Since the 1998/99 campaign, various strategies to manage *H. armigera*'s resistance to pyrethroids have been tested throughout West Africa in reasoned protection programmes. They are based on (i) limiting the use of pyrethroids; (ii) alternation of chemical families; (iii) use of synergizing insecticides in combination with pyrethroids; and (iv) the practice of treatment thresholds to optimize both the cost and effectiveness of insecticides.

Faced with this phenomenon of resistance, major cotton countries of West Africa decided, in 1998, at the meeting of Bobo-Dioulasso (Burkina Faso) to combine their efforts to implement a regional project on the management and prevention of *H. armigera* resistance in West Africa (PR-PRAO). The PR-PRAO has revealed that resistance levels depend on country, year and season. In addition, resistance is crossed to all types of pyrethroids, with the exception of other chemical families. Therefore, a resistance management strategy, inspired from Australian window programme models, was immediately validated in a peasant environment and gradually popularized from 1998 on all cotton areas of the six main cotton-growing countries in West Africa (Benin, Burkina Faso, Côte d'Ivoire, Mali, Togo and Senegal). The window programme strategy is based, firstly, on the definition (setting) of a restriction period for pyrethroids and their replacement by other chemical alternatives (endosulfan, profenofos, indoxacarb, spinosad etc.) before 15 August or 20 August, depending on the cotton-growing regions. This period of no pyrethroid use corresponds to the cotton vegetative stage. In contrast, on the fruiting stage, the use of synergizing pyrethrinoid-organophosphate mixtures are advised, for pyrethroids still prove very effective for controlling endocarpic bollworm species such as *Pectinophora gossypiella* and *Cryptophlebia* (=*Thaumatotibia*) *leucotreta*. The widespread adoption of the window programme, illustrated in Figs 10.4a and 10.4b (Ochou and Martin, 2001) has led for years to lower levels of larval infestations of *H. armigera* and to an increase in seed-cotton production. This reflects the success of the strategy at the regional level.

A survey of 79 farmers, randomly selected over the cotton-growing area of Cote d'Ivoire, was carried out to investigate cotton farmers' knowledge and perception of arthropod pests. This revealed a wide range of ages (20–73 years) among cotton farmers; with 25% being less than 30 years old. Most farmers were illiterate, as

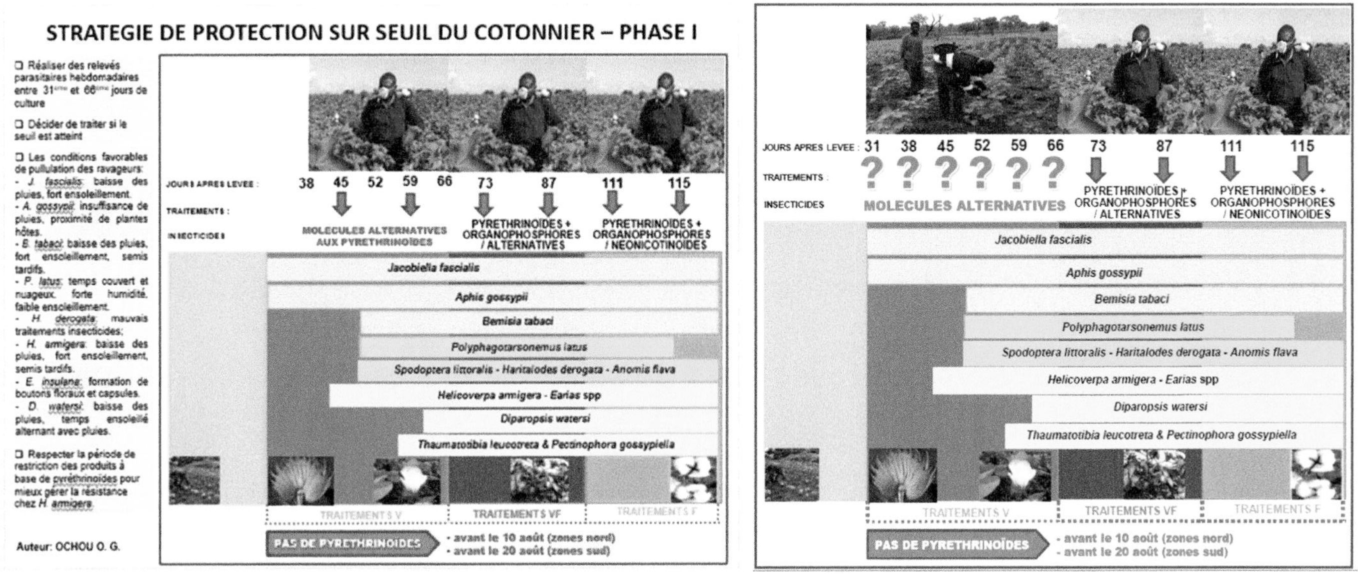

(a) (b)

Fig. 10.4. The strategy of « Window Program » adopted in 1997 for the pyrethroid resistance management in *H. armigera* on cotton crop in Côte d'Ivoire, performed according either to predetermined calendar schedule treatments (a) or to treatment thresholds during cotton vegetative stage (b).

74% had not attended school. Most of the pests which were well known to a high proportion of farmers either were those whose feeding caused very obvious effects (*Syllepte derogata*, *Heliothis armigera*, *Polyphagotarsonemus latus* and *Dysdercus volkeri*) or those found frequently in their fields (*Aphis gossypii*, *Earias insulana*, *Anomis flava*, *Bemisia tabaci*). However, it concluded that it was important to improve farmers' pest-management capabilities by providing them with field diagnostic tools and educational materials along with other relevant facilities. These could be achieved through rural schools appropriate for helping cotton farmers to acquire basic knowledge of key concepts of pest control, and to enable the most receptive farmers to reach the level of independent decision-makers (Ochou *et al.*, 1998a).

The use of treatment thresholds is a solution, if not the only one, to reduce the cost of protection. Another advantage of treatment thresholds is that they allow farmers to monitor major pests during the campaign and to intervene against the early larval stages of *H. armigera*, the most sensitive, based on field observations. Different pest-management approaches aimed at timing and limiting insecticide applications have therefore been investigated since 1985 (Ochou *et al.*, 1998b). Since 1998, the strategy based on the monitoring of major pests during the early cotton stage of cotton helped to reduce cotton-protection costs and improve overall pest control decision-making at the farmer level. Decision tools were adapted through a pest-scouting pegboard and other decision aid tools (Figs 10.5a and 10.5b). The number of treatments and the amount of insecticides applied were reduced while achieving equal or better pest control and yields. The strategies proved to be safer to some beneficial arthropods (Ochou *et al.*, 1998b). At present, this strategy is adopted by a large number of farmers (more than 30,000), and efforts are now being made to improve, at the farmer level, overall pest control decision-making.

Cotton Growing in Burkina Faso

Traders from the Middle East undoubtedly spread cotton growing across Africa and farmers in Burkina Faso and neighbouring countries grew it as a perennial crop rather than on an annual basis as it is today. Despite low yields (Schwartz, 1996), the fibre was spun and woven into fabric to make clothes for local use. They also bartered it with Saharan traders to the north, exchanging cotton for rock salt; or to the south, traded for kola.

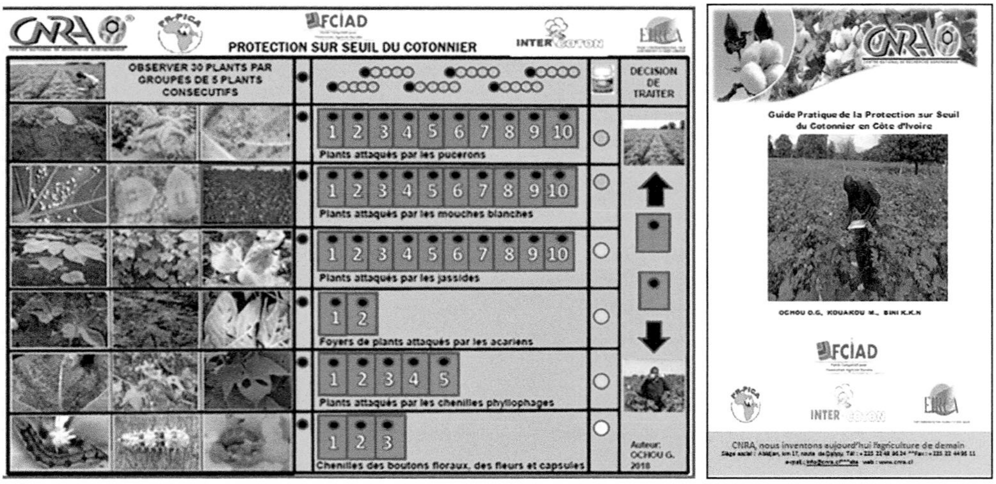

(a) (b)

Fig. 10.5. (a) Decision aid tools to cotton farmers in Côte d'Ivoire illustrated here by a local designed pegboard; (b) A guide for cotton pest scouting and treatment thresholds.

During the French colonial period, growing cotton was encouraged to obtain a reliable and low-cost supply of cotton for the French textile industry, but the smallholder cotton grower was poorly treated under a top-down technocratic system (Bassett, 2001). Early in the 20th century, the French established the Colonial Cotton Association (Association Cotonnière Coloniale, ACC) to promote cotton production, and in 1949 this was transformed into the CFDT (FAO 2018).

In the early 1960s, several governments in francophone West Africa, including that of Burkina Faso, focused their agricultural development programmes on cotton, and yields increased eightfold in Burkina Faso, from 103 to 836 kg per ha, and cotton acreage tripled from 23,000 to 72,000 acres. Spraying cotton had been introduced, but the adoption of ULV spraying in 1975 (Cauquil, 1987) enabled the crop to be treated more easily without needing large volumes of water.

A key contribution was the agricultural development efforts of CIRAD (Centre de Coopération Internationale en Recherche Agronomique pour le Développement), which in conjunction with the national cotton companies extended the use of modern inputs that boosted productivity, including chemical fertilizers, insecticides, herbicides, and improved cottonseeds. The introduction of animal traction in the traditional cotton-growing areas greatly increased labour efficiency in ploughing, planting and weeding operations, and eased early seasonal bottlenecks in labour. As yields and planted area increased in Burkina Faso (as well as neighbouring countries such as Mali and Benin), the West African cotton sector quickly gained a foothold in world markets.

The drop in world cotton prices and the continued need to control insect pests using insecticides, meant production of cotton declined in 2006. In response, Burkina Faso approved the cultivation of GM (Bt) cotton (Héma *et al.*, 2009), which provided control of bollworm in 2008. This helped farmers reduce from an average of eight sprays per season to just three and increase their income. Production then improved again.

By 2012, the Bt cotton with the Cry1Ac and Cry2Ab genes to combat both bollworm and defoliators was grown on over half the area cotton-cultivated, mostly by smallholder farmers on 500,000 ha. The introduction of Bt cultivars helped reduce insecticide use but there was also an increase in the Hemiptera (piercing-sucking) pest population that was not targeted by Bt toxins. Thus Bt cotton farmers had to use a limited number of sprays (Hofs *et al.*, 2013).

Unfortunately, fibre from the new varieties was shorter in length, so processors complained and a decision was taken in 2016 to completely phase out GMO varieties and return to conventional seeds. This was due to the importation of the Bt cottonseeds instead of developing a Bt variety suited to the local climate. The non-Bt cotton

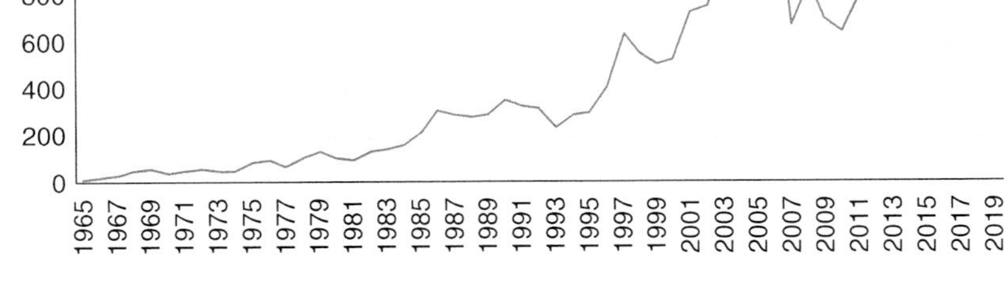

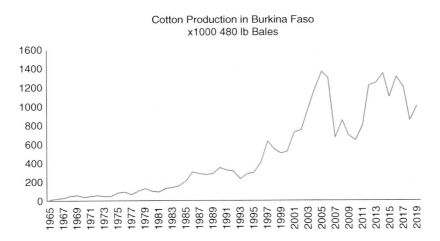

Fig. 10.6. Increased production in Burkina Faso.

Fig. 10.7. GM cotton in Burkina Faso.

had an increased pest infestation which resulted in lower yields, so the government, cotton companies and farmer groups have announced fresh measures to revive the cotton industry, including an increased price for seed cotton and subsidies for inputs for cotton farmers. Clearly indicating the need for a regional system of cotton breeding so that plants adapted to the environment are grown, a variety with the appropriate GM technology has to be combined with disease resistance and pubescent leaves to control jassids within an integrated pest-management programme. With the effects of climate change, cotton farmers are to receive irrigation training and learn better agricultural practices.

Cotton Growing in Cameroon

During the 19th century, traders from Germany were established in Cameroon and at the Berlin Conference of 1884/85, France and Britain ceded their local interests on the coast. Some ten years later, in 1893/94, inland frontiers were agreed with British Nigeria to the west and French Equatorial Africa to the east. When the First World War started in 1914, France and Britain aligned against Germany, so the two German colonies, Togoland and Cameroon on the Gulf of Guinea, were in an impossible position. By early 1916, the British and French controlled both German colonies, and in the Treaty of Versailles, in 1919, Germany renounced sovereignty over all its African colonies. The League of Nations in 1922 confirmed the working division already established in Cameroon between Britain and France; two thin strips on the eastern border of Nigeria became known as the British Cameroons, while most of Cameroon stayed with the French. Much later, in 1956, the French were confronted by a powerful uprising orchestrated by a nationalist party, the UPC (Union des Populations du Cameroun), demanding immediate independence. The uprising was suppressed by

French troops and a vote to remain within the French community was followed by independence in 1960.

No doubt some cotton had been grown through this period, but cotton production for export only began in the north of the country just before independence in Cameroon in the early 1950s. This was under the control of CFDT, a French parastatal, set up a decade earlier to supply the French textile industry with fibre. Difficulties resulted in poor results, so the government set up SODECOTON as a mixed-economy company with the majority of capital belonging to the government and a minority to CFDT (Gergely, 2009). The company was granted a monopoly for developing cotton in the northern part of Cameroon, purchasing seed cotton from farmers, as well as processing and marketing lint cotton. In return, the company assumed an obligation to buy all seed cotton produced anywhere in Cameroon at a fixed price. At the same time, the introduction of ULV spraying in 1975 led to a recovery in production.

SODECOTON remains under the control of the government but has nevertheless expanded its operations despite the major constraints that the cotton-growing area in the north is land-locked and individual farms are small, incurring high extension costs. Chronic soil fertility also has limited production. While other cotton companies in West and Central Africa have abandoned direct oil production, SODECOTON still benefits from the sale of oil and cakes. Cotton is nevertheless the only cash crop cultivated on a large scale in the northern part of the country and is vital to the rural livelihoods, social well-being and political stability in this poor, land-locked region.

SODECOTON made it clear to cotton farmers that they must use agrochemicals, which it supplies together with extension services. Cotton production is the main source of revenue for the peasants in this zone, but with the fall in cotton prices since 1989 and the constant increases in the cost of inputs, SODECOTON tried to diversify their sources of income. Another problem was due to the liberalization of the fertilizer sub-sector, which resulted in an increase in the cost of fertilizers and pesticides. Nevertheless, SODECOTON's involvement contributed to the notable increase in cotton production, which rose from 84,000 tons in 1980 to 165,000 tons in 1988.

Improved varieties helped to increase the yield from 1295 kg/ha in 1980 to 1480 kg/ha in 1988. In 1990, SODECOTON began to experiment with a glandless variety with grains rich in proteins, which the peasants can consume. The idea was to make cotton production more competitive with food crops. The number of farmers increased to over 170,000 in 1991/92 and a more efficient and effective farmer-training programme was provided.

In another in-depth study of the impact of genetic improvement, cotton cultivars were released in Cameroon between 1950 and 2009.

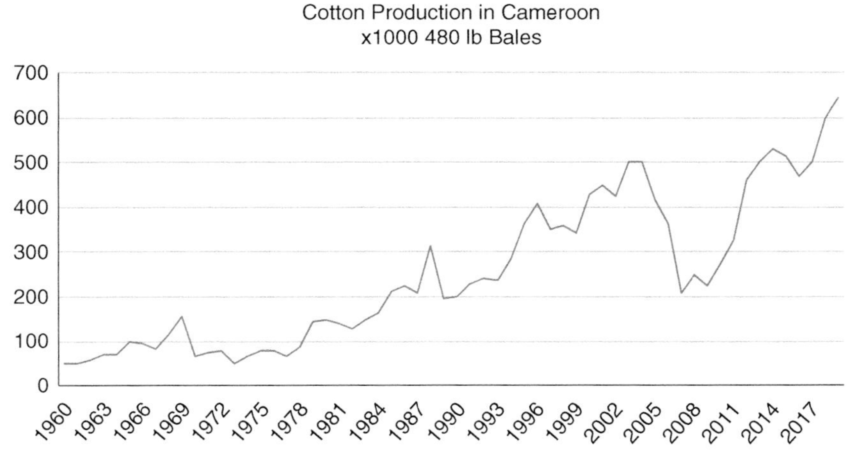

Fig. 10.8. Increased production in Cameroon.

Fig. 10.9. ULV spraying cotton near Garoua, Cameroon.

Loison *et al.* (2017) carried out field experiments in two locations using cotton cultivars released in Cameroon during this period. The study revealed a rate of genetic gain on fibre yield of 3.3 kg/ha/year, due to an increase in the ginning out-turn with a range of 3.9% and 6.2% in 60 years in Garoua and Maroua, respectively. They showed that cotton-breeding efforts in Cameroon had successfully improved cotton fibre yield but further work was needed to improve seed cotton yield improvement.

The insect pest situation is similar to other countries in West Africa, with three bollworm species that need to be monitored to implement an integrated pest-management programme. Deguine and Ekukole (1994) considered crop protection in three areas: cost reduction, conservation of the environment and effective control of pests. They considered these three aspects in a 'staggered targeted control' programme and decided that a scheduled integrated programme was the most suitable for Cameroon. This involved programming spraying dates in advance at two-week intervals from day 45 after seed germination, with doses and active ingredients applied as a VLV spray decided according to which pests were present and how many, based on observations made on the day before spraying. They concluded that this programme enabled cash savings of some 50% and was as effective as a conventional ULV spray programme, which involved up to 14 sprays on a seven-day schedule beginning also on day 45 and taking more account of the environment.

Beyo *et al.* (2004) proposed a pegboard designed to allow small-scale farmers to count the abundance of the three species using a sequential sampling system. With *Helicoverpa* being the most serious pest, detailed studies assessed the consumption of cotton buds and bolls by larvae at different stages of development, with younger larvae preferring buds and the older larvae eating bolls (Nibouche *et al.*, 2007). Following the introduction of transgenic cotton producing the *Bacillus thuringiensis* toxins Cry1Ac and Cry2Ab, a trial in three agricultural landscapes in Cameroon assessed the contribution of non-cotton hosts of *Helicoverpa* as a source of moths ovipositing on Bt cotton. Simulation modelling indicated that planting non-Bt cotton refuges may be needed to significantly delay resistance to cotton producing these two toxins (Brevault *et al.*, 2012).

Cotton Growing in Chad

Arab traders undoubtedly brought cotton to the southern part of Chad long before 1910, when the French colonial administration organized market production on a limited scale under the direction of the military governor. By 1920, the colonial administration was promoting the large-scale production of cotton for export and in 1928 the administration was prescribing a certain area of cotton to be planted in every village. Production then increased from 17 tons to 7900 tons by 1936 (Stürzinger, 1983). Cotton became crucially important to the national economy, both in terms of income generation for farmers and for export revenue. Most cotton is on family-owned farms which cover 1–2 ha. After the Second World War, the acreage of the crop, cultivated under rain-fed conditions in Chad, increased from 270,000 in 1960 to 340,000 in 1978 and thereafter declined to reach 200,000 ha in 1995. Production depends on a number of factors but is primarily due to variations in annual rainfall.

The insect problems on cotton are mainly bollworm and whitefly, which are very severe. ULV spraying was introduced in 1975 (Cauquil, 1987) but was later, in 1995, replaced by very-low-volume (VLV) spraying as the cost of the oil-based formulations was considered too high. A new version of the rotary atomizer, the ULVA+, increased the efficiency of applying the water-based formulation at 10 litres/ha (Clayton, 1992).

The Chadian cotton industry is dominated by *Coton-Chad*, the only cotton processing

Fig. 10.10. Taking cotton to ginnery.

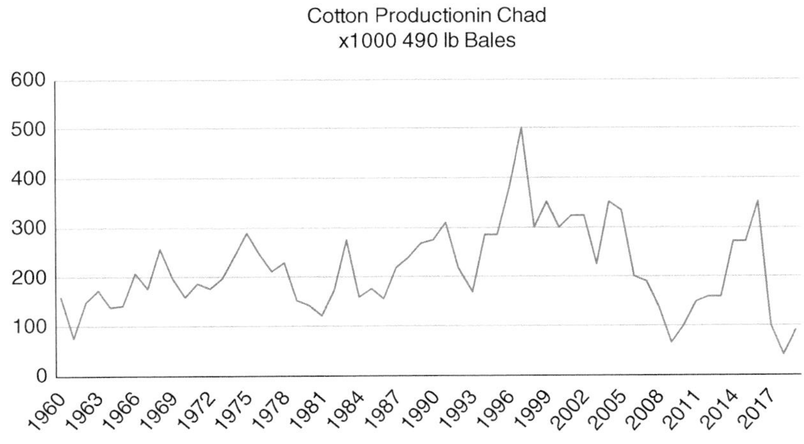

Fig. 10.11. Variable production in Chad.

company largely owned by the government. *Coton-Chad* provides farm inputs to farmers on credit and manages the distribution of such inputs; purchases, collects and transports seed cotton from the villages to its ginneries; and commercializes the lint. Nine ginning factories are active in the south of the country, but farmers have problems with transport. Most cotton grown was with two varieties (A 51, STAM F). During the 1980s, low cotton prices caused the state cotton company to lose money for several years until it modernized its ginning factories and prices recovered.

Cotton Growing in Niger

Helm (1902) confirmed that cotton fibre has, in times far beyond the reach of exact records, been gathered as a spontaneous product, and spun and woven in many parts of Africa; but nowhere, except in Egypt, has it yet become the subject of regular and sustained industry. Attempts, liberally supported, were made during the 'cotton famine' of 1861–1865 to establish the growth of cotton permanently in West Africa, and some fairly considerable success followed, but these experiments must be regarded as failures insofar as lasting results are concerned.

Visiting Niger via Nigeria, it was reported that, 'In every hut is cotton spinning [by hand wheel, presumably], in every town is weaving [Yoruba is famous for its cloth], dyeing, often iron smelting, pottery works by no means despicable, or other useful employments to be witnessed'. Men, women, and children were seen carrying these articles of their production from town to town for many miles for barter or sale (Helm, 1902).

The agricultural economy is based largely upon internal markets, subsistence agriculture and the export of raw commodities: food stuffs and cattle to neighbours. In land-locked Niger, the soil is bad due to drought and the desert, but in the southern area bordering Nigeria, where 15% of Niger's land that is arable is found, the main rain-fed crops are pearl millet, sorghum, rice, maize and cassava, so it is self-sufficient for these crops. Farmers also grow cowpea and onions for commercial purposes. In addition, Niger produces a limited quantity of garlic, peppers, potatoes and wheat. Peanuts are the main source of agricultural export revenue, but a small quantity of cotton is grown, as the crop was introduced by the French in the 1950s

Cotton Growing in Senegal

Some of the earliest cotton was probably in the area near the Senegal and Gambia rivers, going back well over a millennium. It is not known when the people started to make cloth, but by the 10th and 11th centuries it was more widely used. Cotton plants were grown on a small scale with many houses having their own 'cotton

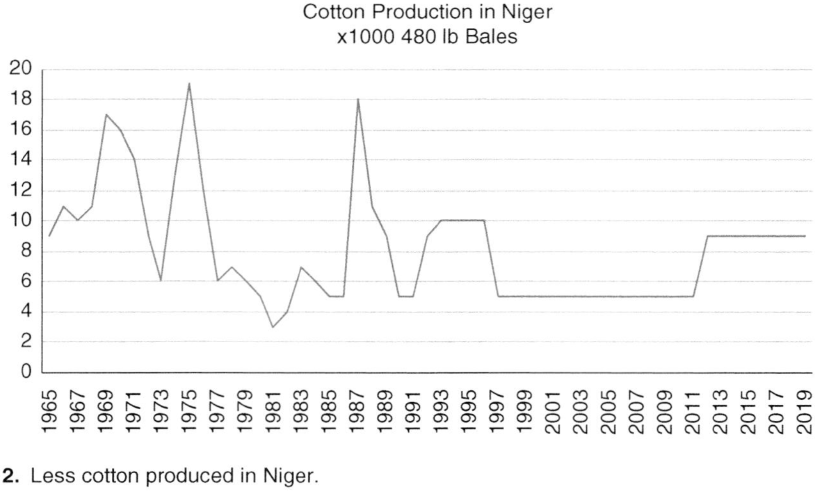

Fig. 10.12. Less cotton produced in Niger.

tree', indicating that it was being grown as a perennial. *G. arboretum* and *G. herbaceum* had been identified in Senegal (Kriger, 2005).

When the Europeans began exploring the coast of Africa, they soon began buying cotton cloth, but when the need for labour in the USA was apparent, ships from Europe would trade goods such as cloth, guns, gunpowder and brandy worth about £3 for a slave captured by local tribal chiefs. These slaves taken to America *c.*1700 were then sold for £20, so despite the risks, a good profit was made. Indian cotton textiles of various types were important for European merchants to purchase slaves along coastal areas of West Africa (Kazuo, 2016). Dark-blue cotton textiles produced in Pondicherry, called *guinées*, were also important in the trade in gum Arabic.

The main development of growing cotton did not occur until after the Second World War, when French scientists began to expand cotton production throughout their West African colonies. As in all these countries, cotton is attacked by many pests, the main ones being the bollworm, *Helicoverpa armigera*, *Diparopsis watersi* and *Earias* spp.; the aphid, *Aphis gossypii*; and the cotton leaf roller, *Haritalodes derogata*; so control of these depended on spraying insecticides, including DDT, initially according to a predetermined calendar schedule.

In practice, only a small acreage was sprayed within the context of threshold-based pest control in cotton. Later, Silvie *et al.*, (2001) introduced a threshold-based scheme in which insecticide was applied at a lower dosage than in the usual calendar-based programme, with scouting six days after spraying. This was to determine if the threshold had reached the level that required a further treatment. The aim was to reduce use of insecticides by 40–50% and lower the cost of crop protection from US$ 50/ha to less than US$ 30/ha. It was realized that accurate timing was needed to increase yields, but growers required training to identify the pests and do the scouting.

Founded in 1974, the Senegalese textile company SODEFITEX sources cotton from smallholder farmers in Senegal to produce textiles, purchasing the cotton at the farm at a market price agreed upon before the planting season. The company, with a Senegalese financial institution, provides cottonseeds and fertilizer on credit and advice to farmers on how to improve cotton yield. It purchases the cotton at the farm level at a market price agreed upon before the planting season. There are about 24,000 cotton farmers in different regions of Senegal who belong to the National Federation of Cotton Producers (FNPC) and are organized in village-level co-operatives to interact with SODEFITEX. Further training has been required to convince farmers to participate actively in their co-operatives, how to set co-operative priorities, how to define and delegate co-operative responsibilities, and how to estimate the profitability of different crops to make informed planting decisions.

Modelling has indicated that adoption of genetically modified cotton varieties would be undeniably useful in sub-Saharan countries

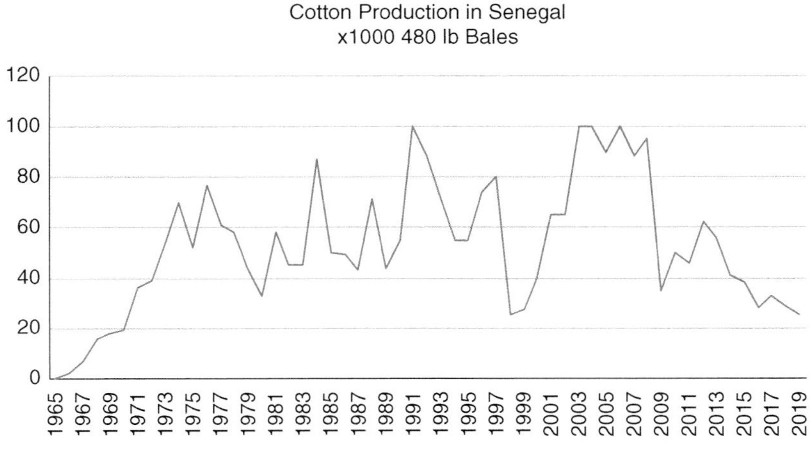

Fig. 10.13. Variability in cotton production in Senegal.

(Bouet and Grouere, 2011), but in using a variety expressing Bt genes or tolerance to a specific herbicide, much depends on selection of a variety that is adapted to the conditions in which it will be grown and produce the right quality of cotton fibre.

Cotton Growing in Benin

Agriculture is the most important sector in the economy of Benin with cotton the major cash crop grown in the country, accounting for up to 80% of export revenues. It is one of the poorest countries in the world with almost half of the population living in extreme poverty. Farmers primarily grow corn, cassava, yam, sweet potato and legumes for their own consumption and the local market. The aim is to boost the output of Benin's agriculture to ensure food security for the population, support economic development and thus more effectively combat poverty. Cotton is typically planted from May through July and harvested from October through December.

In Benin, the introduction of ULV spraying was later than in other West African countries but with its adoption, yields increased to above 1500 kg/ha, although it was usually in the 1100–1200 kg/ha range, which improved production from 1987. Endosulfan sprays were applied on a calendar schedule. A key problem was the high cost of inputs, so two alternative systems were introduced, organic cotton and staggered targeted control, known as LEC (Lutte Etagee Ciblee), to consider the economic threshold of targeted pests (Togbe *et al.*, 2012). One strategy with LEC was to start with full or half the recommended dose of a pyrethroid – cypermethrin – although research suggested replacing endosulfan with a mixture of flubendiamide and spirotetramate (known as 'Tihan O-Teq') for the first two treatments, followed by a mixture of cypermethrin and triazophos. These sprays were determined by using a pegboard designed to indicate the economic threshold (Beyo *et al.*, 2004).

The introduction of LEC to reduce pesticide use generated an estimated benefit of FCFA 48,800 (€74.40) per cotton ha, due to increased productivity with the cost of pesticide reduced. However, the farmer needed more time for pest identification and scouting his crop. A study indicated that the majority of farmers (87%) would be willing to pay for training costs to boost their awareness of improved timing of sprays (Kpade *et al.*, 2016).

Much later, in 2009, criticism of the use of endosulfan resulted in it being banned in Benin as it was a highly hazardous pesticide (HHP) (Glin *et al.*, 2006; Williamson *et al.*, 2008; Ferrigno *et al.*, 2005). Producing organic cotton instead of conventional cotton is favoured due to the absence of adverse health effects. Many farmers – especially those with substantial cattle numbers, such as the Fulani people – also find that it is easier to produce organic cotton. The not-for-profit Organisation for the Promotion of Organic

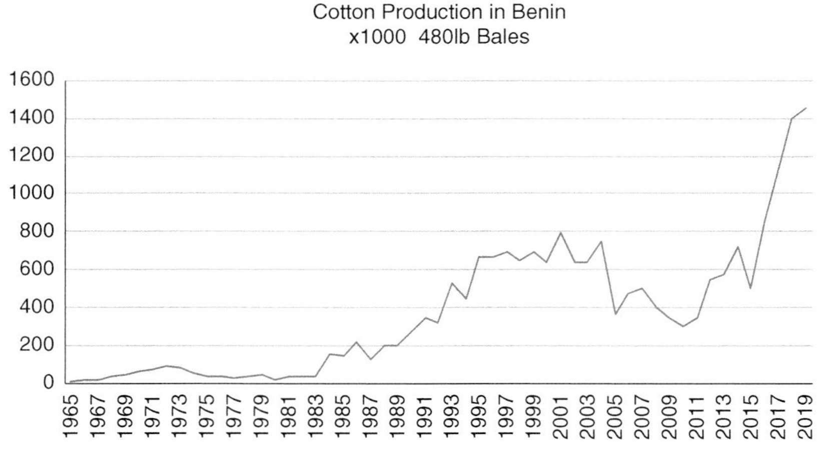

Fig. 10.14. Increased production of cotton in Benin.

Agriculture in Benin (OBEPAB) was founded in 1996 to contribute to the reduction of pesticide use in agriculture.

Organic cotton is considered a high-value crop in Benin, but without controlling the insect pests, yields were low and, in consequence, left cotton farmers with low profit margins. To improve yields of organic cotton, trials examined the application of a novel supplementary food spray to attract and retain beneficial insects on these crops to improve the management of pests. One food spray was Envirofeast used in Australia (See also the studies by Mensah in Chapter 11), with another made in Benin. A neem spray and a baculovirus were also included in the trials (Mensah *et al.*, 2012). An area of maize separated the trials at two sites from an area sprayed with insecticides to avoid any spray drift affecting the organic cotton. The interval between sprays was generally two weeks, but there was no indication on the actual volume of spray applied or the nozzle used on a knapsack sprayer. While the applications of food spray to organic cotton, with and without other biological control agents, attracted and increased the densities of predatory insects, significantly reducing pest insect numbers, higher yields than in cotton treated with neem extract or untreated cotton were achieved; these yields were lower than obtained by four sprays of insecticides. The long interval of two weeks between applications was undoubtedly a factor in minimizing yields, although the distribution of rain during the season, which was not indicated, could have also been a factor. Based on costs and income, Mensah *et al.* (2012) concluded that a supplementary food spray, combined with biological pesticides such as neem extract, sugar and nuclear polyhedrosis virus, could have a positive economic impact on organic cotton production in Benin. All organic cotton is sold to the Beninese government with a guaranteed premium.

Looking to the possibility of using a biopesticide, an arena field experiment coupled with laboratory observations demonstrated that the *Metarhizium anisopliae* isolate (Met 31) has the required characteristics for controlling the cotton bollworm *H. armigera*. However, it showed a more comprehensive assessment of the biological control potential of this isolate as well as studies to confirm its performance under large-scale field conditions in a variety of agro-ecological and farming situations (Kpindou, *et al.*, 2011).

In West Africa, prices paid for cotton lint were based on world prices, but Benin, Burkina Faso and Mali successfully challenged the USA, where farmers had received subsidies for growing cotton. Following involvement of the World Trade Organization, the USDA announced, in 2005, changes to export credit guarantee programmes to comply with WTO findings (Heinisch, 2006).

Cotton Growing in Togo

Archaeological finds indicate that ancient tribes in Togo could produce pottery and, in common with other parts of West Africa, were familiar with making cloth from cotton fibres. Various tribes tended to move to the coast, where the Portuguese and other European powers arrived in the late 15th century. Although the coast of Togo had no natural harbours, the Portuguese did trade at a small fort at Porto Seguro. For the next 200 years, the coastal region was a major trading centre for Europeans in search of slaves, earning Togo and the surrounding region the name 'The Slave Coast'. Following the Berlin Conference in 1884, Germany claimed the coastal area which became the German colony of Togoland in 1905 and was followed by investment in a railway, the port of Lomé and other infrastructure.

In 1900, four sons of slaves from Alabama, James N. Calloway, John Robinson, Allen Burks and Shepherd Lincoln Harris, connected to Booker T. Washington's Tuskegee Normal and Industrial Institute, set sail from New York to Hamburg on their way to new jobs in the German colony of Togo. The Tuskegee graduates began their advisory role to the German colonial administration in 1901. The experiment lasted eight years, when the last of the African Americans, John Robinson, on a mission to spread cotton commerce into the Togolese hinterland, drowned in a 'swift river'. The Tuskegee experts managed to grow a small amount of cotton during their first year in Togo, but the aim was to learn from the experienced African-American cotton farmers and to transfer that knowledge to local growers, so more cotton could be exported to Germany (Beckert, 2005).

Britain and France invaded the area during the First World War and in 1922 the League of Nations mandate stated that the western part of Togo should be governed by Great Britain while France governed the eastern part. After the Second World War, these mandates became UN Trust Territories, but in 1957 the residents of British Togoland voted to join the Gold Coast as part of the new independent nation of Ghana, while French Togoland became the Togolese Republic in 1960 and a year later instituted the National Assembly of Togo as the supreme legislative body. In the same year, the first president, Sylvanus Olympio, dissolved the opposition parties and arrested their leaders. Military coups occurred in 1963 and 1967. Gnassingbé Eyadéma assumed the presidency in 1969 and remained in power for the next 38 years with a one-party system. When he died in 2005, the military installed his son, Faure Gnassingbé as president. Gnassingbé held elections and won, but the opposition claimed fraud. Political violence resulted in around 40,000 Togolese fleeing to neighbouring countries.

Cotton's contribution to the Togolese economic and social development has been considerable. In 1987, 132,000 farmers, about half the country's total, grew cotton as it was their most important source of cash income. The World Bank then supported a cotton development project from August 1989 to June 1997, which saw a further increase in production from 85,000 tons in 1987/88 to 146,473 tons in 1996/97 (84,517 tons in 1993/94), and the number of farmers growing cotton rose to about 200,000. It noted that Togolese farmers received almost the highest price for seed cotton (190 CFAF/kg) in West Africa, plus a 50% share of the profits of the Société Togolaise du Coton (SOTOCO). In addition, SOTOCO was able to reduce significantly its ginning costs (Anon, 1998).

During the World Bank project, instead of spraying on a calendar schedule involving five applications fortnightly, beginning 50 days after sowing, the possibility of applying treatments according to action thresholds was examined on smallholder plots in northern Togo from 1988 to 1991. Thresholds were defined for *H. armigera, D. watersi, Earias* sp., *S. derogate* and *A. gossypii*. A ULV formulation of a mixture of deltamethrin and dimethoate at 3 litres/ha was applied, in both programmes. The results obtained showed that seed cotton yields were significantly different in 1991 (164 kg/ha less in the threshold treatment programme). Calculations shows that a reduction in treatments from five to three was not economical for the producer under the cropping conditions of the experiment. Silvie and Sognigbe (1993) concluded that the implementation of protection programmes based on application thresholds on smallholdings required further, relatively theoretical, studies on threshold determination, training of officials from village groups and of supervisors, and a certain amount of 'flexibility' on the part of the latter, so as to favour farmer acceptance of the method. Later research was reported by Silvie *et al.* (2001).

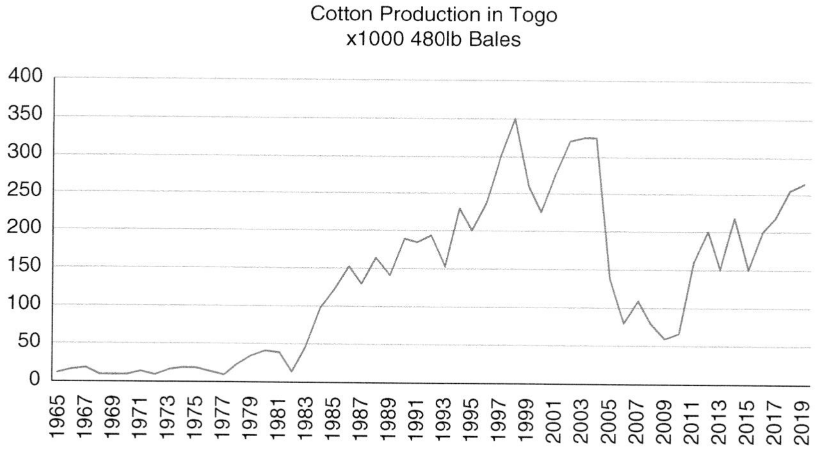

Fig. 10.15. Cotton production in Togo.

Francophone West Africa

In terms of agronomic, biological, chemical and perhaps also varietal control methods (apart from the CGM at Burkina Faso), it seems that it might be redundant to make presentations by country. Indeed in this matter, the basic principles are the same with a few exceptions in all countries; constraints and realities at the level of different countries will be specified. So, for instance, the agronomic control method will address: (i) the rotations implemented to better manage pest infestations at the level of farms, agrosystems, etc.; (ii) the tillage to break the biological cycle of certain predators, for example by bringing pupae to the surface; (iii) the periods of sowing and densities to escape the periods of heavy infestations of such and such pests; (iv) the weeding to destroy target pest habitats other than the cultivated plants; and (v) the management of old cotton plants to break the biological cycles of some pests (*Diparopsis*, *Pectinophora*, *Cryptophlebia*, *Earias*).

It will be the same for the other methods of control except perhaps for the varietal control. In fact, before the GMOs in Burkina Faso, there was a lot of common breeding work (for instance, hairiness cotton, nectariless cotton, coloured cotton, short-cycle cotton varieties), which was achieved by French-speaking scientists from West African research centres in conjunction with the CIRAD in France.

For cotton protection in francophone countries in West Africa, a synthesis can be made on behalf of PR-PICA, which is a platform. Known formally as PR-PRAO between 1998 and 2005, the initiative of PR-PICA was brought out in 1998 by numerous cases of failure of the broadcasted protection programme reported in 1996 in Burkina Faso, Mali, Benin and Côte d'Ivoire. The fault has often been carried over to farmers: poor compliance with treatment instructions, low dosage etc. It must be noted that despite the warnings of research on the decrease in susceptibility of *H. armigera* to pyrethroids, it was necessary to wait for significant cotton-production losses and for a significant increase in the frequency and number of treatments in the field to take curative rather than preventive measures.

Because insects had no borders, thinking and experimentation had to be done not only at the country level but also at the regional and international level in order to benefit from the experience of all partners. The prevention and the sustainable management of insect resistance to insecticides had to involve all partners in the cotton sector: research centres, development organizations, cotton-ginning companies, agrochemical companies, etc.

On the initiative of Burkina Faso, Côte d'Ivoire and Mali, the Regional Project for the Management and Prevention of the Resistance of *Helicoverpa armigera* to pyrethroids in West Africa (PR-PRAO) was created in March 1998 during the Bobo-Dioulasso workshop on this problem. This brought together all partners of the West African cotton industry. It should be noted that the organization of this cotton sector has played a decisive role in the implementation of this project. This project quickly spread to Benin, Togo and Senegal in the following years. It comprises at present six French-speaking countries in West Africa. It has benefitted from financial contributions from cotton companies at the national and regional levels, as well as technical support from international organizations such as the Cirad, CFDT and IRAC (Insecticide Resistance Action Committee). An annual meeting is held in a member country in rotation to compile and to update cotton-pest data.

A bulletin with information concerning the Programme Regional de Production Integrée du Coton en Afrique (PR-PICA) has been published for cotton growers which sets out the effectiveness of chemicals against major cotton insect pests as assessed by PRPICA (Table 10.1) and which pesticides can be applied in each country. The list for the 2020/21 season shown in Table 10.2 is divided into three sections, which cover the first two treatments, then treatments 3 and 4, and 5 and 6, although an extra spray could result in a total of seven treatments. The bulletin provided trade names and type of formulation of different products, some of which had different concentrations of the active ingredient as well as the dose to be applied per ha. Some products were limited to part of the country. Although there is a table listing the pests and whether they are likely to be a low, medium or severe infestation, the need for farmers to monitor crops to determine which pests are present, and their severity, is not indicated and, in consequence, many mixtures of insecticides are recommended to provide control of bollworm and sucking pests.

Table 10.1. Efficiency of chemical against major cotton pest species as assessed by PR-PICA in 2016 and 2017.

Matières actives	Dose g/ha	*Bemisia*	Jassides	Helicoverpa	Thaumatotibia	Pectinophora
Téflubenzuron	15–25					
Abamectin	10.8					
Cyantraniliprole	40					
Diflubenzuron	150–175					
Chlorantraniliprole	20					
Emamectin	10–12					
Bifenazat	120					
Spirodiclofen	120					
Spiromesifen	120					
Spinétoram	18					
Emamectin-Luféneron	10–40					
Emamectin-Acétamiprid	12–16					
Emamectin-Pyriproxifen	10–30					
Emamectin-Bifenthrin	9.5–30					
Cyperméthrin-Abamectin	36–10					
Flubendiamide-Spirotetramat	20–15					
Flubendiamide-Thiacloprid	24–24					
Téflubenzuron-Alphacypermethrin	15–15					
Spinetoram-Acétamiprid	14–16					
Spinétoram-Sulfoxaflor	15–45					
Spinétoram-Métoxyfénosid	18–90					
Deltaméthrin-Pyriproxifen	12–30					
Indoxacarb-Acétamiprid	25–16					
Sulfoxaflor-Lambdacyhalothrin	24–18					
Sulfoxaflor-Chlorpyfos éthyl	30–300					
Spirotétramate-Imidacloprid	30–30					
Lufénuron-Acétamiprid	60–16					
Lambdacyhalothrin-Thiamethoxam	16–21					
Bifenazat-Spirodiclofen	10–59					
Bifenazat-Abamectin	120–10.8					
Emamectin-Abamectin-Acetamiprid	10–10–20					

Good Medium Low Not determined

Table 10.2. Choice of insecticides to be applied at different stages of crop development.

Country	First window Treatments 1 and 2 (or 3)	Second window Treatments 3 and 4 (or 5)	Third window Treatments 5 and 6 (or 6 and 7)
Benin	Emamectine benzoate Acetamiprid	Deltamethrin + chlorpyrifos-ethyl, Cypermethrin + chlorpyrifos-ethyl, Flubendiamide + Thiacloprid	Betacyfluthrin + Imidaclop[rid, Flubendiamide + Thiacloprid, Deltamethrin + chlorpyrifos-ethyl + acetamiprid
Burkino Faso	Indoxacarb, Flubendiamide + Spirotetramat, Chlorantraniliprole, Emamectine benzoate	Indoxacarb, Lambdacyhalothrin + Profenofos-ethyl, Lambdacyhalothrin + sulfoxaflor, Deltamethrin + chlorpyrifos-ethyl, Emamectine benzoate + Abamectin, Chlorantraniliprole	Lambdacyhalothrin + Acetamiprid, Imidacloprid + Acetamiprid, Emamectin benzoate + Acetamiprid, Lambdacyhalothrin + sulfoxaflor, Emamectin benzoate + Pyriproxifen, Alphacypermethrin + Acetamiprid, Lambdacyhalothrin + Acetamiprid, Cypermethrin + Acetamiprid, Spinetoram + Acetamiprid
Cote D'Ivoire	Flubendiamide + Tiaclopride, Abamectin + Acetamiprid + Emamectin benzoate, Indoxacarb, Chlorantraniliprole, Profenofos, Flubendiamide + Spirotetramat	Teflubenzuron + Imidacloprid, + Lambdacyhalothrin, Cypermethrin + Profenofos, Lambdacyhalothrin + Profenofos, Teflubenzuron + Alphacypermethrin	Pyriproxyfen + Deltamethrin, Profenofos + Lambdacyhalothrin, Cypermethrin + Acetamiprid, Cypermethrin + Imidacloprid, Emamectin benzoate + Acetamiprid
Mali	Profenofos, Flubendiamide + Spirotetramat, Emamectine benzoate, Indoxacarb, Teflubenzuron, Hear NPV	Cypermethrin + Imidacloprid, Betacyfluthrin + Imidacloprid, Cypermethrin + Acetamiprid, Emamectin benzoate + Acetamiprid, Emamectin benzoate + Pyriproxyfen, Deltametrin + Chlorpyrifos, Bifenthrin + Acetamiprid	Cypermethrin + Imidacloprid, Betacyfluthrin + Imidacloprid, Cypermethrin + Acetamiprid, Emamectin benzoate + Acetamiprid, Emamectin benzoate + Pyriproxyfen, Deltametrin + Chlorpyrifos, Bifenthrin + Acetamiprid
Senegal	Emamectine benzoate + pyriproxyfene	Cypermethrin + Acetamiprid, Chlorantraniliprole	Cypermethrin + Acetamiprid
Togo	Indoxacarb, Flubendiamide + Spirotetramat, Lambdacyhalothrin + Abamectin, Cypermethrin + Abamectin	Alphacypermethrin + Profenofos, Lambdacyhalothrin + Abamectin, Cypermethrin + Abamectin, Bifenthrin + Abamectin, Deltamethrin + Pyriproxyfen	Deltamethrin + Pyriproxyfen, Bifenthrin + Abamectin Deltamethrin + Acetamiprid, Cypermethrin + Abamectin

Cotton Growing in Ghana

Silas W. Avicor

In Ghana, cotton production is regarded as the cocoa of the northern regions and can be traced back to the pre-colonial and colonial era as one of the areas in West Africa that grew cotton around 1500 AD (Kriger, 2005). Textiles, especially cotton, were a major merchandise in high demand in ports along the Guinea coast. Around 1900, the failure to propagate American cotton in Ghana was traced to the infestations of insect pests, particularly stainers (*Dysdercus*), which punctured and discoloured developing bolls early in the season, and several varieties of bollworms (Dumett, 1975). During this period, an attempt to commercially produce cotton began and the British Cotton Growing Association established cotton farms in 1901 along the Volta river at Anum (Obeng-Ofori, 2007). In global perspective, the extent of growing cotton in Ghana was minimal, but in 1968 the government established the Cotton Development Board (CDB) to oversee activities of the cotton industry. The mandate of the board was to develop the cotton industry in relation to production and ensure adequate supply of raw materials to local textile companies (Asinyo *et al.*, 2015). It also undertook research on improved varieties and by 1977 the production of cotton witnessed a massive but short-lived increase.

New companies arriving during the 1980s faced more difficulties and conflicts that were largely because each company was struggling for space. Subsequently, with the global price of cotton falling in the 1980s together with industrial financial problems in Ghana, the Cotton Development Board was dissolved. As a result, the government zoned the cotton-growing areas and restricted farmers from dealing with any company outside their jurisdiction. The idea initiated by the Ministry of Food and Agriculture (MOFA) was to inject discipline into the

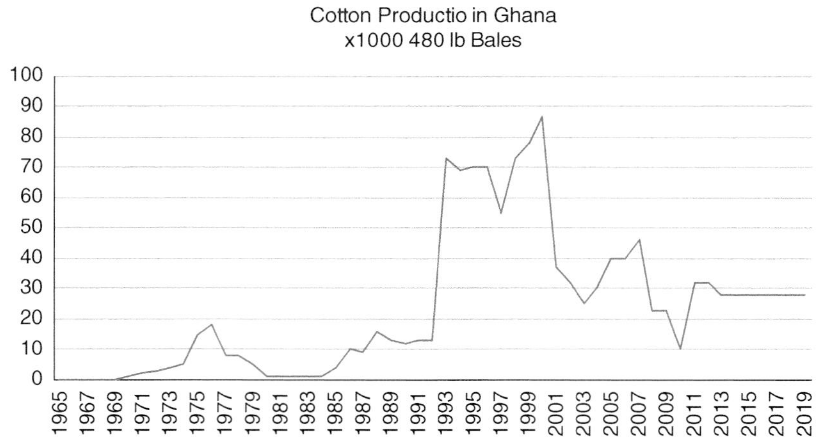

Fig. 10.16. Low production of cotton in Ghana.

Address for correspondence: wintuma@live.com

activities of the companies. The zoning also had problems as some of the companies did not have the capacity to buy cotton from the areas assigned to them. Currently, the cotton companies provide inputs and seeds on loans together with extension services. The value of the loan is deducted from the sales of the cottonseed to the companies during harvesting period. During the 1993–2001 period, when there was a direct injection of funds by government and a ready market by local textile companies, production was much improved. With ginneries not functioning, the companies move the seed cotton to a ginnery in a neighbouring country, but it is hoped that the new Cotton Development Authority will revamp them.

There was a period when the cotton companies had cotton production assistants (CPAs) who would assemble the ultra-low volume (ULV) sprayers in a community for their farmers to use for a period. Afterwards, the equipment was moved to another community for the farmers there to use. This 'nomadic' movement of equipment meant some farmers could not access the equipment when they needed to spray and the equipment had been moved out of their locality. As a result, the farmers resorted to several measures to spray their crops. However, unlike previous years, most of the farmers now have knapsack sprayers so they do their own application.

Planting of cotton usually starts in early June or to coincide with the onset of the rainy season in the northern part of the country so that when the boll begins to open, it coincides with the dry season or a period of less rain. The first application of insecticide is usually done about four weeks after emergence and subsequently every two weeks for a total of six applications. Insecticide application ends when the boll begins to open and is left to dry on the field until harvesting is done.

Insecticides are measured in milk tins (150 ml) and used per filling (15 litres) of knapsack sprayer depending on the type of product. Some insecticide products also come with measuring cups/containers to enable farmers to correctly measure the amount of pesticide needed. Usually, four fillings (60 litres) of knapsack is used per acre (*c.*150 l/ha); however, this is dependent on the insecticide product/type. Averagely, 1.2 litres of insecticide product is used per ha. However, relatively smaller volumes (200–800 ml/ha) of some insecticide products are also used (Abudulai *et al.*, 2017).

A survey of 337 cotton farmers in northern Ghana in 2002/03 revealed that over 70% of the farmers with an average of nine years of cotton-farming experience were illiterate. Farmers were familiar with cotton insect pests, especially those whose feeding caused obvious symptoms in the field such as bollworm, leafrollers and cotton stainer. Farmers relied on chemical insecticides supplied by their contract cotton companies to control insect pests on their fields. Insecticides from three classes, namely organochlorines (e.g. endosulfan), pyrethroids (e.g. lambda-cyhalothrin), organophosphates (e.g. chlorpyrifos), or their mixtures, were used, either as emulsifiable concentrates (EC) or

Fig. 10.17. First spray of a cotton crop using a knapsack spray with cone nozzle. (Photo: Silas W. Avicor)

ultra-low-volume (ULV) concentrations. An average of five applications were made during the season. Farmers indicated that the control practice was largely ineffective against the pests and attributed the cause to the use of expired and inadequate quantities of chemicals. The study showed that only 28% of the respondents had some knowledge of alternative pest-control measures, indicating that training in integrated pest management (IPM) was required to improve their management of insect pests on their fields (Abudulai *et al.*, 2007).

A later survey involving vegetable and cotton farmers reported 25% of farmers exceeded the recommended dose and the frequency of insecticide application was generally high (weekly or less). Most farmers lacked safety and application equipment with a few improvising some of these items, such as applying insecticides with 'brooms' which were being dipped in the liquid and then used to 'sprinkle' it over the crops. Farmers changed insecticides mostly as a result of their availability on the market rather than perceived ineffectiveness or cost. They considered that the storage and disposal of insecticides by farmers was risky and could adversely affect their health and the environment (Avicor *et al.*, 2011).

Nboyine *et al.*, (2013) examined the effect of neem sprayed with a knapsack sprayer, as neem seed oil and neem seed kernel using different concentrations in comparison with alternate use of chlorpyrifos and lambda cyhalothrin to control cotton pests. Aphids were the most important pests, but few bollworms were recorded on all treatments. The seed kernel formulation gave a higher yield than seed oil, but the chemical treatment gave the highest yields. No details were given on the volume of spray or how it was actually applied.

In 2009, the government considered the rejuvenation of the cotton industry as key to the alleviation of poverty in the northern part of the country, due to an increase in cotton price on the international market. The government under the Ministry of Trade and Industry (MOTI) dissolved the zones and brought in multi-national companies, such as OLAM, to help revamp the cotton industry. Production, however, dropped to 2500 tonnes in 2010. This was considered to be due to an unfavourable policy environment, poor sector organization, lack of professionalism of stakeholders and weak farmer organization.

In recent years, the sector is characterized by low productivity levels (±500 kg/ha) and inefficient contractual agreement between farmers and the companies. Cotton farmers grow staple crops on fertile lands and cotton on marginal lands because they claim they cannot farm cotton and use the revenue generated to buy food crops to feed their families, thereby resulting in low cotton yields. Some insecticides and other agrochemicals provided by cotton companies are also diverted for use on other crops (maize and other staples) by farmers contributing to low cotton yields.

The possibility of using *Bacillus thuringensis* (Bt)-engineered cotton in Ghana has been explored (Abudulai *et al.*, 2018) and it was observed that while bollworm populations on Bt-cotton were lower, the population of natural enemies (ladybird beetles and lacewings) were higher compared to non-Bt cotton. In addition, higher seed cotton yield and net profit were obtained from the Bt cotton. However, the use of engineered cotton is a contentious issue and currently Bt cotton is not commercially grown in Ghana.

Currently, all activities in relation to, and in, the cotton industry in Ghana are regulated by the Cotton Development Authority. The authority is tasked with co-ordinating and revitalizing the cotton sector and making cotton farming a lucrative venture.

Cotton Growing in Nigeria

The Portuguese navigator Ramusio recorded that cotton thread and cloth were sold in Nigeria in 1450, presumably introduced from across the Sahara. In 1900, it was reported that cotton grows wild in many parts of West Africa. Although the author mentions *Gossypium barbadense*, a footnote suggests that it could be *G. herbaceum* or *arboreum* and is extensively cultivated in some districts, but more for the purpose of supplying local demands than for export (Muckler-Ferryman, 1900). Small quantities were exported to the UK at the end of the 19th century, but trade in this article has proved a disappointment; great hopes were entertained at one time that Manchester would be able to draw largely on West Africa for supplies, but so far the total annual value of the cotton imported into Great Britain from West Africa has never exceeded a few thousand pounds.

Subsequently, there was a will to promote West African cotton, but was there a way? A number of projects were promoted from the UK, but with limited funding and knowledge these failed (Ratcliffe, 1982). Production continued to be primarily to provide for local traditional making of cloth.

According to Duggan (1922), Lord Palmerston prophesied that the west coast of Africa would outstrip all other countries in the production of cotton, except the USA. The cotton was described as a bushy plant from three to six feet high, with spreading branches, covered with fruits bearing masses of white, wool-like substance – the raw cotton. This wool, or more properly lint, adheres to the seeds of the fruit, there being some 27–45 seeds to each fruit, or boll, as it is called. The lint is composed of fibres, the length of which varies according to the type of cotton.

The British Cotton Growing Association decided that the actual growing of the cotton should be a native industry as African farmers did not take kindly to farm work as hired labourers in the plantation system, which was impossibly expensive due to the cost of labour, poor results and the expense of European supervision. However, it was realized that any great increase in bale output depended almost entirely on the progressive development of transport facilities. It was also important that the quality of the cotton grown in areas needed to be improved and the African farmer educated in better methods of farming and handling the crop to provide the firm foundation, if the ever-increasing possibilities of cotton cultivation presaged by the extension of transport facilities in Nigeria were to be ever adequately realized (Duggan, 1922).

Nigeria's cotton exports to the UK remained insignificant because the domestic market for cotton was proportionately larger than that for the other major export crops, such as cocoa. Prior to 1941, they endeavoured to export cotton, but then export of Nigerian cotton was suppressed until after the Second World War when the policy of export maximization was restored, in 1945, with the maximization of Nigeria's cotton exports adopted to assist in the UK's economic recovery (Hinds, 1996).

The Empire Cotton Growing Corporation established a research unit at *Nigerian College of Arts, Science and Technology at Samaru*, just outside *Zaria*, which was absorbed within the Amadu Bello University in 1962. In 1951/52, a record yield of 113,510 bales of cotton was produced. The application of DDT was started and achieved good control of *Campylomma* and a marked reduction in leaf tattering at Samaru by late spraying with 1% DDT emulsion, while at Daudawa, treatment with DDT emulsion gave a 50% increase in the yield of clean seed cotton (Engledow, 1961). Early spraying trials used a Falcon compression sprayer or Lancet – a simple syringe-type sprayer (manufactured by Cooper Pegler & Co.) – to apply the sprays (de B. Lyon, 1971; Usenbo, 1976). This sprayer would have given very irregular spray over the top of cotton plants, but was the sprayer used by farmers.

The average yield obtained by approx. 75% of farmers was 600 kg seed cotton per ha, whereas in research trials implementing recommended practices, the yields were between 1.5 and 2.5 tons. Adeniji (2007) pointed out that farmers' low yields were due to late planting, as they gave priority to sowing and then weeding food crops. They also failed to apply adequate fertilizer. Farmers have to take crucial decisions in allocation of limited resources in relation to land, labour, cash and the need to support their food crops.

The cotton growing was concentrated in the northern part of the country, especially in the north-west region, which contributed nearly 80% of the product. The majority are small and marginal farmers with an average landholding of 2 ha, so the operations are seldom mechanized and the seed cotton is hand-picked. Cottonseed was frequently sown in the same patch alongside maize and cowpea, so to apply an insecticide, farmers had to start sowing in rows.

With subsequent separation of cotton to sow it alone in rows, trials enabled the use of insecticides to be evaluated.

The bollworm, *Diparopsis watersi*, was considered to be an important pest. It also occurs in the Sudan and is similar to another species in Southern Africa. Cotton can also be damaged by *Helicoverpa armigera*, the spiny bollworm (*Earias insulana* and *E. biplaga*.) and the leafroller (*Syllepta derogata*). Early studies at Samaru showed that early-sown cotton may suffer more damage from the red bollworm, *Diparopsis watersi*, than later-sowing dates, but earlier sowing produced

Fig. 10.18. Cotton on a small farm north of Zaria, 1960, intermixed with maize and other food crops. (Photo: Graham Matthews)

Fig. 10.19. ULV spraying, Nigeria. (Photo: Graham Matthews)

better yields due to a longer growing season. By eliminating ratoon and stand-over cotton, it would prevent early development of the bollworm and consequent red bollworm attack on early sowings of cotton (Geering and Baillie, 1954).

In 1971, two trials by John Hayward compared spraying with mistblower and using a spinning-disc sprayer (ULVA) at Kadawa near Kano; and a knapsack sprayer and ULVA at Daudawa to the north-west of Samaru (Hayward, 1973). During the three seasons (1971–1973), ultra-low-volume spraying was examined. In 1972, two trained ULVA operators contracted for six routine sprays treated cotton grown by 70 farmers at four villages (Beeden, 1974; Beeden *et al.*, 1977).

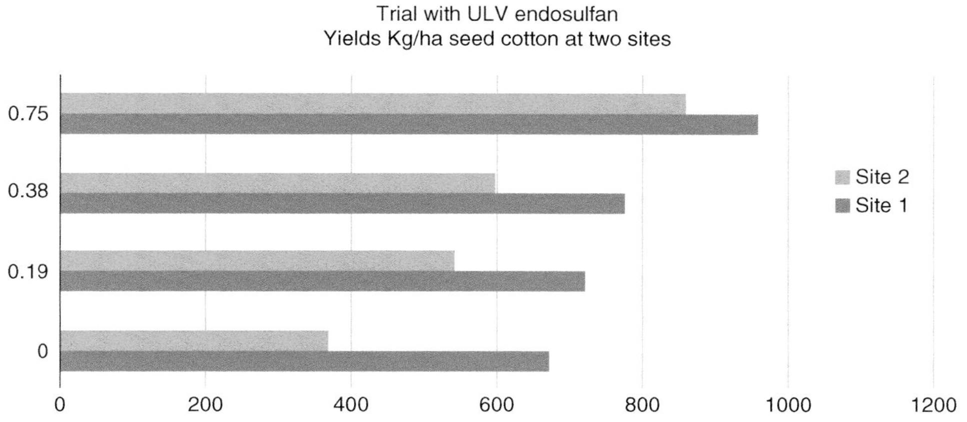

Fig. 10.20. Trial in northern Nigeria reported by Usenbo (1976).

In a trial reported by Usenbo (1976), cotton at two sites was treated. Site 1 had the higher yield with all treatments, but the plant population was higher at that site. The much lower yield at Site 2 indicated heavier pest infestation or the soil could have been less fertile.

Deposition and drift of ULV sprays of small spray droplets (70–130 μ vmd) at ULV rates (3 l/ha with UL formulations, and at very low-volume rates (6–12 l/ha) using water-based formulations were examined in determining practical recommendations (Johnstone and Huntington, 1977). ULV spraying was considered to be much more suitable for farmers as the conventional spraying with a knapsack sprayer demanded too much water.

The World Bank supported development projects in Gombe to the east of Zaria, Funtua near Daudawa to the north-west of Samaru and in Gusau of Sokoto state to the west of Daudawa. These all promoted ULV applications to cotton. ULV spraying of cotton was also evaluated for rain-fed cotton in the Niger state, Nigeria, where endosulfan and monocrotophos were considered the most effective insecticides applied at 3 l/ha for an economical crop (Chaudhry, 2008).

Control of the insect pests, especially when the ULV spraying technique was introduced, enabled yields and production to increase. For a short period, the Electrodyn sprayer was also used in Nigeria.

In 2011, Nigeria passed the National Biosafety Management Act, which sets out the regulations scientists and corporations need to follow to allow GM crops to be grown. Subsequently, approval for the commercialization of Bt cotton was given in a bid to revive its textile industry by the Nigerian government following guidance from the National Biotechnology Development Agency (NABDA). In the 2019 season, the seed company Mahyco demonstrated the value of two newly released Bt cotton hybrids, MRC 7377 BG 11 and MRC 7361 BG 11, which were developed by Mahyco in collaboration with the Institute for Agricultural Research (IAR) by supplying seed and technology to selected Nigerian farmers so they could understand the crop. They trained the farmers on appropriate product use with the right agronomy to optimize the yield in their own fields. In addition to the pest-resistant traits, the Bt varieties provide early maturity, fibre length of 30.0–30.5 mm, fibre strength of 26.5–27.0 g/tex (tenacity) and micronaire (strength) of 3.9–4.1, and replace the local conventional variety, which is no longer accepted in the international market. The hope is that by involving an Indian company in the plant selection, Nigeria would escape the problem experienced by Burkina Faso when the quality of lint deteriorated.

Nigeria has 50 saw gins but only 17 ginneries are active due to a reduction in production and obsolete equipment. A new variety increased ginning out-turn ratio from 36% to 43% in 2003–2008. As farmers have used polypropylene bags when picking and to deliver their seed cotton, Nigeria's cotton has high levels of polythene

Fig. 10.21. Electrodyn sprayer in Nigeria. (Photo: Graham Matthews)

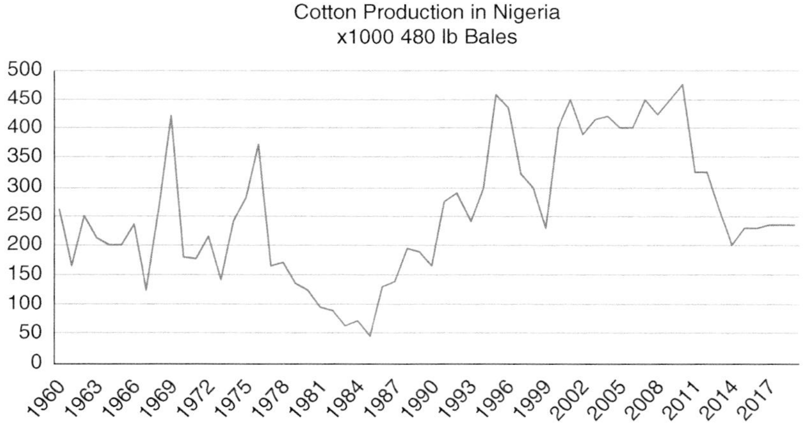

Fig. 10.22. Changes in cotton production in Nigeria.

contamination affecting the returns to farmers, operating cost of ginners, quality of yarn and also the export value. Farmers are encouraged to use other packaging material to earn a premium price for polythene-free cotton. The Nigerian textile industry uses up to 50% of the total production and the rest is exported to the EU, China, South Korea and Taiwan.

References

Abudulai, M., Abatania, L. and Salifu, A.B. (2007) Farmers' knowledge and perceptions of cotton insect pests and their control practices in Ghana. *Journal of Science and Technology* 26, 34–40.

Abudulai, M., Seini, S.S., Nboyine, J.A., Seidu, A. and Ibrahim, Y. Jr (2017) Field efficacy of some insecticides for control of bollworms and impact on non-target beneficial arthropods in cotton. *Experimental Agriculture* 54, 315–322.

Abudulai, M., Chamba, E.B., Nboyine, J.A., Adombilla, R., Yahaya, I., Seidu, A. and Kangben, F. (2018) Field efficacy of genetically modified FK 95 Bollgard II cotton for control of bollworms, Lepidoptera, in Ghana. *Agriculture and Food Security* 7, 81.

Adeniji, O.B. (2007) Constraints to improved cotton production in Katsina State, Nigeria. *Journal of Applied Sciences* 7, 1647–1651.

Ajayi, O.C., Akinnifesi, F.K., Sileshi, G. and Ajayi, A.O. (2009) Agricultural policies and the emergence of cotton as the dominant crop in northern Côte d'Ivoire: Historical overview and current outlook. *Natural Resources Forum* 33(2), 111–122. DOI: 10.1111/j.1477-8947.2009.01215.x

Anon (1998) *Implementation Completion Report Republic of Togo Cotton Sector Development Project*. World Bank.

Asinyo, B., Frimpong, C. and Amankwah, C. (2015) The state of cotton production in Northern Ghana. *International Journal of Fiber and Textile Research* 5, 58–63.

Avicor, S., Owusu, E.O. and Eziah, V.Y. (2011) Farmers' perception on insect pests control and insecticide usage pattern in selected areas of Ghana. *New York Science Journal* 4, 23–29.

Bassett, T.J. (2001) *The Peasant Cotton Revolution in West Africa: Cote D'Ivoire, 1880–1995*. Cambridge University Press, New York.

Beckert, S. (2005) From Tuskegee to Togo: the problem of freedom in the empire of cotton. *The Journal of American History* 92, 498–526.

Beeden, P. (1974) *ULV Techniques on Cotton in the Northern States of Nigeria: Pesticide Application by ULV Methods*. Proceedings of the Cranfield Symposium. BCPC Monograph No. 11.

Beeden, P., Hayward, J.A. and Norman, D.W. (1977) A comparative evaluation of ultra low volume insecticide applications on cotton farms in the north central state of Nigeria. *Nigerian Journal of Crop Protection* 2, 23–29.

Beyo, J., Nibouche, S., Goze, E. and Deguine, J.-P. (2004) Application of probability distribution to the sampling of cotton bollworms (Lepidoptera: Noctuidae) in Northern Cameroon. *Crop Protection* 23, 1111–1117.

Bingen, R.J. (1998) Cotton, democracy and development in Mali. *The Journal of Modern African Studies* 36, 265–285.

Bouet, A. and Gruere, G.P. (2011) Refining opportunity cost estimates of not adopting GM cotton: an application in seven sub-Saharan African countries. *Applied Economic Perspectives and Policy* 33, 260–279.

Brevault, T., Nibouche, S., Achaleke, J. and Carriere, Y. (2012) Assessing the role of non-cotton refuges in delaying *Helicoverpa armigera* resistance to Bt cotton in West Africa. *Evolutionary Applications* 5, 53–65.

Cauquil, J. (1987) Cotton-pest control: a review of the introduction of ultra-low volume (ULV) spraying in sub-Saharan French-speaking Africa. *Crop Protection* 6, 38–42.

Chaudhry, A.B. (2008) Field evaluation of some new ultra-low-volume (u.l.v.) insecticides for rainfed cotton in the Niger State of Nigeria. *Tropical Pest Management* 28, 122–125.

Clayton, J. (1992) A new generation hand held spinning disc sprayer the Micron ULVA+ for small farmer crop protection. *IRCT Cotton Conference, N'Djanena, Chad.*

De B. Lyon, D.J. (1971) *The timing of insecticide applications on cotton in the Northern States of Nigeria. Proceedings of the Cotton Institute Conference,* Blantyre, Malawi, 92–107.

Deguine, J.-P. and Ekukole, G. (1994) A new cotton crop protection programme in Cameroon. *Agriculture et Developpement. Special issue*, 41–45.

Dembele, B., Bett, H.K., Kariuki, I.M., Le Bars, M. and Ouko, K.O. (2018) Factors influencing crop diversification strategies among smallholder farmers in cotton production zones in Mali. *Advances in Agricultural Sciences* 6, 1–16.

Duggan, E. De C. (1922) The cotton growing industry of Nigeria. *African Affairs* 21, 199–207.

Dumett R.E. (1975) Obstacles to government-assisted agricultural development in West Africa: cotton-growing experimentation in Ghana in the early twentieth century. *The Agricultural History Review* 23, 156–172.

Engledow, F.L. (1961) Cotton-growing in Nigeria. *Nature* 192, 1248.

FAO (2018) *Economic Importance of Cotton in Burkina Faso.* Report by the Food and Agriculture Organization of the United Nations.

Ferrigno, S., Ratter, S.G., Ton, P., Vodouhê, D.S., Williamson, S. and Wilson, J. (2005) *Organic Cotton: A New Development Path for African Smallholders.* International Institute for Environment and Development.

Fok, M. (1993) *Cotton Development in Mali by Analysis of Contradictions: Actors and Crises from 1895 to 1993*. CIRAD-CA, Montpellier, France.

Geering, Q.A. and Baillie, A.F.H. (1954) The biology of red bollworm, *Diparopsis watersi* (Roths.) in northern Nigeria. *Bulletin of Entomological Research* 45, 661–681.
Gergely, N. (2009) *The Cotton Sector of Cameroon*. Paper prepared for the World Bank. Africa region working paper 126.
Glin, L., Kuiseu, J., Thiam, A., Vodouhê, D., Dinham, B. and Ferrigno, S. (2006) *Living with Poison: Problems of Endosulfan in West African Cotton Growing Systems.* PAN UK. Available at: https://issuu.com/pan-uk/docs/living_with_poison (accessed 7 August 2021).
Hau, B. (1988) Histoire de la sélection du cotonnier en Côte d'Ivoire. *Coton et Fibres Tropicales* 43, 177–193.
Hayward, J.A. (1973) *Northern States, Nigeria*. CRC Annual Research Report.
Heinisch, E.L. (2006) West Africa versus the United States on cotton subsidies: how, why and what next? *Journal of Modern African Studies* 44, 251–274.
Helm, E. (1902) The cultivation of cotton in West Africa. *Journal of the Royal African Society* 2, 1–10.
Héma, O., Some H.N., Traore, O., Greenplate, J. and Abdennadher, M. (2009) Efficacy of transgenic cotton plant containing the Cry1Ac and Cry2Ab genes of *Bacillus thuringiensis* against *Helicoverpa armigera* and *Syllepte derogata* in cotton cultivation in Burkina Faso. *Crop Protection* 28, 205–214.
Hinds, A. (1996) Colonial policy and Nigerian cotton exports, 1939–1951. *International Journal of African Historical Studies* 29, 25–46.
Hofs, J.L., Gozé, E., Cene, B., Kioye, S. and Adakal, H. (2013) Assessing the indirect impact of Cry1Ac and Cry2Ab expressing cotton (*Gossypium hirsutum* L.) on hemipteran pest populations in Burkina Faso (West Africa). *IOBC-WPRS Bulletin* 97, 49–54.
Johnstone, D.R. and Huntington, K.A (1977) Deposition and drift of ULV and VLV insecticide sprays applied to cotton by hand applicator in Northern Nigeria. *Pesticide Science* 8, 101–109.
Kazuo, K. (2016) *Indian Cotton Textiles and the Senegal River Valley in a Globalising World: Production, Trade and Consumption, 1750–1850.* London School of Economics and Political Science (UK). ProQuest Dissertations Publishing.
Kpade, C.P., Mensah, E.R., Fok, M. and Ndieunga, J. (2016) Cotton farmers' willingness to pay for pest management services in northern Benin. *Agricultural Economics* 48, 105–114.
Kpindou, O.K.D., Djegui, D.A., Glitho, L.A. and Tamo, M. (1011) Dose transfer of an oil-based formulation of *Metarhizium anisopliae* (Hypocreales: Clavicipitaceae) sprays to cotton bollworm in an arena trial. *International Journal of Tropical Insect Science* 31, 262–268.
Kriger, C. (2005) Mapping the history of cotton textile production in pre-colonial Africa. *African Economic History* 33, 87–116.
Loison, R., Audeberts, A., Chopart, J.-L., Debaeke, P., Dessauw, D., Gourlot, J-L., Goze, E., Jean, J. and Gerardeaux, E. (2017) Sixty years of breeding in Cameroon improved fibre but not seed cotton yield. *Experimental Agriculture* 53, 202–209.
Martin, T., Ochou, G.O., Hala-N'Klo, F., Vassal, J.-M. and Vaissayre, M. (2000) Pyrethroid resistance in the cotton bollworm, *Helicoverpa armigera* (Hubner), in West Africa. *Pest Management Science* 56, 549–554.
Mensah, R.K., Vodouhe, D.S., Sanfillippo, D., Assogba, G. and Monday, P. (2012) Increasing organic cotton production in Benin, West Africa with a supplementary food spray product to manage pests and beneficial insects. *International Journal of Pest Management* 58, 53–64.
Muckler-Ferryman, A.F. (1900) *British West Africa.* Swannerschein, London.
Nboyine, J.A., Abudulai, M. and Opare-Atakora, D.Y. (2013) Field efficacy of neem (*Azadirachta indica* A. Juss) based biopesticides for the management of insect-pests of cotton in Northern Ghana. *Journal of Experimental Biology and Agricultural Sciences* 1, 321–327.
Nibouche, S., Goze, E., Babin, R., Beyo, J. and Brevault, T. (2007) Modeling *Helicoverpa armigera* (Hubner) (Lepidoptera: Noctuidae) damages on cotton. *Environmental Entomology* 36, 151–156.
Obeng-Ofori, D. (2007) Arthropod pests of cotton, *Gossypium* spp. (Malvaceae). In: Obeng-Ofori, D. (ed.) *Major Pests of Food and Selected Fruit and Industrial Crops in West Africa*. The City Publishers Limited, Accra, Ghana, 179–192.
Ochou, G.O. and Martin T. (2001) Une stratégie pour gérer la résistance de *Helicoverpa armigera* (Hübner) aux pyréthrinoïdes en Côte d'Ivoire: une fenêtre en phase végétative. Actes 3ème Réunion Bilan du PR-PRAO, Yamoussoukro, Côte d'Ivoire, 3–6 April, pp. 58–67.
Ochou, G.O., Matthews, G.A. and Mumford, J.D. (1998a) Farmers' knowledge and perception of cotton insect pest problems in Cote d'Ivoire. *International Journal of Pest Management* 44, 5–9.
Ochou, G.O., Matthews, G.A. and Mumford, J.D. (1998b) Comparison of different strategies for cotton insect pest management in Africa. *Crop Protection* 17, 735–741.
Ratcliffe, B.M. (1982) Cotton imperialism: Manchester merchants and cotton cultivation in West Africa in the mid-nineteenth century. *African Economic History* 11, 87–113.

Schwartz, A. (1996) Attitudes to cotton growing in Burkina Faso: different farmers, different behaviours. In: Benoit-Cattin, M., Griffon, M. and Guillaumont, P. (eds) *Economics of Agricultural Policies in Developing Countries*. Revue Française d'Economie, Paris.

Silvie, P. and Sognigbe, B. (1993) Use of action thresholds on cotton crops in northern Togo. *International Journal of Pest Management* 39, 51–56.

Silvie, P., Deguine, J.P., Nibouche, S., Michel, B. and Vaissayre, M. (2001) Potential of threshold-based interventions for cotton pest control by small farmers in West Africa. *Crop Protection* 20, 297–301.

Stürzinger, U. (1983) The introduction of cotton cultivation in Chad: the role of the administration, 1920–1936. *African Economic History* 12, 213–225.

Togbe, C.E., Zannou, E.T., Vodouhe, R., Haagsma, R., Gbehounou, G., Kossou, D.K., and van Huis, A. (2012) Technical and institutional constraints of a cotton pest management strategy in Benin *NJAS-: Wageningen Journal of Life Sciences* 60–63, 67–78.

Usenbo, E.I. (1976) *Approaches to Integrated Control of Cotton Pests in the Mid-Western State of Nigeria.* PhD thesis, University of London.

Williamson, S., Ball, A. and Pretty, J. (2008) Trends in pesticide use and driver for safe pest management in four African countries. *Crop Protection* 27, 1327–1334.

11 Cotton Growing in Australia

Graham Matthews and Paul Grundy*

The cotton industry in Australia is now well established, having undergone rapid expansion since the 1960s. Cotton lint is a major agricultural commodity in Australia and generates AU$2 billion per annum in export earnings, and cottonseed has become an important feed source for the livestock industry. The success of the modern Australian cotton industry has been enabled by significant investment in research that has developed solutions to a range of challenges. Current Australian cotton production is underpinned by locally developed cultivars that incorporate several transgenic traits, and growing practices that are focused on maintaining longer-term production sustainability.

Early History

According to Craven *et al.* (1994), 17 *Gossypium* species were endemic to Australia, long before the first European explorers reached Australia in February 1606. Upland cottonseeds, *Gossypium hirsutum*, were first brought to Australia by the British with the first fleet that arrived at Botany Bay in January 1788; however, interest in growing cotton did not happen until the American Civil War stimulated a search for alternate sources of cotton. Production was temporary, and by 1886 only 15 acres remained under cultivation. In 1926, the Queensland Cotton Marketing Board was established to promote industry growth in central Queensland. By 1934, cotton production had reached 17,000 bales, but it was not until after the Second World War that the modern era of Australian cotton began.

An Industry Develops

The completion of the Keepit Dam in 1958 on the Namoi river near Narrabri, New South Wales, provided for the option of developing irrigated cropping on land that had been used for sheep grazing. Several growers from California took the opportunity to test farm cotton in the region, and early success with the crop soon led to the emergence of an industry and construction of ginning facilities. Similarly, cotton production established in the Macquarie Valley at Warren following the construction of the Burrendong Dam in 1966. The Australian Cotton Research Institute was established near Narrabri and supported interest in growing cotton as a better-managed, higher-input irrigated crop. Cotton was subsequently established in the Gwidyr Valley at Moree in 1976 with the construction of the Copeton Dam, and by 1977, cotton-production

*Corresponding author: paul.grundy@daf.qld.gov.au

DOI: 10.1079/9781800620216.0011

capacity in southern Queensland had also increased following the construction of the Pindari and Glenlyon dams (de Garis, 2013).

In 1963, cotton was introduced in the Ord Valley, situated in the semi-arid tropics of north-western Australia, with production peaking at 4800 ha in 1967. However, frequent and increasing applications of insecticide for caterpillar control resulted in resistance to DDT by the bollworm, *Helicoverpa armigera*, with upwards of 45 applications made in the final growing season (Wilson, 1974; Hearn, 1975). The industry collapsed in 1974, and commercial cotton production in northern Australia was abandoned (Basinski and Wood, 1987).

The Australian cotton industry is highly mechanized and mostly irrigated. The average Australian cotton farm is family-owned and operated and grows 570 ha of cotton. Many cotton producers grow a range of crops in rotation with cotton including wheat, chickpea and sorghum. There are up to 1500 growers in some seasons with production acreage reflecting variations in seasonal rainfall and conditions (Fig. 11.1). Until 1983, only US-bred varieties, predominantly Deltapine, were grown (Fitt, 1994). However, an extensive breeding programme led by the CSIRO at Narrabri has resulted in cultivars that outperformed imported lines in both yield and lint quality. These cultivars now account for 100% of Australian production.

During these early industry phases, pests caused damage to cotton crops throughout the entire growing period, with over 40 species of insects and seven species of mites recorded. Bollworm (*H. armigera* and *H. punctigera*) caused economic damage in every crop. A range of sucking pests including *Aphis gossypii* (cotton aphid), *Creontiades dilutus* (green mirid), *Thrips tabaci* (onion or cotton seedling thrips) and *Tetranychus urticae* (two-spotted spider mite), were also frequent pests.

Insecticide-resistance Problems

In the late 1970s, pyrethroids were introduced, delivering high efficacy against *Helicoverpa* spp., especially *H. armigera*, which had become resistant to other insecticides (Forrester *et al*., 1993).

Despite what had happened in the Ord Valley with DDT, pyrethroids were extensively used and *H. armigera* acquired resistance (Fig. 11.2). This led to the development of an insecticide resistance management strategy (IRMS) in an effort to better preserve the efficacy of these and other products (Forrester, 1990). The IRMS limited the use of endosulfan and pyrethroids to defined periods, referred to as 'windows', and ensured that rotation between insecticides with different modes of action occurred in a strategic manner. End of season cultivation of cotton residues during winter was added to the insecticide resistance strategy to reduce the survival of diapausing *H. armigera* pupae (termed pupae

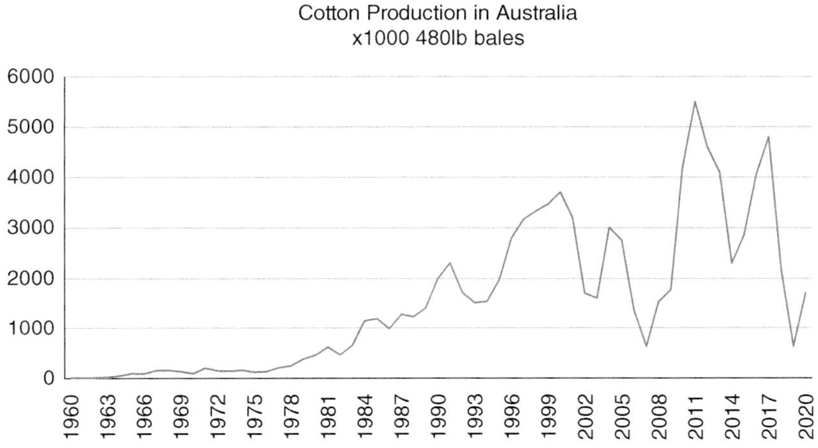

Fig. 11.1. Changes in cotton production, 1960–2020. (Figure courtesy of Graham Matthews)

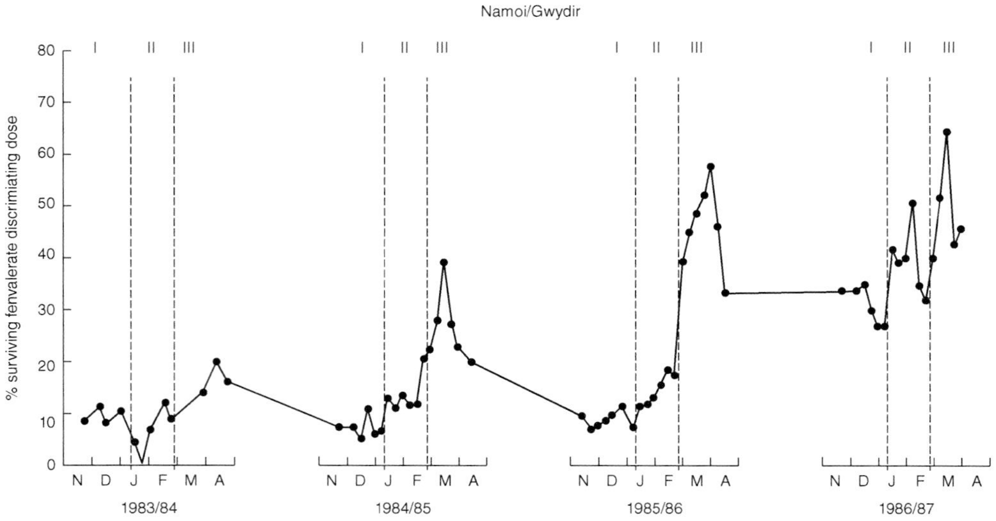

Fig. 11.2. Changes in level of resistance to pyrethroid (fenvalerate) insecticides. (From Forrester and Cahill, 1987)

busting), and the carryover of resistance alleles between cropping seasons (Duffield and Dillon, 2005). Resistance was monitored to assess the strategy's effectiveness and provide a basis for modification of the programme. Despite these efforts, control of *H. armigera* became more difficult, especially late in the season, as resistance to pyrethroids, endosulfan and other insecticides increased both during and across seasons.

The insecticide sprays were applied with tractor-mounted equipment whilst the crop was still relatively small. As the crop developed, aerial spraying generally superseded ground-based equipment to enable crop clearance and overcome difficulties traversing fields that are wet due to surface irrigation. The use of aircraft raised concerns about spray drift into sensitive areas near to cotton fields. To address this issue, studies found that large droplet placement (LPD) application (VMD <250 um) when applying water-based sprays and the implementation of crop buffer distances could be used to effectively mitigate spray drift (Woods *et al.*, 2001).

Pest control during the early industry stages was largely dependent on advice from agrochemical company agronomists. This subsequently changed to the use of independent consultants, who charge a fee per ha for pest-sampling and management advice (Fitt, 1994). A computerized decision support system, SIRATAC, was developed that used simple pest and crop models along with thresholds to determine whether control was required (Hearn *et al.*, 1981). Growers that used SIRATAC applied fewer sprays, demonstrating the importance of using pest thresholds in deciding when to spray. Despite SIRATAC having achieved a 25% market share over five years, the service was discontinued as the company required a 50% market share to remain solvent (Hearn and Bange, 2002).

Bt Cotton Consolidates IPM

Bt cotton (cotton transformed with insecticidal gene(s), *cry* isolated from soil bacterium, *Bacillus thuringiensis* (Bt)) has been commercially grown in Australia since 1996. The first version of Bt cotton (Ingard®) was introduced to the Australian marketplace by the Monsanto Company (Wilson *et al.*, 2018). This product expressed the Cry1Ac protein which was toxic to *H. armigera* and *H. punctigera*. Anticipating the potential for resistance, a Transgenic and Insecticides Management Strategies Committee (TIMS) was established with the regulators for the purpose of managing the use of the technology in a way that would minimize the risk of resistance. A key strategy was capping the planting of Ingard®

varieties at 30% of the crop area together with the use of other tactics such as pupae busting as part of a comprehensive resistance management plan (RMP), which was supported by annual Bt resistance monitoring. INGARD® technology provided effective control of *Helicoverpa* spp. Although expression of the toxin declined during late flowering and boll fill, the level of control achieved was sufficient to reduce insecticide use by 40–50% in those crops (Fitt *et al.*, 1998).

To further improve pest management, research was conducted on a wide range of tactics including area-wide management involving trap crops, the conservation and use of beneficial insects, development of new thresholds and sampling strategies, and plant breeding to select for traits that may reduce pest damage (Mensah *et al.*, 2013).

One tactic included examining the spraying of a supplementary food product, 'Envirofeast', onto cotton to modify predator/prey ratios through the attraction and in-crop retention of beneficial insects. This evolved to examine the growing of unsprayed lucerne strips to attract green mirids, *Creontiades dilutus*, away from cotton as well as provide a potential reservoir for beneficial insects that may enhance biological control when a food spray was applied to the crop (Fig. 11.3) (Mensah and Harris, 1995; Mensah, 1999; Mensah, 2002). Later a biopesticide was studied to control *Creontiades dilutus* (Mensah and Austin, 2012). With the arrival of GM cotton achieving good control of bollworm, Envirofeast® or similar products such as Mobate®, Aminofeed® and Pred Feed® did not continue to be used commercially, although beneficial insects could be attracted into the cotton fields when using these products. This was partly due to the requirement for lucerne strips to be grown near the cotton as a refuge from which the beneficials could be drawn, which was difficult to achieve within large-scale cotton farms. Difficulties were also encountered with the application of some powder formulations, including clogging of nozzles. The introduction of herbicide-tolerant cotton cultivars that utilized over-the-top glyphosate sprays also posed challenges for the incorporation of lucerne strips within cotton fields.

Despite these efforts, insecticide use on non-Bt cotton increased from 10–12 applications per season in the early 1990s to 15–18 applications per season by 1997/98. However, an increased uptake of Ingard® Bt cotton, combined with the increased availability of more selective insecticides against *Helicoverpa* spp. (notably spinosad, indoxacarb, and emamectin, which provided effective control without decimating beneficial populations), improved the situation for a period of time.

Bollgard II® varieties expressing Cry1Ac and Cry2Ab proteins were commercialized and replaced INGARD® in 2004/05. The introduction of two Bt toxins provided a much more robust option, considered to lower the likelihood of *Helicoverpa* spp. evolving resistance. Plant expression of the Cry2Ab toxin throughout the growing period provided season-long control of

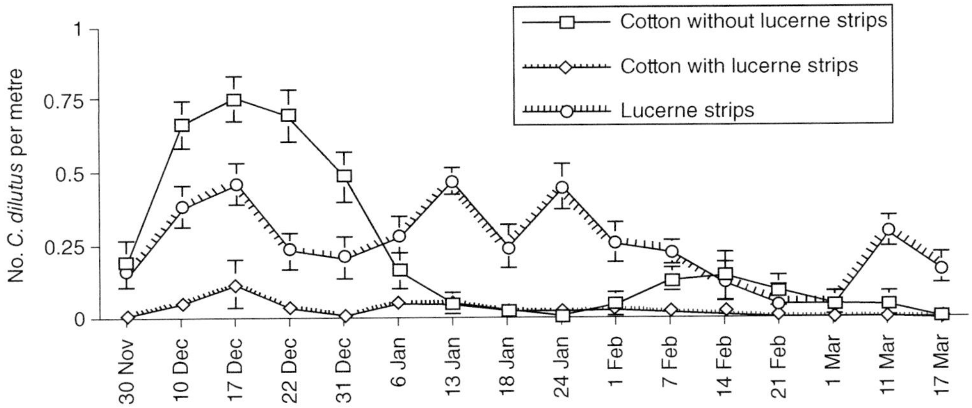

Fig. 11.3. Populations of *Creontiades dilutus* (green mirids) in cotton inter-cropped with lucerne strips. (From Mensah and Khan, 1997)

Fig. 11.4. View of cotton field. (Photo: P. Grundy)

Helicoverpa, greatly reducing the need for insecticide application.

A resistance-management plan was implemented together with the release of Bollgard II® technology for the Australian marketplace. The new plan centred upon three basic principles:

1. minimizing the exposure of *Helicoverpa* spp. to the Bt proteins Cry1Ac & Cry2Ab;
2. providing a population of susceptible individuals that can mate with any resistant individuals, hence diluting any potential resistance; and
3. removing resistant individuals at the end of the cotton season.

How these principles were implemented varied across production areas depending on the number of *Helicoverpa* generations per season and the likelihood of diapause, with warmer production regions such as central Queensland having both a more restrictive planting window to limit crop pest exposure and the inclusion of end-of-season trap crops (Sequeira, 2001). Adoption of Bollgard II was dramatic, growing from ~16% in 2003/04 to ~96% by 2013 (Wilson *et al.*, 2013).

The deployment of Bt cotton in Australia has been within a regulatory framework that required technology users to adopt practices that aimed to reduce the likelihood of pest resistance. This framework has included the implementation of refuge requirements that are amongst the most stringent in the world (Knight *et al.*, 2021). The ratio of structured refuge to Bt cotton as well as naturally occurring non-Bt host crops and native vegetation that contribute additional refugia has likely been a significant factor in the success of resistance management for *Helicoverpa* budworm and bollworm species in Australia (Tabashnik *et al.*, 2013).

Growers planting Bt cotton sign a legally binding contract to follow a comprehensive resistance-management plan (RMP). It is considered that growing Bt cotton provides strong support for IPM, as the reduction in spraying has enabled the natural enemies of other pest species to exert meaningful levels of biological control that has greatly reduced the need to spray for secondary pests.

Downes *et al.* (2016) concluded that compliance with mandatory resistance-management plans for Bt cotton necessitated a shift from pest control at the level of individual fields or farms towards a co-ordinated area-wide landscape approach and that control of *H. armigera* and resistance management is essential in genetically modified crops and must be season-long and area-wide to be effective.

However, while Bollgard II® was well protected from *Helicoverpa*, there were sucking pests that were formerly controlled by insecticides targeted at *Helicoverpa* spp., notably mirids,

Creontiades dilutus, that became a more important pest in the absence of broad-spectrum insecticides (Mensah and Khan, 1997). Without proper sampling protocols and insecticide-application thresholds, control measures often relied on broad-spectrum insecticides with a high reliance on fipronil (Regent®) to control mirids (Whitehouse, 2011), despite the risk of secondary outbreaks of aphids and mites (Wilson *et al.*, 1998).

Later, Bollgard® 3 (which contains Cry1Ac, Cry2Ab, and Vip3A) was registered to further minimize the risk of resistance to Bt cotton increasing in *Helicoverpa* spp., and in 2016/17 was >90% of the area planted (Mahon *et al.*, 2012). The introduction of Bollgard II® and subsequently Bollgard® 3 cotton has enabled significant industry growth. The ability to rapidly set fruit uninterrupted by insect damage has allowed expansion into shorter-season southern regions. Likewise, the protection afforded by Bt cotton has enabled cotton production to be considered in northern Australia with test farming well underway in the Ord Valley and parts of the Northern Territory. So far, the resistance-management measures for Bt cotton seem to have delivered well with no discernible change in background resistance levels from when the technology was first released in 1996 (Knight *et al.*, 2021).

Weed Management

Prior to the introduction of transgenic varieties, weed management in cotton systems was based on frequent cultivation, residual herbicides, and applying some post-emergent herbicides. This diverse range of tactics reduced the likelihood of selecting for herbicide resistance (Norsworthy *et al.*, 2012). Within this system, glyphosate was used as a pre-planting knockdown treatment, with any within-crop use being applied through shielded sprayers directed at the inter-row spaces.

Roundup Ready Flex® technology was released in combination with Bollgard II®. This genetic trait allowed the season-long application of glyphosate as an over-the-top crop spray for weed control. Some weeds, including *Hibiscus verdcourtii*, *Hibiscus tridactylites* and *Echinochloa colona*, had tolerance to glyphosate and were considered to be able to survive glyphosate applications or increase in dominance over time within a glyphosate-based system (Werth *et al.*, 2013). The presence of resistant weeds has also been encouraged by the zero-tillage system that is widely adopted throughout Australia's rain-grown cropping regions where glyphosate is used extensively to control weeds in fallows. Farmers prefer no-till or minimal tillage as it retains crop residues at the soil surface, conserving soil moisture and reducing topsoil exposure to wind and water erosion.

Studies with a mimic weed, common sunflower, sown with the cotton and removed at different times during the season was used to determine the critical period for weed control in high-yielding cotton. It indicated that season-long interference by five or more weeds per m^2 resulted in no harvestable cotton (Charles *et al.*, 2019a, b).

The presence of glyphosate-resistant weeds and volunteer glyphosate-tolerant plants has necessitated that growers utilize an integrated weed management approach that utilizes multiple herbicides deployed at different stages, cover and rotation crops, and mechanical cultivation. Other tactics under investigation include organic herbicides, RNAi technology and robotics (Iqbal *et al.*, 2019). The industry will soon have access to multiple herbicide tolerance traits (glufosinate and dicamba) as a triple stack which will enable greater in-crop herbicide rotation.

Industry Productivity Gains

In Australia, average irrigated cotton lint yields have increased by 33.6 kg ha^{-1} y^{-1} over five decades with a recent five-year average of 2483 kg ha^{-1} for irrigated crops (Conaty and Constable, 2020); new cultivars and improved management have accelerated the rate of gain three-fold since the mid-1990s (Liu *et al.*, 2013).

With the introduction of Bollgard® varieties, the primary focus of crop managers was able to move beyond the prevention of insect-related crop loss to the optimization of agronomic inputs to maximize yield potential. A comparison of conventional non-Bt and Bollgard II® cultivars found that Bollgard cultivars were able to maintain yield potential with altered sowing times, which was attributed to the higher retention of early fruit and shorter fruiting cycle (Bange *et al.*, 2004). The ability to maintain yield potential over a much broader sowing time window has provided a range of farming system and yield advantages for cotton growers in Australia. Being able to sow cotton later in the spring and early summer period has allowed many growers to grow cotton as a dry-land crop and take advantage of rainfall that generally increases during November in northern NSW and southern Queensland, a time of year when the sowing of conventional cotton varieties is traditionally riskier. Similarly, the rapid fruiting accumulation cycle of Bollgard varieties enabled the crop to prosper in shorter-season, cooler environments and has been a major factor underpinning significant industry expansion and the construction of new ginneries around the Lachlan, Murrumbidgee and Murray river irrigation regions during the last decade (Knight *et al.*, 2021).

The adoption of Bollgard varieties also helped to advance industry-wide changes to how irrigation water was both scheduled and applied to cotton crops in Australia. For traditional flood-irrigation techniques, there has been a major emphasis on changing field grades and application rates to ensure that water is moved on and off fields quickly so that, despite more frequent irrigation, less water is used by minimizing deep drainage and water-logging (Wigginton *et al.*, 2012). The increase in yields from newer Bollgard varieties combined with more efficient water-management systems has underpinned a 40% improvement of water-use productivity by Australian cotton growers between 2003 and 2013 (Roth *et al.*, 2013). With Bollgard cotton crops typically subject to no more than 0–3 insecticide treatments per crop (Wilson *et al.*, 2013), cotton is now produced much more efficiently and sustainably compared to previous conventional methods.

References

Bange, M.P., Deutscher, S.A., Larsen, D., Linsley, D. and Whiteside, S. (2004) A handheld decision support system to facilitate improved insect pest management in Australian cotton systems. *Computers and Electronics in Agriculture* 43, 131–147.

Basinski, J.J. and Wood, M. (1987) Kimberly Research Station. In: Basinski, J.J., Wood, M. and Hacker, J.B. (eds) *The Northern Challenge: A History of CSIRO Crop Research in Northern Australia*. Research Report no. 3. CSIRO, Brisbane, Australia.

Charles, G.W., Sindel, B.M., Cowie, A.L. and Knox, O.G.G. (2019a) The value of using mimic weeds in competition experiments in irrigated cotton. *Environmental Entomology* 33, 601–609.

Charles, G.W., Sindel, B.M., Cowie, A.L. and Knox, O.G.G. (2019b) Determining the critical period for weed control in high-yielding cotton using common sunflower as a mimic weed. *Environmental Entomology* 33, 800–807.

Conaty, W.C. and Constable G.A. (2020) Factors responsible for yield improvement in new *Gossypium hirsutum* L. cotton cultivars. *Field Crops Research* 250, 107780.

Craven, L.A., Stewart, J.M., Brown, A.H.D. and Grace, J.P. (1994) The Australian wild species of *Gossypium*. In: Constable, G.A. and Forrester, N.W. (eds) *Challenging the Future*. Proceedings of the World Cotton Research Conference 1. CSIRO, Brisbane, Australia, pp. 278–281.

de Garis, S.A. (2013) The cotton in Australia: an analysis. In: 19th Annual Pacific-Rim Real Estate Society Conference, Melbourne, Australia, 13–16 January. Available at: http://www.prres.net/papers/DeGaris_THE_COTTON_INDUSTR_%20IN_AUSTRALIA%202013.pdf (accessed 7 August 2021).

Downes, S., Kriticos, D., Parry, H., Paull, C., Schellhorn, N. and Zalucki, M.P. (2016) A perspective on management of *Helicoverpa armigera*: transgenic Bt cotton, IPM, and landscapes. *Pest Management Science* 73, 485–492.

Duffield, J. and Dillon, M.L. (2005) *Helicoverpa armigera* emergence in southern New South Wales. *Australian Journal of Entomology* 44, 316–320.

Fitt, G.P. (1994) Cotton pest management: part 3: an Australian perspective. *Annual Review of Entomology* 39, 543–562.

Fitt G.P., Daly, J., Mares, C. and Olsen K. (1998) Changing efficacy of transgenic Bt cotton – patterns and consequences. Paper presented at 6th Australian Applied Entomology Research Conference, 29 Sept.–2 Oct., Brisbane, Australia.

Forrester, N.W. (1990) Designing, implementing and servicing an insecticide resistance management strategy. *Pest Science* 28, 167–179.
Forrester, N.W. and Cahill, M. (1987) Management of insecticide resistance in *Heliothis armigera* (Hubner) in Australia. In: Ford, M.G., Hollomen, D.W., Khambay, P.S. and Sawicki, R.M. (eds) *Biological and Chemical Approaches to Combatting Resistance to Xenobiotics*. Elsevier.
Forrester, N.W., Cahill, M., Bird, L.J. and Layland, J.K. (1993) Management of pyrethroid and endosulfan resistance in *Helicoverpa armigera* (Lepidoptera: Noctuidae) in Australia. In: Walker, A. (ed.) *Buletin of Entomological Research,* Supplement Series 1. CAB International, Wallingford, UK.
Hearn, A.B. (1975) Ord valley cotton crop: development of a technology. *Cotton Growing Review* 52, 77–102.
Hearn, A.B. and Bange, M.P. (2002) SIRATAC and CottonLOGIC: persevering with DSSs in the Australian cotton industry. *Agricultural Systems* 74, 27–56.
Hearn, A.B., Ives, P.M., Room, P.M., Thomson, N.J. and Wilson L.T. (1981) Computer-based cotton pest management in Australia. *Field Crops Research* 4, 321–332.
Iqbal, N., Manalil, S., Chauhan, B.S. and Adkins, S.W. (2019) Glyphosate-tolerant cotton in Australia: successes and failure. *Archives of Agronomy and Soil Science* 65, 1536–1553.
Knight, K., Grundy, P., Kauter, G. and Ceeney, S. (2021) 20 years of successful Bt cotton production in Australia. In: Gujar, G., Trisyono, Y.A. and Chen, M. (eds) *Genetically Modified Crops in Asia Pacific*. CSIRO Publishing.
Liu, S.M., Constable, G.A., Reid, P.E., Stiller, W.N. and Cullis, B.R. (2013) The interaction between breeding and crop management in improved cotton yield. *Field Crops Research* 148, 49–60.
Mahon, R.J., Downes, S.J. and James, B. (2012) Vip3A resistance alleles exist at high levels in Australian targets before release of cotton expressing this toxin. *PLOS ONE* 7, e39192.
Mensah, R.K. (1999) Habitat diversity: implications for the conservation and use of predatory insects of *Helicoverpa* spp. in cotton systems in Australia. *International Journal of Pest Management* 45, 91–100.
Mensah, R.K. (2002) Development of an integrated pest management programme for cotton, part 2: integration of a Lucerne/cotton interplant system, food supplement sprays with biological and synthetic insecticides. *International Journal of Pest Management* 48, 95–110.
Mensah, R.K. and Austin, L. (2012) Microbial control of cotton pests. Part 1: Use of the naturally occurring entomopathogenic fungus *Aspergillus* sp. (BC 639) in the management of *Creontiades dilutus* (Stal) (Hemiptera: Miridae) and beneficial insects on transgenic cotton crops. *Biocontrol Science and Technology* 22, 567–582.
Mensah, R.K. and Harris, W.E. (1995) Using envirofeast Ò spray and refugia technology for cotton pest control. *Australian Cotton Grower* 16, 30–33.
Mensah, R. and Khan, M. (1997) Use of *Medicago sativa* (L.) interplantings/trap crops in the management of the green mirid, *Creontiades dilutus* (Stal) in commercial cotton in Australia. *International Journal of Pest Management* 43, 197–202.
Mensah, R.K., Gregg, P.C., Del Socorro, A.P., Moore, C.J., Hawes, A.J. and Watts, N. (2013) Integrated pest management in cotton: exploiting behaviour-modifying (semiochemical) compounds for managing cotton pests. *Crop and Pasture Science* 64, 763–773.
Norsworthy, J.K., Ward, S.M., Shaw, D.R., Llewellyn, R., Nichols, R., Webster, T. *et al.* (2012) Reducing the risks of herbicide resistance: best management practices and recommendations. *Weed Science* 60, 31–62.
Roth, G., Harris, G., Gillies, M., Montgomery, J. and Wigginton, D. (2013) Water-use efficiency and productivity trends in Australian irrigated cotton: a review. *Crop & Pasture Science* 64, 1033–1048.
Sequeira, R. (2001) Inter-seasonal population dynamics and cultural management of *Helicoverpa* spp. in a central Queensland cropping system. *Australian Journal of Experimental Agriculture* 41, 249–259.
Tabashnik, B.E., Brévault, T. and Carrière, Y. (2013) Insect resistance to Bt crops: lessons from the first billion acres. *Nature Biotechnology* 31, 510–512.
Werth, J., Boucher, L., Thornby, D., Walker, S. and Charles, G. (2013) Changes in weed species since the introduction of glyphosate resistant cotton. *Crop & Pasture Science* 64, 791–798.
Whitehouse, M.E.A. (2011) IPM of mirids in Australian cotton: why and when pest managers spray for mirids. *Agricultural Systems* 104, 30–41.
Wigginton, D., Carter, M. and Grabham, M. (2012) Surface irrigation performance and operation. Cotton Research and Development Corporation, Narrabri, Australia.
Wilson, A.G.L. (1974) Resistance of *Heliothis armigera* to insecticides in the Ord irrigation area, North-western Australia. *Journal of Economic Entomology* 67, 256–258.

Wilson, L.J., Bauer, L.R. and Lalley, D.A. (1998) Effect of early season insecticide use on predators and outbreaks of spider mites in cotton. *Bulletin of Entomological Research* 88, 477–488.
Wilson, L.J., Downes, S., Khan, M., Whitehouse, M., Baker, G. *et al.* (2013) IPM in the transgenic era: a review of the challenges from emerging pests in Australian cotton systems. *Crop and Pasture Science* 64, 737–749.
Wilson, L.J., Whitehouse, M.E.A. and Herron, G.A. (2018) The management of insect pests in Australian cotton: an evolving story. *Annual Review of Entomology* 63, 215–237.
Woods, N., Craig, I.P., Dorr, G. and Young, B. (2001) Spray drift of pesticides arising from aerial application in cotton. *Journal of Environmental Quality* 30, 697–701.

12 Cotton Growing in South America and the Caribbean

Simone Silva Vieira* and Graham Matthews

Cotton Growing in Brazil

Introduction

The use of cotton fibres in South America had been noted in Peru long before Europeans sailed across the Atlantic and was discovered in Brazil after Pedro Álvares Cabral had reached Brazil's shores in April 1500. However, it was not until the 1930s that Brazilian growers began planting a long-staple cotton in the north-east regions, which was later spread south to São Paulo in the 1940s. However, competition from sugar, oranges and cattle prompted reduced interest in cotton because it was labour-intensive and the costs of inputs and transportation remained high. It was then cheaper and easier for Brazil to import cotton than to grow it.

In the 1970s, some cotton was still grown on approximately 2.5 million ha and then the boll weevil arrived and created new problems. It was first reported in the state of Sao Paulo in February 1983. The federal government of Brazil attempted to eradicate the boll weevil, but efforts failed because the programme was not applied completely or at the correct time (Ramalho, 1994). Despite introducing the use of insecticides, the area of cotton declined to about 650,000 ha by the 1996/97 season, when farmers were not as organized, educated or technologically advanced as they are today. Now Brazil occupies a prominent position in relation to the production and commercialization of cotton. According to the Department of Agriculture of the United States (USA), the country was the second largest exporter in the world, behind the USA, and is fourth in terms of production.

It is a world leader in certification of good labour and environmental practices, about 86% of production is certified by auditors' international standards. With high technology and extremely organized producers, through the Brazilian Association of Cotton Producers (Abrapa), the sector invests in quality, traceability and marketing so that it can continue growing and seeking positive results. In addition to the recognized quality, cotton production has reached above 2.5 million tons of fibre per year, and the country has become a regular exporter to markets around the world for delivery throughout the year.

In the early 1990s, the cotton cultivation and production areas in Brazil were concentrated in the southern, south-eastern and north-eastern regions (Stadler and Buteler, 2007). After this period, the cultivation of this fibrous plant shifted significantly to the Cerrado areas, in the mid-western region of Brazil. In 1990, this region had a total cotton cultivation area of 123,000 ha, representing only 8.8% of the total

*Corresponding author: sisilvavieira@gmail.com

(eds G.A. Matthews and T. Miller)
DOI: 10.1079/9781800620216.0012

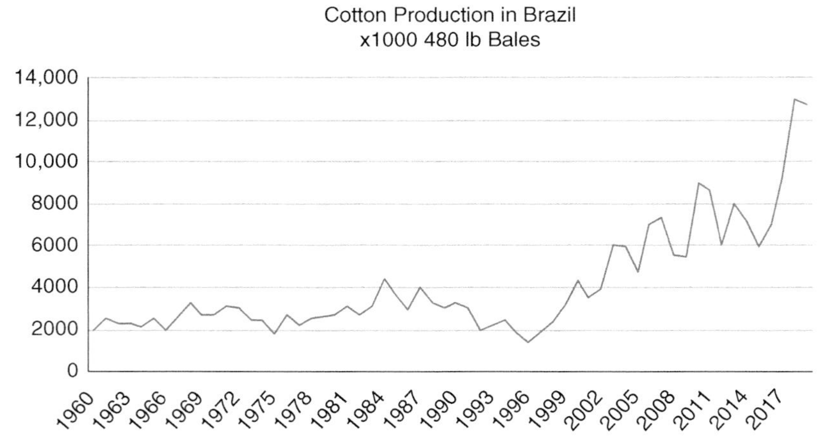

Fig. 12.1. Increase in cotton production in Brazil.

cotton cultivation area in the country. Cotton has been planted in integrated production systems in the Cerrado (Brazilian savannah) ecosystem and is an agronomically feasible and sustainable option for corn and soybean farmers (Lima Jr *et al.*, 2013).

However, to remain competitive in the world market, Brazil needed to produce in quantity and quality, in a sustainable and economically viable way. The occurrence and damage of various pests in the crop brings great challenges to producers. These pests have high reproductive capacity and wide dispersion, which infest quickly, causing damage, either in falling productivity or adversely affecting seed and fibre (Santos, 2007).

The first commercial transgenic cotton (Bt) cultivar, known as Bollgard, was generated in 1989 and released for sale in the USA and Australia in 1996/97. In Brazil, it was released in 2005. Bollgard has the Cry1Ac gene, which gives it resistance to the larvae of lepidopterans that attack cotton in Brazil, namely the bollworm, *Heliothis virescens*, pink bollworm, *Pectinophora gossypiella* and the cotton leafworm, *Alabama argillacea* (Barroso and Hoffmann, 2007). The growing of transgenic cotton has made a significant increase in cotton production since 2005.

According to Silva *et al.* (1968), it was estimated that the entomofauna associated with growing cotton includes about 259 species of insects in Brazil. Of these, 12 are considered important pests together with three species of phytophagous mites (Gallo *et al.*, 2002). In the Brazilian Cerrado region, there has been an expansion of crops, in which frequent presence of pests has the potential to cause significant damage. In addition to the pests noted above, the arrival of the boll weevil (*Anthonomus grandis*) in Brazil in 1983, despite the apparent barrier of the Amazon rainforest, added another major problem, which later spread into Paraguay and Argentina (Santos, 2007). It infested more than 90% of the total cotton cultivation area in Brazil (Stadler and Buteler, 2007). Management of boll weevil populations required the destruction of the cotton stalks after harvest to prevent any regrowth of cotton plants on which the boll weevils could survive during the off season (Lu *et al.*, 2010; Bianchini and Borges, 2013; Grigolli *et al.*, 2015). Destruction of cotton stalks also affected other pests including *Bemisia tabaci*, *Aphis gossypii* and *Frankliniella schultzei*. However, the way cotton stalks were removed initially did not have sufficient impact on the boll weevil in 2015, but there was a reduction in the population of larvae of *Spodoptera frugiperda*, *Alabama argillacea*, *Chrysodeixis includens*, *Helicoverpa* sp., and *Pectinophora gossypiella* during the fallow period (Silva *et al.*, 2018). Other pests include *Spodoptera* spp., whitefly, *Bemisia argentifolii* and *Bemisia tabaci*; as well as migrant soya bugs, *Nezara viridula*, *Euschistus heros* and *Piezodorus guildinii*, with some of these being more evident following changes in cultivation (Silva *et al.*, 2013).

Boll Weevil *Anthonomus grandis* Boheman (Coleoptera: Curculionidae)

The boll weevil was first observed in Brazil in 1983, in two separate areas (Degrande, 2006), in the region of Campinas-Sorocaba, state of São Paulo, and in the region of Campina Grande-Ingá, state of Goiás. The weevils devastated the economies of much of the southern, south-eastern, and north-eastern cotton-producing states in the country.

Boll weevils can attack several parts of the cotton plant (Bleicher, 1991a,b; Degrande, 1998), but their population increase is dependent on environmental factors, such as temperature and humidity, with most infestations concentrated along the perimeters of cotton fields (Rummel *et al.*, 1977). Losses from floral bud ('square') abscission and fruit damage can reduce production by as much as 100% if no control measures are taken (Degrande, 1991; Gallo *et al.*, 2002).

Controlling the boll weevil has resulted in the largest number of insecticide sprays against a single target pest species in cotton fields (Lima Jr *et al.*, 2013; Belot *et al.*, 2016). Cost estimations of cotton pest control indicated the boll weevil increased costs; as its presence in cotton fields determines the spray frequency and choice of insecticide (Showler, 2012; Belot *et al.*, 2016). When boll weevil infestations reached the economic threshold, at least five sequential sprays were carried out to attain control of adult weevils (Barros *et al.*, 2019). As seen in the USA, the management of the boll weevil requires an area-wide plan to achieve eradication (Smith, 1998).

Integrated Pest Management

Decision-making aimed at increasing and preserving populations of natural enemies in the cotton agri-ecosystem are promising, technically and ecologically viable actions that could result in great savings for cotton farmers, in improving the quality of the environment and the reduction of health problems resulting from the indiscriminate use of chemicals.

The cotton grower has been able to recognize pests and their natural enemies that may occur during the crop cycle, performing periodic sampling on the farms and making appropriate and economical decisions. Generally, sampling should be done at intervals of five days, examining 100 plots randomly in a 100 ha of homogeneous area, by walking through the crop to observe plants in different areas of the crop. The sample must be completed taking into account the critical period, sampling location and the levels of pest control (Table 12.1). The studies were conducted in Maracaju, MS, Brazil, in two seasons with cotton cultivar FM 993 on a 10,000-m^2 area of cotton, which was subdivided into 100 ten-metre plots. Five plants per plot were evaluated weekly, recording the number of squares with feeding plus oviposition punctures of *A. grandis* in each plant. A sequential sampling plan by the maximum likelihood ratio test was developed (Grigolli *et al.*, 2017).

A large-scale field-testing of a Boll Weevil Suppression Program (BWSP) was implemented to assess its technical and operational feasibility for boll weevil suppression in the state of Goiás, Brazil. The pilot plan focused on 3608 ha of cotton during the 2006/07, and 6011 ha in the 2007/08, growing seasons; the areas were divided into four inner zones with an outer buffer zone. Data on boll weevil captures using pheromone traps installed in the BWSP fields were analysed from the detection of the first weevil and subsequently during the period of crop damage in relation to the number of insecticide applications.

Fourteen pheromone-baited trapping evaluations were used to compare the weevil populations from 2006/07 to 2007/08 growing seasons. The BWSP regime reduced in-season boll weevil captures from 15- to 500-fold compared to pre-suppression levels in the preceding year. The low capture rates were related to delays in infestation and damage by weevils. The smaller population size measured by trapping and field monitoring reduced the number of required insecticide treatments. The BWSP strategy was efficient in suppressing populations of this pest and is a viable programme for cotton production in subtropical and tropical areas (Lima Jr *et al.*, 2013).

Ten insecticide formulations were studied with five different active ingredients in mixtures, namely lambda-cyhalothrin + thiamethoxam, lambda-cyhalothrin + chlorantraniliprole, thiamethoxam + chlorantraniliprole, and fenitrothion + esfenvale separaterate, and singly as five formulations. Adult mortality was assessed 48 hours after caging

Table 12.1. Pest, critical time, sampling place and control level recommended to cotton IPM.

Pest	Critical time	Sampling place	Control level
Brazilian cotton borer *Eutinobothrus*	Emergence of seedlings to the appearance of first floral bud	Root and stem of plant	
Black cutworm *Ipsilon agrotis*	Emergence of seedlings to the appearance of the first true leaves	Root and stem of plant	
Root stink bugs	Emergence of seedlings to the appearance of first boll ready for harvest	Root of plant	
Thrips	Emergence of seedlings up to 20 days after germination	New leaves of apical region of plant	70% of plants attacked
Aphid	Emergence of seedlings to the appearance of first boll ready for harvest	New leaves of apical region of plant	70% of plants with colonies
Cochineal	Emergence of seedlings to the appearance of first boll ready for harvest	New leaves of apical region of plant	10% of plants attacked
Cotton leafworm *Alabama argillacea*	Emergence of seedlings to the appearance of first boll ready for harvest	Third leaf expanded at apical region	22% or 53% of attacked plants by caterpillars > or < 15mm
Boll weevil *Grandis anthonomus*	Appearance of first buttons on the plant until the appearance of first boll ready for harvest	Larger floral bud more than 3 mm and less that 6 mm in diameter, half top of the plant	10% of plants with flower buds damaged (hole for oviposition and/or food)
Budworm tobacco *Heliothis virescens Helicoverpa armigera*	Appearance of first buttons on the plant until the appearance of first boll ready for harvest	Leaves located in the upper third or in the bracts of floral buds of the plants	13% of plants with caterpillars small *H. virescens*; 10% of plants with caterpillars small *H. armigera*
Fall armyworm *Spodoptera frugiperda*	Appearance of first buttons on the plant until the appearance of first boll ready for harvest	Leaves located in the upper third or in the bracts of floral buds	10% of damaged plants
Pink bollworm *Gossypiella pectinophora*	Emergence of first firm boll until the appearance of boll ready for harvest	Bigger firm square than 2.5 cm and less that 3.0 cm diameter, next to the apex of the plant	11% of plants with ripening bolls damaged
Whitefly *Bemisia tabaci*	Emergence of seedlings to the appearance of first boll ready for harvest	Ventral face of third leaf expanded region apical	40% of plants with nymphs or 60% of plants with adults
Mite	Appearance of flower buds to the appearance of first boll ready for harvest	Ventral face of region leaves apical (mite white) and median (red mites and brindle) of the plant	40% of plants with colony
Stink bugs of structures reproductive	From flowering to fruiting	Floral bud and square (less than 2.0 cm in diameter)	20% of plants attacked

adults on treated and untreated plants. The LC50s concentrations varied from 0.004 to 0.114 g a.i./L, while the relative potency between a single and mixture of insecticides varied from 1.37- to 29.59-fold. Furthermore, lambda-cyhalothrin and thiamethoxam in single formulation were the most toxic insecticides to boll weevil. Among insecticide mixtures, only lambda-cyhalothrin + chlorantraniliprole resulted in a synergic effect, as the remaining mixtures did not enhance toxicity against the boll weevil and should be recommended only when aimed at different pests (Barros *et al.*, 2019).

In crops of agricultural importance, like cotton, the adequate technology for the application of phytosanitary products has been fundamental for the successful phytosanitary treatment. Bearing in mind that efficiency in agricultural production is increasingly demanded, it is essential that investments in treatment result in greater efficiency, without meaning extra impact on production costs and the environment. The efficiency of phytosanitary treatment depends not only on the choice of the insecticide but also on factors related to the application technology, which can make a difference when accounting for the cost of production and profitability (Miranda *et al.*, 2008).

To optimize decision-making in integrated cotton aphid management programmes, the vertical and horizontal distribution of *A. gossypii* in non-transgenic and transgenic Bt cotton were examined during two cotton seasons. The number of apterous and alate aphids found per plant was lower on the Bt cotton than on its isoline and there were significant effects on the vertical, horizontal, spatial and temporal distribution patterns of *A. gossypii*, so distribution behaviour inside the plant canopy changed as the cotton crop developed (Fernandes *et al.*, 2012).

As a part-insecticide resistance management, studies on the susceptibility of the bollworm *Helicoverpa armigera* and *H. zea* to Bt cotton with Vip3Aa20 were conducted from 2014 to 2015, as transgenic maize with this insecticide protein had been increasingly adopted. Based on LC50 data, *H. armigera* was more tolerant to Vip3Aa20 protein than was *H. zea* (40- to 75-fold). The baseline susceptibility data for Vip3Aa20 in these bollworm populations was aimed to help in IRM programmes in Brazil (Leite *et al.*, 2018).

Recommended Strategies

Traits handling: Where cotton growers opt for the use of transgenic cotton cultivars, it is essential that farmers comply with areas of refuge as recommended by companies with Bt technology. The aim is to promote the crossing between pest populations exposed and not exposed to Bt toxin. The refuge must represent at least 20% of the cultivated area with insect-resistant transgenic cotton and be located not more than 800 m away from the area with Bt plants to favour and facilitate mating among insects. In the refuge areas, control of pests is by biological control, using natural enemies and baculovirus-based bioinsecticides. Therefore, the agricultural sector must be aware of how to manage the transgenic events available, by avoiding planting of transgenic cultivars whose events express the same toxins in different fields simultaneously and successively. Cotton varieties in the refuge area should be cultivars of short cycle.

Sowing time: The sowing of cotton within a region should be within a period of 3–4 weeks and whenever possible in areas and periods of time proven to have a lower incidence of pests, to break the synchrony between the availability of food and the occurrence of insects and thus anticipate the harvest and the practice of destroying the remains of culture.

Soil conservation and fertilization: Correct land use, based on recommendations, preparation and fertilization techniques, for maintaining their fertility and structure, contribute directly to the formation of vigorous plants and, therefore, less vulnerable to pest attack.

Planting density: The planting density should avoid an excessive number of plants to facilitate penetration of sunlight and the distribution of spray droplets to the biological target.

Collection of flower buds and bolls fallen on the ground: Removal of flower buds and ripening bolls that fall on the ground, to eliminate any weevils and other pests that develop inside these structures areas.

Biological control: Carry out extensive releases of the wasp, *Trichogramma* spp., and/or the predatory bug *Podisus nigrispinus* at the time of appearance of pest lepidoptera in the field. Another option is to apply Bt as a spray. Particular attention should be paid to the presence of predators (ladybugs, syrphids, garbage bugs and spiders) and aphid parasitoids (braconid parasitoid wasp, *Lysiphlebus testaceipes*) and lepidopteran egg parasitoids (*Trichogramma* spp.) in the field, obeying the action level of these natural enemies (71% of plants with predators and/or parasitoids).

Destruction of culture remains: Immediately after harvesting, the cotton plants must be destroyed. This requires the roots, stems, flowers, ripening bolls and bolls not harvested to be destroyed and/or incorporated into the soil. The destruction of the remains of the crop aims to interrupt the biological cycle of certain root pests, like the drill, and pests of aerial parts, like the weevil, aphids, whitefly and caterpillars. It is also important to eliminate the voluntary plants that appear in the off-season, as those that germinate from the fallen seeds by the side of the highway. Cutting cotton stalks immediately after harvest effectively reduces *M. incognita* reproduction, and may lead to a lower initial population density of this nematode in the following year (Lu *et al.*, 2010).

Chemical control: Chemical control should only be carried out when necessary, based on monitoring the crop. In areas with a history of early pest infestation, such as the borer and the brown stink bug, it is suggested to observe the presence of seedlings or volunteers and carry out sampling of soil 20 days before sowing, with the aid of an auger or shovel, in order to detect the presence of these organisms. If the presence of these pests is detected, the infested area should be avoided for the cultivation of cotton, as treatment of the soil and /or insecticide applications in the planting furrow for the control of these insects has shown little effect. Sucking-pest insects such as aphids, whitefly and mealybug should be preventatively controlled by means of seed treatment with systemic insecticides. The use of defoliants after 60% of the ripening bolls are open is practical and recommended as it also suppresses populations of aphid and whitefly by eliminating food, while preserving the fibre quality by preventing contamination with honeydew deposition.

Small-scale farmers in two areas of Brazil were able to use the hand-carried Electrodyn sprayer to apply ULV insecticide sprays (Fig. 12.2). Farmers liked the equipment as they did not have to collect water and the insecticide formulation in the Bozzle container was simply screwed onto the sprayer avoiding measuring out and mixing the insecticide. Yields were good using less insecticide, but the manufacturer withdrew the availability of the sprayer.

However, few small-scale farms now grow cotton, due to the expansion of cotton in the states of Mato Grosso and Bahia, where it is a crop for big farmers (usually 5000–10,000 ha or more). Some groups are larger with more than 20,000 ha. As these areas have extensive flat areas, insecticides are applied on cotton crops, using large self-propelled sprayers operated

Fig. 12.2. The electrostatic sprayer Electrodyn was used successfully in Brazil. (Photo: Graham Matthews)

at high speeds, often over 20 km h^{-1} (Figs 12.3 and 12.4). It is also common to use aircraft as an alternative and/or complement to ground applications, using low-volume or ultra-low-volume application, specially to control *A. grandis* with insecticides (Figs 12.3 and 12.4).

São Paulo State University and the Agriculture and Forestry Studies and Research Foundation (FEPAF) developed 'Periodic Inspection of Sprayers' (IPP), which indicated that the spray volume applied was reduced from an average of 118.6 l/ha^{-1} in 2008 to 79.9 l/ha^{-1} in 2016. Information collected in that programme about the agricultural sprayers showed that 40% of the sprayers had less than two years of usage, and 75% of the spray booms were 28 metres wide or wider, as they invested in new technologies. These sprayers were fitted with flat-fan or hollow-cone nozzles with many farmers now using 50 l/ha^{-1} (Carvalho *et al.*, 2017)

Spray drift is a major concern, due to the meteorological conditions during spray application. Hence the need to select spray nozzles carefully (Carvalho, 2016). There is a major concern in Brazil to instruct farmers to use nozzles with drift-reducing technologies (DRTs), such as air-induction and pre-orifice nozzles, especially for spraying systemic herbicides and when applications are next to sensitive areas, such as water supplies. With the need to adopt integrated pest management, the timing and accuracy of any application of a pesticide or biopesticide has to be more precise to minimize adverse effects on non-target species (Carvalho *et al.*, 2020).

The development of UAVs (unmanned aerial vehicles) provides a means of providing a more targeted application of the correct dose, especially by using formulations that are more persistent, thus minimizing loss of spray in areas subject to rain (Figs 12.5 and 12.6). Avoiding use of too high a dosage allows greater survival of natural enemies and reduces the selection pressure for pests becoming resistant to specific modes of action. The downward flow of air from

Fig. 12.3. Sprayer with 32-m wide boom equipped with air assistance and electrostatic spraying on cotton. (Photo: Fernando Carvalho)

Fig. 12.4. Aerial spraying with fixed-wing aircraft using (a) hydraulic nozzles and (b) rotary atomizers. (Photos: Fernando Carvalho)

a UAV should also provide better distribution and impaction of droplets within a crop canopy, reduce soil impaction caused by taking heavy loads of spray applied with tractor sprayers, and allow treatments when fields are too wet to access with ground equipment. In Asia, many smallholder farmers are using a drone in preference to using a knapsack sprayer. According to Matthews (2018), it has been shown that ULV spraying can be effective, but it needs a narrow droplet spectrum with the droplets remaining stable and not shrinking to become too small. Formulation research can reduce the volatility of the spray, hence the success of oil-based sprays. However, instead of petroleum-based oils, there is a chance to develop vegetable oil carriers with micro-sized particle suspensions to deliver low-toxicity pesticides in droplets that can be deposited within the crop and not drift beyond the crop boundary. Oil deposits will be less prone to loss after rain so less should be lost in neighbouring ditches and water.

Perspective and Research

The integrated pest management in cotton is still a challenge, mainly due to the difficulty in controlling some pest species such as boll weevil. The need to solve problems in the short term ends up hampering the solution in the long term. Therefore, it is necessary that the concept of sustainability management be considered when making a decision. Thus, the main points to work on immediately are: monitoring and management strategies for insect resistance, for Bt plants resistant varieties, or to insecticides. For the management of insect resistance to insecticides, the first step is to use them, within the technical recommendations of good agricultural practices, seeking to obtain the maximum efficiency of control of the available molecules. In this scenario, the use of the recommended doses and the appropriate application technology, are the first steps to obtaining greater efficiency of pest control in the cotton culture.

Cotton Growing in Peru

The origin of *Gossypium barbadense* was in the Ñanchoc Valley in Peru when around 5000 years ago, or earlier, the Ñanchoc people were growing cotton. So far, no earlier instances of the farming of these crops are known, although the early history of cotton indicates the presence of cotton cloth in the tombs of Egyptian pharaohs and in the Indus Valley, also about 6000 years ago. The north-west of the Andes should be considered the centre of origin of cotton classified as New World cottons (Hutchinson

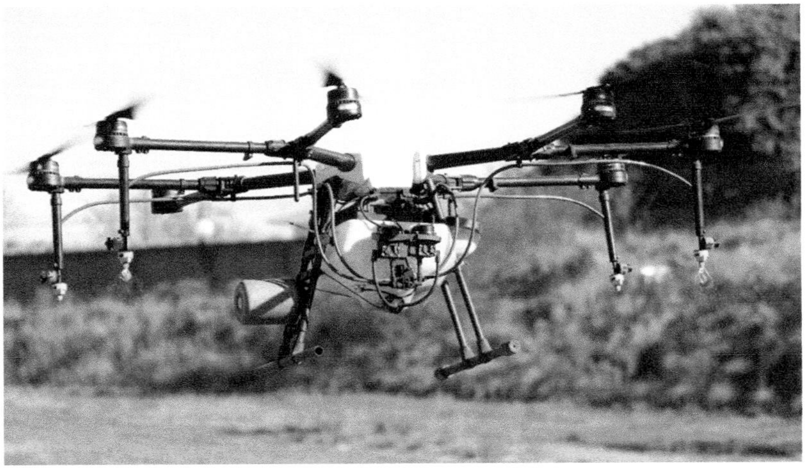

Fig. 12.5. DJI Agras MG1-S drone, with eight rotors equipped for spraying. (Photo: Fernando Carvalho)

Fig. 12.6. Cotton plants growing well. (Photo: Fernando Carvalho)

et al., 1947). The Arawak people in Venezuela were also using cotton and were thought to have spread the cottonseeds to the Caribbean.

Cultivation of *G. barbadense* in Peru continued domestically until 1830, and then until 1874, commercial production was directed towards export. Towards the end of the 19th century, native cottons were replaced by introduced varieties, and later the cultivation of the improved varieties Tangüis and Pima has since occupied an important place in the economic development of the country (Eguren Lopaz *et al.*, 1981). About 154 tons of cotton were exported to Liverpool in 1862.

Despite this early development and the use of cotton by the Incas, the cotton industry of Peru became important commercially only at the end of the 19th century. Although the Andean highlands are too cold and the eastern slopes of the mountains in general are too humid for cotton, it thrived in the sunny, arid coastal area. Exports of cotton increased from 10 million pounds in 1896 to 90 million in 1924; ranking 26% of the total exports with more than 12,500 bales consumed annually by domestic mills (Rosenfeld and Jones, 1927).

With the availability of DDT in the late 1940s, farmers in the Canete Valley began to use large quantities of dust such as 5% DDT, 3% HCH and mixtures of both with sulfur in a 3-5-40 formula and 20% toxaphene. Scientists proposed that ratoon cotton should be banned, and Tanguis cotton should be sown in rotation of crops, but farmers failed to take adequate notice of these recommendations. In consequence, by 1956, there was the worst crop despite 15 applications of insecticides, so farmers agreed to accept the recommendations made seven years earlier (Barducci, 1973).

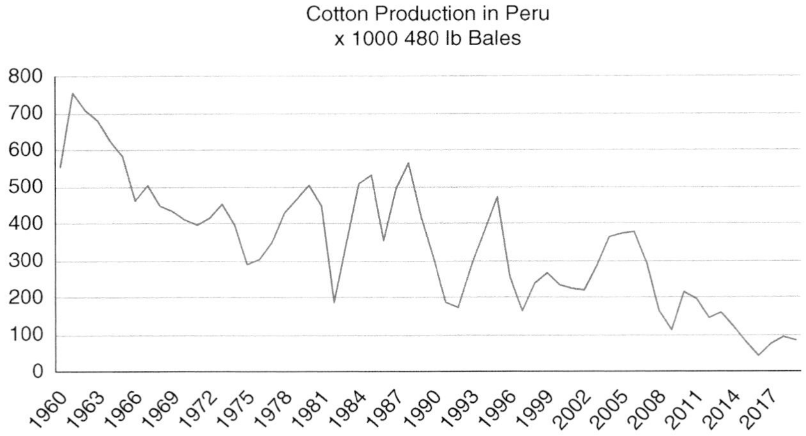

Fig. 12.7. Decline in cotton production in Peru since 1962.

In 1970, only 46,900 ha were sown with cotton, producing a volume of over 37,000 tons. By 1985, however, 385,900 ha were covered with cotton, yielding almost 159,000 tons, but dropped to 275,000 ha and 84,000 tons during the drought of 1986. Commercial large-scale production in the eastern border region began to surpass the small-scale grower production in the central region production in the late 1980s. Despite the advances in cotton production, cotton cultivation in the 1980s was still characterized by low yields and a low technological level. Production declined from a peak of 89,000 MT produced in 2006 (Fig.12.7). A major factor was the failure to select plants with higher yields and an unwillingness of famers to establish associations. A typical cotton producer had less than 5 ha, which made it very difficult and expensive to buy inputs and increase mechanization. The Peruvian government therefore initiated credit lines and technical assistance for farmers that formed an association, but such measures did not have a significant effect on the traditional cotton producer. Thread and textile dumping from India and China and better profit opportunities in other crops also played an important role in reducing Peruvian cotton output.

Cotton Growing in Paraguay

Grown since the time of the Jesuit missions, cotton was once a principal crop for Paraguay, as the Paraguayan climate and soils are suited to growing cotton primarily by small farmers in the central region. After the 35-year dictatorship under Alfredo Stroessner from 1954 to 1989, the Paraguayan economy struggled to grow. However, much of the 77% of the land in Paraguay, owned by only 1% of the population was sold to foreign companies, from Brazil and Europe, to produce soy beans, as these companies argued that their farms are the way forward for sustainable agriculture in Paraguay. Highly mechanized, these farms efficiently used the land to boost the economic value of Paraguay's agriculture industry. Soy bean production took over from cotton, but the government has tried to revive the cotton industry. Some Brazilian and Italian companies decided to invest in cotton factories in Paraguay, after the fall of the Stroessner regime and with the profit margin improved by processing and spinning the cotton in Paraguay, while creating jobs, both the companies and the Paraguayan farmers have benefitted. Nevertheless, the environmental diversity in Paraguay has been hurt by the expansion of soy bean production as forests that once covered 85% of eastern Paraguay now cover less than 8%. The arrival of the boll weevil also made a major impact, increasing the problems of crop protection, as the toxins due to the Bt gene are not effective against it.

More recently, this has involved the PIEA (Cotton Research and Experimentation Program) cotton-improvement programme in collaboration

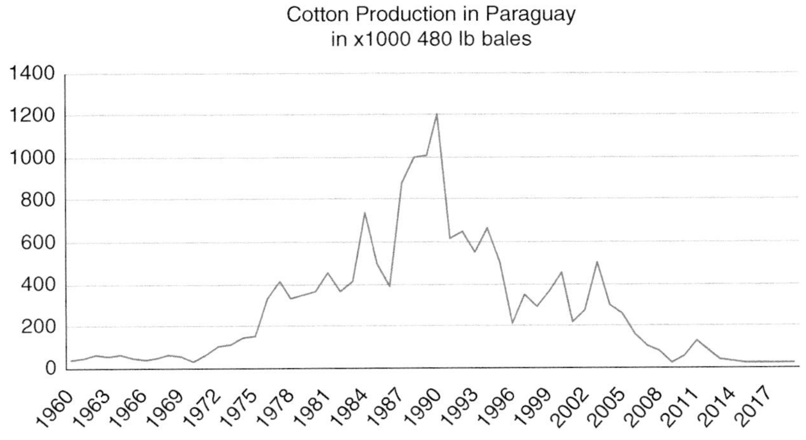

Fig. 12.8. Cotton production in Paraguay 1960–2018.

with the French research organization CIRAD, particularly to examine and understand the similarities and differences in agronomic behaviour of selected varieties in different environments in order to guide the breeding programmes towards varieties better adapted to the prevailing environmental conditions and cultural practices. The fibre analysis laboratory of AGUSA (Algodonera Guarani SA, group) with Ecom Agroindustrial Corporation Ltd, Switzerland, analysed 80 samples for PIEA in 2009. Paraguay then doubled the area sown to cotton in the 2010/11 season to 30,000 ha but production has remained very low (Fig. 12.8).

Cotton Growing in Argentina

Cotton production began more than 400 years ago in lands which are now part of Argentina. The Spanish missionaries introduced cotton in the province of Santiago del Estero in 1556 and it quickly spread to the neighbouring provinces of Tucumán and Catamarca. The product was the first ever export that left the port of Buenos Aires, in 1587, and was a batch of cotton fabric manufactured on looms in the north of the country. After an extensive period of growth, growing cotton declined and disappeared almost entirely around 1860, despite British attempts to promote production, due to the war in the USA, in 1861, which encouraged the English industry to seek new fibre-supplying markets. In 1862, the first great official cotton-promotion campaign began in Argentina, which included the importation of seeds and two saw gins, but managed by hand. Once the war in the US ended, Americans soon became competitive again.

Cotton did not return as a cash crop in Argentina until about 1890 and in 1904 the Ministry of Agriculture of Argentina carried out another campaign to increase cotton planting, providing seeds from North America and training producers. After the First World War, cotton expanded rapidly as a cash crop with the appearance of the first mechanical ginning.

A historic production record was achieved in the 1997/98 season, when a maximum area of 1,133,500 ha was planted, but subsequently a period of decline began, reaching 160,000 ha in the 2002/03 season (Fig. 12.9). The decline was due to several reasons, including unfavourable prices, competition with soybean and adverse environmental conditions (Paytas and Ploschuk, 2013).

In Argentina, cotton growing is concentrated now in the northern provinces, predominantly in the state of Chaco, and covers an estimated 500,000 ha. Planting begins in early October and continues through to the end of the year. The harvest can begin in mid-February and continue to mid-July, depending on location. The introduction of genetically modified cotton has enabled most insects to be controlled, except plant bugs, notably cotton stainers (*Dysdercus* spp.) and the red-shouldered bug (*Jadera* spp.)

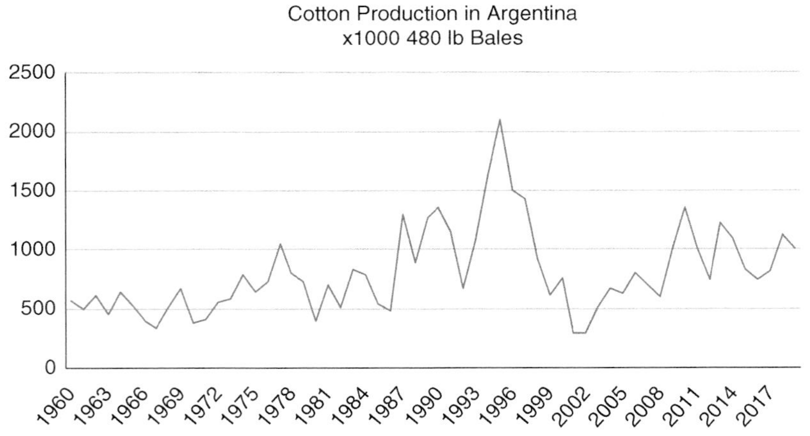

Fig. 12.9. Cotton Production in Argentina 1960–2018.

and various sucking pests, especially aphids (*Aphis gossypii*) and thrips (*Frankliniella* spp.). The boll weevil, *Anthonomus grandis*, arrived in Argentina in 1994, when boll weevils were captured in cotton fields in Formosa province on the border between Argentina and Paraguay. The pest has subsequently moved to new areas, and in 2006 it was reported in Argentina's main cotton-growing region (Grilli *et al.*, 2012). However, cotton farming in Argentina is currently living a revival, with some cotton sown at higher plant densities with a narrower row spacing and minimum tillage followed by using a stripper harvester.

GM Bt cotton technology has significantly reduced insecticide applications and increased yields, but these advantages are curbed by the high price charged for genetically modified seeds (Qaim and De Janvry, 2005). The cost of Bt seeds is significantly higher than conventional varieties. Gross benefits could be highest for smallholder farmers who are not currently using the technology. Rapid-resistance build-up and associated pest outbreaks appear to be less likely if minimum non-Bt refuge areas are maintained.

Many of 19,000 cotton growers in Argentina are small, family-based operations as they enable employment at ginneries and support the local textile industry, which accounts for more than 10% of Argentina's total industrial employment. The main current and future challenges for the Argentine cotton sector are: (i) the development of varieties well adapted to different production regions; (ii) the reduction of the negative impact of boll weevil and other pests that increase production costs; and (iii) the implementation of support, training and policies for the sector. This will require greater collaboration between the different organizations involved in the cotton industry.

Cotton Growing in Bolivia

Commercial production of cotton in Bolivia began in 1950 and 52,000 ha were cultivated in the 1990s, but production now does not exceed 4000 ha with 150 farmers in the Department of Santa Cruz, as the majority now mainly grow crops, such as soybean and corn (Fig. 12.10). Currently, the socioeconomic impact of cotton is minimal and practically does not appear in the country's economic indices (Carvalho and Aguirre, 2017).

Cotton is now suitable for family farmers in areas with the lowest rainfall and is profitable for famers with 2 ha. Conventional and transgenic varieties from Argentina, the USA and Brazil are sown. Bolivia is facing challenges to activate the cotton sector mainly related to protective policies for small and medium-sized farmers.

Cotton Growing in Venezuela

The Arawak people inhabited the northern and western areas of the Amazon basin, where they

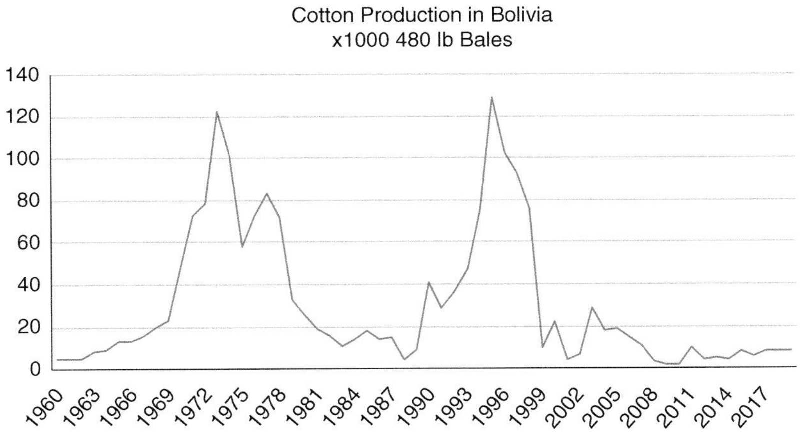

Fig. 12.10. Cotton production in Bolivia.

shared the means of livelihood and social organization of other tribes of the tropical forest. They were sedentary farmers who hunted and fished, lived in small settlements, but spread to the Caribbean, where Arawakan speakers were also historically known as the Taíno. Although Arawak people were isolated from the Andean civilizations, cottonseeds, presumably from Peru, were grown by them and they spread cotton growing to the Caribbean.

In more recent times, cotton in Venezuela has been mainly in the flood zones of the Orinoco river and its branches, specifically in the south of the Guárico state (Navarro *et al.*, 2010). With the development of varieties with uniform growth and the government's effort to provide the sector with machinery, the cotton-planting system has been mechanized, dating back to 1949 (León Díaz, 1980).

In the 1990s, imports of raw materials and fabrics from both China and India affected the domestic market, such that in 2010, only 10,446 ha were harvested and most of the spinning mills closed (Fig. 12.11).

The main challenge for Venezuela is to consider policies that promote the competitiveness of cultivation and its domestic processing in parallel with the incorporation of technologies.

Cotton Production in Colombia

No doubt, some cotton had been grown on a small scale in Colombia before the first settlers arrived on the Caribbean coast and on the banks of the Magdalena river. These settlers began the commercial production of cotton and the first exports were made between 1834 and 1891 from the regions of the Magdalena, Santander and Bolívar department. The cultivation of *G. barbadense* cottons was important and the export of fibre to Europe reached its maximum level between 1858 and 1873. Then the farmers changed to varieties of *G. hirsutum*.

Between 1950 and 1977, the production of raw cotton increased almost 25-fold, due to its profitability and because its market was outside Colombia. The great increase in cotton production occurred when the Peso was devalued and made the export of fibre profitable in 1977; 380,000 ha were planted, more than 75% of the planted area on the Atlantic coast and the rest in the 'interior' of the country, providing employment for more than 480,000.

However, between 1978 and 1992, there was a strong crisis in the cotton sector, generated by inaccurate government policies, the explosion of pests in the crop due to agronomic mismanagement in the use of pesticides, the decrease of international prices, the reduction of taxes for imports, and the increase in the costs of pesticides and agricultural machinery, among others (García, 1995).

The area sown to cotton decreased from 260,000 ha in 1992 to only 50,000 ha in 1999. This was due to the importation of subsidized

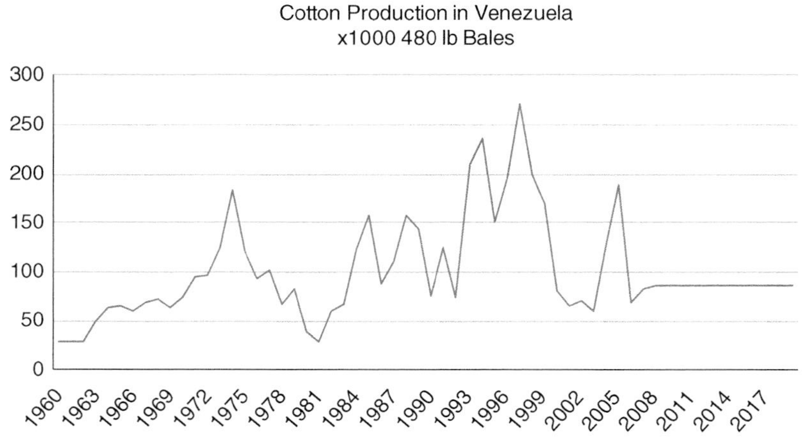

Fig. 12.11. Cotton production in Venezuela.

fibre from the USA in 1993; so that, currently, more than 60% of national consumption is imported cotton. By 2007, only 54,497 ha were sown and in 2008 the cotton area barely exceeded 15,000 ha (Espinal *et al.*, 2005) and has continued to decrease until today (Fig.12.12).

Cotton cultivation in Colombia is one of the most expensive in the world. Add to this the low availability of irrigation districts, lack of soil preparation, lack of good agricultural practices, obsolete machinery, lack of harvesting equipment, and insufficient companies offering seed in the market (Conalgodon and Ruiz, 2015).

The Ministry of Agriculture authorized the commercial planting of Bt cotton since 2002 and the area cultivated with transgenic cotton has increased in recent years, but not at the rate expected by organizations in the sector. GM cottonseed costs more than three times as much as conventional cotton. In addition, for transgenic technology to work, farmers have to incur new expenses that increase production costs, such as: implementation of efficient irrigation systems, precision sowing machines, adequate soil fertilization and comprehensive management of the crop (pests and refuges), among others, which prevent small and medium-sized cotton farmers adopting this new technological package.

The main cotton pest is the boll weevil (*Anthonomus grandis*), which is not controlled by Bt, and for its control farmers apply about 70% of the total pesticides of the season, as they have to use up to six insecticide applications to control it. In Tolima, the most important pests are *Spodoptera* spp., and pink bollworm *Pectinophora gossypiella*, for which farmers have to apply additional pesticides for their control. In some areas of Tolima, a resurgence of the whitefly reached critical injury levels with increased costs for control.

Bt technology requires the use of refuges, areas with a variety of non-transgenic cotton within the transgenic crop to decrease the probability that pests will not become resistant to Bt toxins. However, many farmers are not managing refuges effectively.

Despite the significant reduction in the area sown with cotton, the industry continues to be an important route for the country's agricultural economy. According to the Colombian Cotton Confederation, by 2020, a planting area of more than 20,000 ha is expected throughout the country. To recover its competitiveness, the industry needs to invest in technology and high-quality improved seeds adapted to local conditions.

Among the current and future challenges of the Colombian cotton sector are: (i) adjust production costs and variable yield between production regions; (ii) implement technologies related to irrigation, mechanization and seeds adapted to each productive region; (iii) improve the competitiveness of the national cotton sector through sectorial policies to mitigate the negative impacts related to prices; and (iv) improve the organization of the national cotton chain.

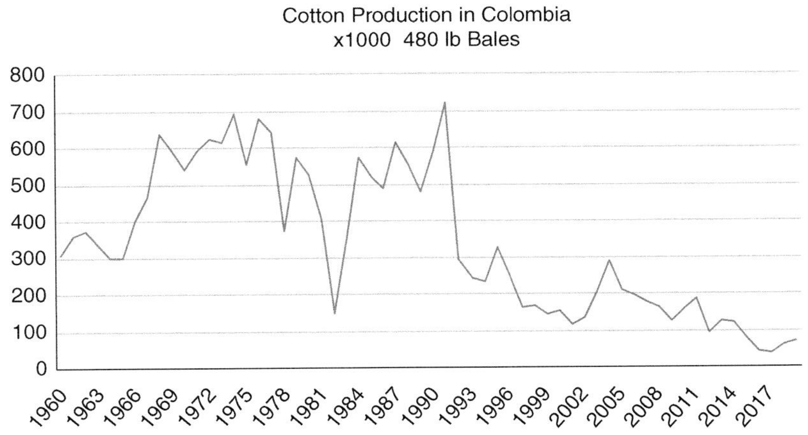

Fig. 12.12. Decline in cotton production since 1993 in Colombia.

Cotton Growing in Ecuador

Historically, small-scale cotton growing probably occurred deeply rooted in society over the centuries, but growing the crop more recently was initially for the purposes of extracting oil from the seed. It then played a major role in the agricultural sector, mainly from the late 1960s to the 1990s. Cotton is cultivated on land in mountainous areas with a dry to semi-humid climate from January to harvest between June and August (Carvalho and Aguirre, 2017). Most cotton is on small family farms.

Due to the permanent deficit of national production (Fig. 12.13), the textile industry is supported by imports, which registered a growth rate of approximately 9% between 1990 and 2012 and a rapid increase in imports of clothing of all different textile fibres.

Cotton Growing in the Caribbean

The first cottonseeds to reach the Caribbean were taken there over 2000 years ago by the Arawaks from South America and were derived from the wild cotton that initiated using the cotton fibres and making cloth as early as 6000 BC in Peru. Cotton cloth was seen in the Caribbean islands when Christopher Colombus arrived at the end of the 15th century and undoubtedly took seeds back to Europe, but soon the Caribbean was invaded by other European nations, with the British establishing colonies in Barbados in 1625 and in Jamaica in 1655. Sea Island cotton with a long fibre then became a commodity with demand by the British textile industry increasing into the 18th century. Most farmers in the Caribbean were growing cotton and tobacco in the 17th century, but large landowners decided to switch to growing sugarcane and purchased slaves to provide the labour for this work.

As the UK relied on the supply of cotton from the USA, as soon as there was a disruption in supply there was an immediate search for new sources of cotton. The British used mainly Barbados and Jamaica as one source of cotton in the 18th and 19th centuries, but documentation of early cotton growing in the Caribbean is scarce. According to Jaquay (1997), the amount exported to the UK rose from very little in 1800 to 2 million pounds of fibres by 1919. Cotton growing was attempted on Jamaica in 1632. St Lucia was cultivating cotton by 1651 and in Antigua, as it was mentioned in a petition in 1656. However, Jamaica cotton '...here hath an especial fineness, and is by all preferred before that of the Caribbee Isles' (Beer, 1959). It must be remembered that by this time, England had secured the bulk of the Caribbean trade.

With an increasing proportion of cotton grown in the USA being used by its own textile companies, there was less available for Lancashire mills, and other sources were not of sufficient quality for the Lancashire mills, which

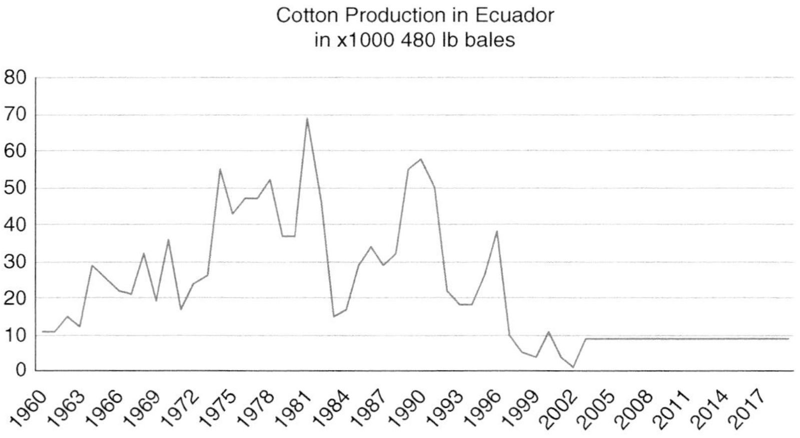

Fig. 12.13. Cotton Production in Ecuador.

needed the longer staples, so in 1902, Sea Island cotton was reintroduced into the West Indies. A thousand bales were produced in 1903 and by 1907 this had increased to 6000 bales. The West Indian planters soon realized that some of the islands were more suited to lower grades of cotton, but with the islands limited in size and population, the opportunities for cotton production were naturally limited. Nevertheless, within a few years, the West Indies were producing sufficient Sea Island cotton to meet British demand.

In the 1920s, an internationally registered trade name was established – West Indian Sea Island cotton – that represents cotton of a particular lineage and quality, grown in a specific geographic location, as required by the Trade Related Intellectual Property Rights Agreement (TRIPS) of the World Trade Organization. Compliance with strict quality standards, climatic conditions and cultivation practices are critical, in addition to the genetic material, all of which prevail in Jamaica. The trademark was presumably to protect the special high-grade West Indian Sea Island Cotton (WISIC), which is long, strong and silky, making it more expensive than the next-best grade cotton available. This species of historic significance, *Gossypium barbadense*, is the source from where all other plants in the Eastern Islands of the Caribbean are sown from. Of the different islands, Barbados had an advantage as it benefitted from annual rainfall, bright sunshine and trade winds that bring to life this astonishingly soft and rare cotton, a cotton that carries a consistency of purity and softness quite unlike any other in the world.

The West Indies Sea Island Cotton Association (WISICA) was formed in 1933 in Trinidad and Tobago to safeguard the interests of West Indian cotton producers and to regulate pricing and production. There have been several technical, financial and legal challenges to the Association. In Barbados, new initiatives considered to establish a viable reconfigured cotton industry indicated that there was a need to incorporate value-added, maintain lint quality and achieve yields that sustain profitability. Thus, the acreage planted under cotton needed to be increased, supported by a cotton-breeding programme to improve link quality with the appropriate machinery to create a state-of-the-art industry (Bell, 2004).

Sea Island cotton (*Gossypium barbadense L.*) is attacked by various phytophagous insects which reduce productivity and quality of the final product. Primary and secondary pests attacking the cotton crop have been identified in Barbados, including the pink bollworm, *Pectinophora gossypiella* (Lepidoptera: Gelechiidae), *Helicoverpa zea* (Lepidotera: Noctuidae), *Heliothis virescens* (Lepidotera: Noctuidae), *Spodoptera* spp. (Lepidotera: Noctuidae), *Alabama argillacea* (Lepidoptera: Noctuidae), *Scirtothrips dorsalis* (Thysanoptera: Thripidae), *Thrips palmi* (Thysanoptera: Thripidae) and *Bemisia tabaci* (Hemiptera: Aleyrodidae) (Bell, 2004).

Of these, the pink bollworm, *P. gossypiella*, and other bollworms, *H. zea* and *H. virescens*, are

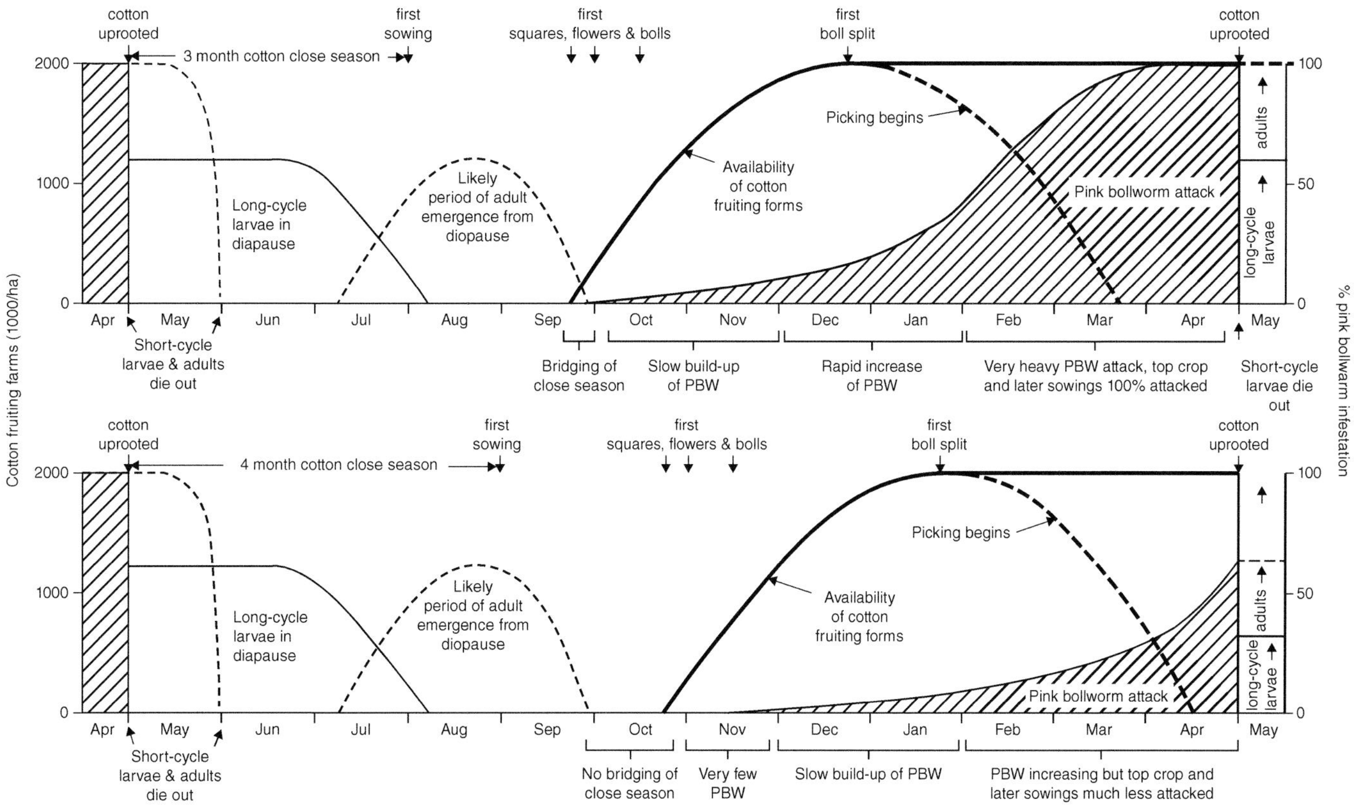

Fig. 12.14. Diagram illustrating the effect of a three- or four-month close season on the survival and infestation of pink bollworm in the Caribbean. (Figure: Ingram, 1980)

responsible for high yield losses in cotton in Barbados. More than 50% crop loss was caused by *P. gossypiella* in some years (Thrakika, 2007). Until the accidental introduction of pink bollworm in the 1920s, falling prices and rising costs led to the collapse of the industry. Falling sugar prices subsequently led to a revival of interest in the crop but pink bollworm still presented a problem and warranted investigation (Ingram, 1980).

Moth catches using pheromone traps, with gossyplure, showed that moths were caught throughout the close season. However, numbers in traps were correlated with boll damage, so catches of 8–9 moths in a night represented a 10% level of boll damage some ten days later. This trapping level was used to advise cotton growers when to begin insecticidal treatment.

Sea Island cotton is grown in Barbados on a relatively small scale, with smallholders having small plots (0.1 ha compared to estates with 280 ha under cotton and individual fields ranging from 3–20 ha). Smallholders do not spray and very few, if any, are currently growing cotton. Some estates spray with tractor-mounted equipment when they can get onto the land and before the crop has grown too tall, but most of the fields are sprayed with hand-held, battery-powered, spinning-disc ULV sprayers. This method appears to be the only way possible in the very tall (>1.75 m) rank growth commonly found in Sea Island cotton fields by the time they are about 12 weeks old in Barbados. Carbaryl, which could be mixed with molasses or a pyrethroid, was recommended (Ingram, 1980). A parasite, *Perisierola nigrifemur*, was detected on pink bollworm larvae towards the end of the season but this was too late to be of economic value. Diapausing larvae were also attacked by the mite *Pyemotes ventricosus*. After harvesting, all plants were slashed and the debris stacked and burnt and the stumps ploughed in.

Ingram (1980) also pointed out that the cotton close season in Barbados should start as early as possible as pink bollworm damage is liable to be very severe in the top crop and late sowings. The starting date can be varied, but should be no later than mid-May and run for at least three months, and planting should not be allowed until good rains have fallen to help break the larval diapause. Ideally, a four-month close season would ensure that no cotton was planted until all diapausing insects had emerged (Fig. 12.14).

More recently, in 2003/04 crops in the Caribbean, Barbados was the principal producer that cultivated 90 ha (Fig. 12.15), while Jamaica grew 80 ha. Antigua is also attempting to resuscitate its industry. The resuscitation of the West Indian Sea Island cotton industry has been assisted by Japanese interests investing just under US$ 107,000 to it.

Colmenárez *et al.* (2016) monitored 17 farms planted with West Indian Sea Island cotton weekly during two cotton seasons (September 2009–February 2010 and September 2010–February 2011) to determine occurrence of natural

Fig. 12.15. *G. barbadense* in West Indies. (Photo: Graham Matthews)

enemies associated with cotton pests in Barbados. They concluded that the natural enemies identified needed further study to determine their efficacy in order to use this information for establishing an integrated pest management programme in cotton. The implementation of a proper IPM programme, using biocontrol as a strategy, can help to reduce the indiscriminate use of pesticides in the island and reach, in this way, a sustainable production of cotton.

Two other islands in the Caribbean were known for growing cotton, namely Puerto Rico and Haiti, both of which have suffered the consequences of being in the Caribbean hurricane belt. Both countries share a common history of cotton production that stopped in the late 1980s in Haiti and the early 1940s in Puerto Rico. Both nations are now trying to rebuild their agricultural sectors to reduce dependence on imports.

References

Barducci, T.B. (1973) Ecological consequences of pesticides used for the control of cotton insects in Canete Valley, Peru. In: Farvar, M.T. and Milton, J.P. (eds) *The Careless Technology*. Stacey, London, pp. 423–438.

Barros, E.M., Rodrigues, A.R.S., Batista, F.C., Machado, A.V.A. and Torres, J.B. (2019) Susceptibility of boll weevil to ready-to-use insecticide mixtures. *Archives of the Biological Institute* 86, 1–9.

Barroso, P.A.V. and Hoffmann, L.V. (2007) Genetically modified bakers. In: Freire, E.C. (ed.) *Cotton in the Cerrado of Brazil. ABRAPA*, Brasilia, pp. 141–173.

Beer, G.L. (1959) *The Origins of the British Colonial System 1578–1660*. Peter Smith, Gloucester, Massachusetts. [Quoted by Jaquay, 1997].

Bell, P. (2004) *A Guide to Cotton Growing*. Barbados Ministry of Agriculture and Rural Development.

Belot, J.L., Barros, E.M. and Miranda, J.E. (2016) Risks and opportunities: the umdo-do-umdo. In: Ministry of Agriculture Livestock and Supply (ed.) *Cerrado Challenges*. MAPA, Cuiabá, pp. 77–118.

Bianchini, A. and Borges, P.H. de M. (2013) Evaluation of cotton stalks destroyers. *Agricultural Engineering* 34, 965–975.

Bleicher E. and Almeida T.H.M. (1991a) Horizontal dispersion of the cotton boll weevil, *Anthonomus grandis* (Coleoptera, Curculionidae). *Anais da Sociedade Entomologica do Brasil*, 20(1), 75–80.

Bleicher, E. and Almeida, T.H.M. (1991b) Spray systems for the cotton boll weevil, *Anthonomus grandis* (Coleoptera, Curculionidae). *Anais da Sociedade Entomologica do Brasil* 20(1), 81–87.

Carvalho, F.K. (2016) Viscosity, surface tension and droplets size on spray solutions with formulations of insecticides and fungicides. Universidade Estadual Paulista, Faculdade de Ciências Agronômicas, Botucatu.

Carvalho, F.K., Chechetto, R.G., Mota, A.A.B. and Antuniassi, U.R. (2017) Characteristics and challenges of pesticide spray applications in Mato Grosso Brazil. *Outlooks on Pest Management* 31, 4–6.

Carvalho, F.K., Chechetto, R.G., Mota, A.A.B. and Antuniassi, U.R. (2020) Challenges of aircraft and drone spray applications. *Outlooks on Pest Management* 31, 83–88.

Carvalho, J. and Aguirre, G. (2017) El estado de arte del sector algodonero en países del Mercosur y asociados.

Colmenárez, Y., Gibbs, I.H., Ciomperlik, M. and Vásquez, C. (2016) Biological control agents of cotton pests in Barbados. *Entomotropica* 31, 146–154.

Conalgodon and Ruíz, L. (2015) Algodón: Revista de la Situación Mundial Comité Consultivo Internacional del Algodón. 68 – No. 3 – Enero-Febrero.

Degrande, P.E. (1991) Bicudo-do-algodoeiro: manejo integrado. Dourados, UFMS/Embrapa-Uepae, Dourados.

Degrande, P.E. (1998) Guia prático de controle das pragas do algodoeiro. UFMS, Dourados.

Degrande, P.E. (2006) Ameaça do bicudo exige organização e empenho de todos. *Visão Agrícola* 6, 55–58.

Espinal, C., Martínez, H., Pinzón, N. and Barrios, C. (2005) La cadena de algodón en Colombia: Una mirada global de su estructura y dinámica, 1991–2005. Ministerio de Agricultura y Desarrollo Rural – *MINAGRICULTURA. Documento de trabajo* No. 53. Colombia.

Fernandes, F.S., Ramalho, F.S., Nascimento Junior, J.L., Malaquias, J.B., Nascimento, A.R.B., Silva, C.A.D. and Zanuncio, J.C. (2012) Within-plant distribution of cotton aphids, *Aphis gossypii* Glover (Hemiptera: Aphididae), in Bt and non-Bt cotton fields. *Bulletin of Entomological Research* 102, 79–87.

Gallo, D., Nakano, O., Neto, S.S. *et al.* (2002) *Entomologia Agrícola*. FEALQ, Piracicaba, Brazil.

García, J. (1995) El cultivo de algodón en Colombia entre 1953 y 1978: una evaluación de las políticas gubernamentales. Conalgodon, Colombia.

Grigolli, J.F.J., Crossariol Netto, J., Izeppi, T.S., Souza, L.A. de, Fraga, D.F. and Busoli, A.C. (2015) *Anthonomus grandis* infestation (Coleoptera: Curculionidae) in birch regrowth. *Tropical Agricultural Research* 45, 200–208.

Grigolli, J.F.J., Souza, L.A., Mota, T.A., Fernandes, M.G. and Busoli, A.C. (2017) Sequential sampling plan of *Anthonomus grandis* (Coleoptera: Curculionidae) in cotton plants. *Journal of Economic Entomology* 110, 763–769.

Grilli, M.P., Bruno, M.A., Pedemonte, M.L. and Showler, A.T. (2012) Boll weevil invasion process in Argentina. *Journal of Pest Science* 85, 47–54.

Hutchinson, J.B., Silow, R.A. and Stephens, S.G. (1947) The evolution of *Gossypium*. Oxford University Press, London.

Ingram, W.R. (1980) Studies of the pink bollworm, *Pectinophora gossypiella*, on Sea Island Cotton in Barbados. *Tropical Pest Management* 26, 118–136.

Jaquay, B.G. (1997) The Caribbean cotton production: an historical geography of the region's mystery crop. Dissertation. Texas A&M University.

Leite, N.A., Pereira, R.M., Durigan, M.R., Beloved, D., Fatoretto, J., Medeiros, F.C.L. and Omoto, C. (2018) Susceptibility of Brazilian populations of *Helicoverpa armigera* and *Helicoverpa zea* (Lepidoptera: Noctuidae) to Vip3Aa20. *Journal of Economic Entomology* 111, 399–404.

León Díaz, J. (1980) Repercusión de la tecnología en el desarrollo de los principales rubros de producción en Venezuela: Sexto Caso Algodón. Maracay, VE, Centro Nacional de Investigaciones Agropecuarias (CENIAP). Oficina de Análisis de Proyectos. Venezuela.

Lima Junior, L.S., Degrande, P.E., Miranda, J.E. and Santos, W.J. (2013) Evaluation of the boll weevil *Anthonomus grandis* Boheman (Coleoptera: Curculionidae) suppression program in the state of Goiás, Brazil. *Neotropical Entomology* 42, 82–88.

Lu, P., Davis, R.F. and Kemerait, R.C. (2010) Effect of mowing cotton stalks and preventing plant re-growth on post-harvest reproduction of *Meloidogyne incognita*. *Journal of Nematology* 42, 96–100.

Matthews, G.A. (2018) Getting the right droplet size: requirement for food production. *Aspects of Applied Biology* 137, 229–235.

Miranda, J.E., Bettini, P.C. and Gusmao, L.C.A. (2008) Deposição de gotas por pulverizações terrestre e aérea na cultura do algodoeiro. Campina Grande, Embrapa Algodão, (*Embrapa Algodão, Comunicado Técnico, 350*).

Navarro, R., Gutiérrez, M., Alfonzo, N. and Piñango, L. (2010) Cultivo del algodón en las zonas de vega del río Orinoco y sus afluentes. Maracay, Venezuela. Instituto Nacional de Investigaciones Agrícolas. Venezuela.

Paytas, M. and Ploschuk, E. (2013) Capitulo Algodón. In: De La Fuente, E., Gil, A. and Kantolil, A. (eds) *Cultivos Industriales*. Editorial Facultad de Agronomía Universidad de Buenos Aires.

Qaim, M and De Janvry, A. (2005) Bt cotton and pesticide use in Argentina: Economic and environmental effects. *Environmental and Development Economics* 10, 179–200.

Ramalho, F.S. (1994) Cotton Pest Management: Part 4. A Brazilian Perspective. *Annual Review of Entomology* 39, 563–578.

Rosenfeld, A.H. and Jones, C.F. (1927) The cotton industry of Peru. *Economic Geography* 3, 507–523.

Rummel, D.R., Jordan, L.B., White, J.R. and Wade, L.J. (1977) Seasonal variation in the height of boll weevil flight. *Environmental Entomology* 6, 674–678.

Santos, W.J. (2007) Management of cotton pests with emphasis on the Brazilian cerrado. In: Freire, E.C. (ed.) *Cotton in the Cerrado of Brazil.* ABRAPA, Brasilia, Chapter 12, pp. 403–478.

Showler, A.T. (2012) The conundrum of chemical boll weevil control in subtropical regions. In: Parveen, F. (ed.) *Insecticides – Pest Engineering*. Tech Europe, Croatia, pp. 437–448. Available at: https://doi.org/10.5772/27981 (accessed 11 August 2021).

Silva, A.G.A., Goncalves, C.R., Galeo, D.M., Goncalves, A.J.L., Gomes, J., Silva, M.N. and Simoni, L. (1968) *Quarto Catálogo dos Insetos que Vivem nas Plantas do Brasil: Seus Parasitos e Predadores, Parte II, tomo 1o, Insetos, hospedeiros e inimigos naturais*. Ministério da Agricultura, Rio de Janeiro, Brasil.

Silva, C.A.B., da Ramalho, F.S., Miranda, J.E., Almeida R.P., Rodrigues, S.M.M. and Alburqueque, F.A. (2013) Technical suggestions for integrated pest management. Brazilian Agricultural Research Company Ministry of Agriculture, Livestock and Supply, EMBRAPA Cotton. Available at: https://ainfo.cnptia.embrapa.br/digital/bitstream/item/101745/1/Folder-2013.pdf (accessed 11 August 2021).

Silva, R.A. Da, Degrande, P.E., Gomes, C.E.C., Souza, E.P. De and Leal, M.F. (2018) Phytophagous insects in cotton crop residues during the fallow period in the state of Mato Grosso do Sul, Brazil. *Pesquisa Agropecuária Brasielira, Brasília*, 53, 875–884.

Smith, J.W. (1998) Boll weevil eradication: area-wide pest management. *Annals of the Entomological Society of America* 91, 239–247.

Stadler, T. and Buteler, M. (2007) Migration and dispersal of *Anthonomus grandis* (Coleoptera: Curculionidae) in South America. *Revista de la Sociedad Entomological Argentina* 66, 205–217.

Thrakika (2007) Barbados still struggling to make cotton farming viable. *Thrakika Ekkonistiria S.A.* August, 2015. Available at: http://thrakika.gr/en/ news/detail.php?NEWS_ID=6519 (accessed 11 August 2021).

13 Cotton Growing Around the Mediterranean

Turkey

Feza Can*, Cafer Mart, Berkant Ödemiş and Yaşar Akişcan

Cotton is a product of strategic importance for Turkey because it has an 18.1 billion-dollar value for textile products outfit exports (14.4% of Turkey's total exports) (Anon, 2020a). According to ICAC, Turkey's global cotton ranking in the 2019/20 season was eleventh for plantation area, fifth for yield, sixth for total amount of production, fourth for consumption and fifth for import (Anon, 2020b).

However, cotton-planting areas in Turkey are shrinking due to various reasons. Most producers are shifting to alternative products such as corn due to high input costs resulting from agricultural struggles, and labour-related problems. High yields of cotton are obtained by farmers, but Turkey has to import cotton almost at the same level as the amount produced yearly due to the fact that the quantity produced locally is not sufficient to meet the demands of the nation's textile industries.

Figure 13.1 shows Turkey as one of the top five countries in terms of cotton consumption (Anon, 2020b).

The pattern of cotton production in Turkey over the last 30 years with regards to the cultivation area, production amount and yields per unit area according to Turkish Statistical Institute is shown in Fig. 13.2 (Anon, 2020c).

Cotton cultivation is mostly conducted in the south-eastern Anatolia region, Aegean region, Çukurova and Antalya. There has been an increase of 41.6% in cotton cultivation areas across the south-eastern Anatolia region in the last 25 years, while significant decrease has occurred in cotton-cultivation areas in the other cotton-growing regions according to Turkish Statistical Institute data. The reduction in cotton-cultivation areas in the Aegean and Çukurova regions, which were the most important cotton regions in the past, is 65%. While producers who have moved away from cotton farming are switching to corn, fruit, onions, watermelon and soybean agriculture in the Çukurova Region, in the Aegean, it tends towards corn, tomato, fruit and vegetables.

The area of cotton grown in Turkey has decreased over the years but the amount of production has increased due to increased efficiency. Turkey's cotton-production fields for the 2020/21 season are expected to be about 420,000 ha. This value is the same as that of the 2009/10 season, which was the lowest in the production history. However, the increase in cotton yield in Turkey has reached 82.8% in the last 30 years (Fig. 13.2).

*Corresponding author: fezacan@mku.edu.tr

DOI: 10.1079/9781800620216.0013

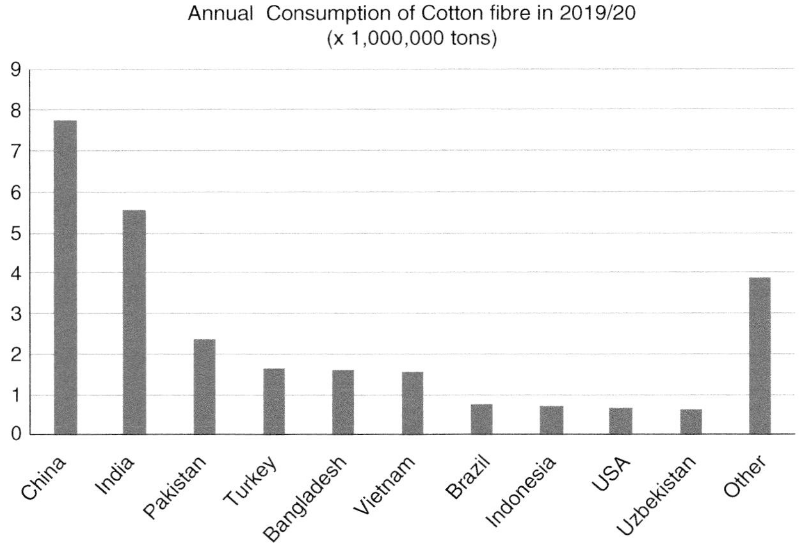

Fig. 13.1. Comparison of fibre consumption by textile industry in ten countries.

The transition to the use of certified delinted seeds and the significant increase in their use are the result of the additional support given to their use, as well as the introduction of high-yielding cotton varieties suitable for the requirements of modern cotton agriculture, developments in cultivation techniques and the transition to mechanical harvesting. Within the scope of the Five-Year Development Plans, adequate production levels and ensuring less impact of adverse conditions of production, the subsidies to cotton production were for the purpose of raising the self-sufficiency rate (Fig. 13.3) (Anon, 2012, 2020b). The first support was implemented in the 1993/94 season and was planned to be made directly to production with applications that lasted until 2001.

Producers that used certified seeds got more than 10% of the determined pay in the 2001/02 season. The pay difference that was paid to the producers using certified seeds was increased to 20% in the 2004/05 season. In parallel with this practice, the rate of certified 'delinted' seed usage, which was extremely low in the 1990s, approached 90% in recent years (Fig. 13.3).

As of the 2012/13 season, the support given to cotton producers has been completely attributed to the use of certified quality seeds free from disease, which increased yields; and the support given to certified seed and its utilization rate resulted in the increase in cotton production in Turkey and has contributed to an increase in the product per unit area.

Types of Cultivated Cotton

Cotton varieties developed by research institutes of government were widely used in Turkey until a certain period. The balance between the public and private sector in developing varieties has changed with the release of seed imports with the legal regulations made in the field of seeds in 1984, the enabling of the private sector for seed growing activities and the support of the private sector by the state. Until the mid-1990s, up to 90% of cotton varieties developed in public research institutes were used with different breeding methods, while cotton varieties registered by the private sector have become widespread as of this date with the effect of the stated legal regulations. A significant portion of the registered cotton varieties in the market of Turkey (approximately 65%) have been developed in foreign countries and consist of cotton varieties registered as the result of adaptation attempts by the private sector; about one-third is made up of local varieties developed by institutes, private firms and public research.

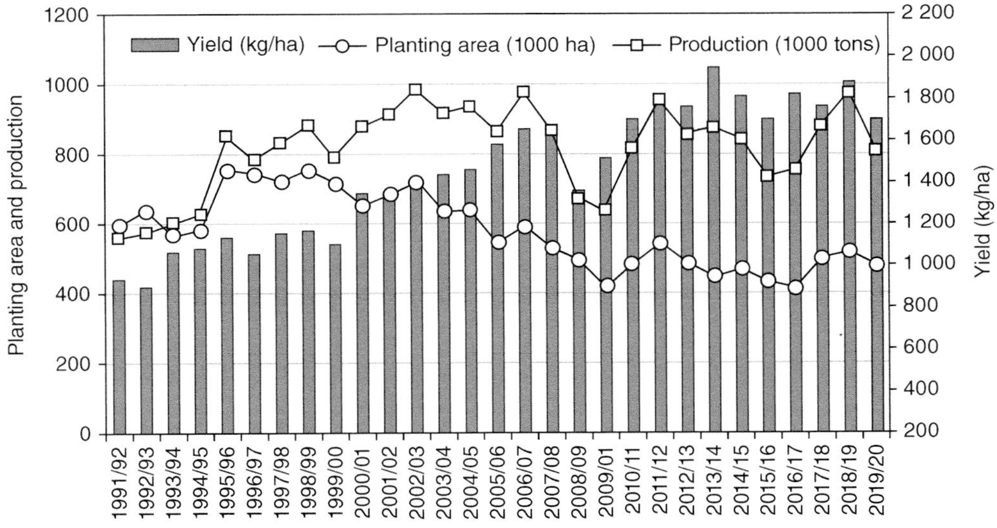

Fig. 13.2. Cotton-planting area, production values and yearly yields in Turkey.

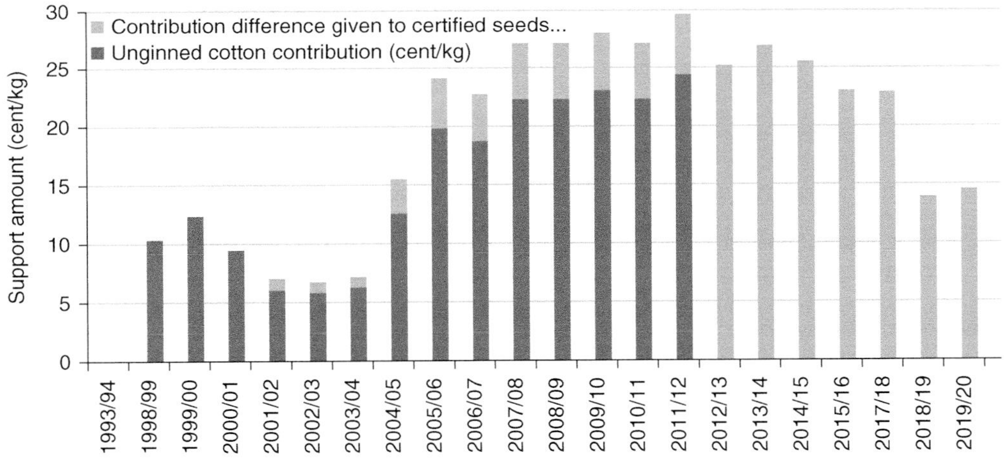

Fig. 13.3. Rate of contributions in cotton production and contribution difference given to certified delinted seed (cent/kg).

Turkey is exporting cottonseeds to certain countries every year. Almost all cottonseeds exported are from local varieties. The number of registered cotton varieties available in the National Variety List is 106 (Anon, 2020d). However, the number of registered cotton varieties with a market share of 2% and above changes within 10–14 years. Cotton varieties that are cultivated in Turkey are members of *Gossypium hirsutum* L. sp. with an average fibre length of 29.1 mm, fibre tensile strength averaged 30.9 g/tex, fibre fineness (micronaire) has an average 4.7 level (Özbek, 2017).

Seed activities in Turkey are carried out according to the Law of Seed. The law includes acts on variety breeding, seed production and marketing. Private sector organizations have a significant dominance in the marketing of cottonseeds. Almost all of the annual cottonseed production and marketing is carried out by private sector organizations. Cultivation of genetically modified cotton varieties is prohibited by law in Turkey.

The use of conventional varieties in both seed and cotton production is considered as an advantage in the world cotton market, where 76% is GMO cotton varieties, and a branding study has been carried out under the name of GMO Free Turkish Cotton by the National Cotton Council.

Technique of Cultivation

The main objective in cotton farming is high yield and quality as in other cultivars. A good field preparation is important in terms of healthy emergence and root development of planted seeds. Field preparations for cotton begin with the breaking down of the plant residues and stalks in the field right after the harvest of the pre-plant in autumn. The shredded stalks and plant residues are mixed into the soil by making a deep ploughing. This practice is important for the protection of soil fertility in soils with low organic matter content. In regions where the climate permits, after the first rain, the field is ploughed again with plough discs and similar tools to break up the clods formed after the plough. The ridges are prepared in autumn if sowing is to be done on the ridge. After the winter, soil preparation starts immediately in spring when the climate and soil factors of the production region allow. A light ploughing is made with a disc harrow at a depth of 8–10 cm if there are excessive weeds in the field during this period.

Seed-bed preparation starts approximately 8–10 days before the planned planting date. Sometimes, due to sudden and extreme temperatures, soil tilth can be lost. In such cases, irrigation is done before sowing, and when soil tilth is achieved, seed-bed preparation should be done. If sowing will not be done with a combined seeder, half of the nitrogen fertilizer and all of the phosphorus and potassium are sprinkled on the soil with fertilizer distributors and taken about 10 cm under the soil with a disc harrow. In addition, in regions where weeds are a problem, herbicide is applied just before the last disc harrow or harrow. After the seed bed is prepared, the soil is pressed with a harrow. The situation is not different in ridge sowing. Special sliders are passed over the ridges prepared in autumn and the edges of the ridges are pressed. Then, one day is allowed for the soil tilth to stabilize before planting.

The physical and chemical structure of the soil, together with the climatic conditions, are effective in determining the planting time. In Turkey, cotton planting is usually done in the last week of March in the Çukurova region, in the first week of April across the Aegean and the Antalya regions while in the south-eastern Anatolia region it begins in the second week of April and continues until the end of May. Sowing is usually done if the soil temperature remains stable between 15°C and 18°C for a few days. Cultivation is widely performed in Turkey using a sowing machine. While flat sowing is generally preferred in the Aegean, Antalya and south-eastern Anatolia regions, ridge planting with a combined seeder is widely preferred in the Çukurova region. The ridge sowing enables cotton planting to be done earlier and minimizes the negative effects of rainfall after sowing.

The yield and quality of the product produced are determined by the genetic potential of the variety used, environmental factors and their interactions. Delinted certified seed with maximum genetic purity and germination capacity are the main reasons for high cotton production in Turkey. Diseases and pests that can be transmitted by seed, soil-borne pests and various fungal diseases can be controlled using certified seeds. The appropriate agronomic practices for the variety should be applied in addition to having high yield, quality traits and adaptation to the particular production regions. When favourable environmental conditions for cotton cultivation occur late or when the sowing needs to be renewed due to insufficient emergence, it is important to choose earlier cotton varieties in order to minimize yield loss.

In Turkey, 20–30 kg/ha of certified seed is enough under normal environmental conditions, although it varies according to production regions. Approximately 20–25 kg/ha of seeds are used in the south-eastern Anatolia region, while 25–30 kg/ha are used in Çukurova and Antalya regions and 40–50 kg/ha in the Aegean region. Sowing depth varies between 2 cm and 4 cm according to planting time, annealing depth, soil structure and temperature. The planting frequency varies between 70 cm and 75 cm between rows and 4–20 cm above rows, depending on the morphological characteristics of the variety, the fertility of the soil and the planting time. After the seedling is established, thinning is done to adjust the planting density with appropriate plant to plant distance. Hoeing in cotton is done by hand and machine. The places in the cotton field where mechanical hoeing is not possible are generally hoed manually. Hoeing is done two to four times depending on plant growth, weed density, rainfall after sowing and the severity of root-rot disease.

Fertilization is one of the most important issues in cultivation technique. To increase the yield and quality of the harvested from the unit area, a conscious fertilization is required together with all agricultural practices. Conscious fertilization is very important in terms of economics as well as for protection of the physical, chemical and biological structure of soil. Since the soils are very different in terms of physical and chemical structure, there is no standard fertilization formula to be applied in all conditions. For this reason, the physical and chemical structure of the soil should be revealed with the soil analysis for the most appropriate fertilizer recommendation, and the necessary fertilization recommendations should be made by combining the results obtained from the field experiments. Fertilization is done by sprinkling or tape method in Turkey. The amount of fertilizer varies depending on factors such as climatic conditions, soil structure, cultivar and irrigation conditions. In addition, product and fertilizer prices may affect the amount of fertilizer used. Nitrogenous fertilizers are applied in two split doses as it is easily washed out from the soil and the rapid increase in nutrient need of cotton in the generative period (Crozier and Cleveland, 2014). Generally, the first half of nitrogen is given to the band with phosphorus and potassium fertilizers before planting, and the second half before the first water. As a result of various studies carried out in Turkey 80–180 kg/ha of pure nitrogen and 60–90 kg/ha of phosphorus fertilizer is recommended for cotton (Özer and Dağdeviren, 1986; Tozan, 1990; Özer, 1992; Haliloğlu and Oğlakçı, 2000; Karademir *et al.*, 2005; Görmüş, 2014; Basal *et al.*, 2020). However, the amount of nitrogen applied by the farmer exceeds 200 kg/ha, and the amount of phosphorus exceeds 115 kg/ha. Turkey's soils are rich in potassium overall (Sonmez *et al.*, 2018). For this reason, potassium fertilizer is not recommended unless potassium deficiency is found in soil analysis. In the generative period, potassium is applied in the foliar form by the farmers in recent years, if the plants need it.

Methods of Irrigation

Irrigation water requirement

Cotton has three development periods from planting to harvest. These are vegetative development, flowering and boll formation, and opening of boll. In Turkey, almost no precipitation falls in all of the periods indicated. This situation reveals the necessity of irrigation in the Mediterranean, Aegean and south-eastern Anatolia regions of Turkey where cotton is grown extensively.

Since the total temperature values of these regions are different from each other, significant differences can be observed in the amount of irrigation water to be applied to cotton during the growing season (Table 13.1).

Evapotranspiration

Each of the cotton-growing areas in Turkey have different macro-climate. The differences in each average temperature, vapour pressure deficit, wind and sunshine hours and intensity cause the differentiation of evapotranspiration values in cotton agriculture. Evapotranspiration in the main cultivation regions varies between 570 and 800 mm in the Aegean region, 630 and 1000 mm in the Mediterranean region and 1100 and 1300 mm in the south-eastern Anatolia region (Kanber, 1982). Compared to other regions, the relative humidity is low in the south-eastern Anatolia region, and the vapour pressure deficit is high, causing more evapotranspiration.

Evapotranspiration varies between 210 and 240 mm in the July when the water need is highest in the Mediterranean region, while it varies between 280 and 300 mm in the south-eastern Anatolia region (Şanlıurfa and Diyarbakır). The similarity of the climatic characteristics of the Mediterranean and Aegean

Table 13.1. Studies conducted in different regions on cotton irrigation in Turkey.

City	Irrigation method	Water (mm)	Evapotranspiration (mm)	Yield (kg/da^{-1})	Water use effiiency (kg/da^{-1}mm$^{-1)}$	References
Hatay	Drip	1097–91	1096–247	556–163	0.49–0.75	Ödemiş *et al.* (2017)
Hatay	Drip	1136–91	1070–256	564–168	0.83–0.43	Kazgöz Candemir and Ödemiş (2018)
Hatay	Drip	682–128	678–283	599–306	0.78–1.26	Can and Ödemiş (2017)
Aydın	Drip	557	723–496	649–537	0.84–1.17	Sarı and Dağdelen (2010)
Kahram-anmaraş	Drip	887	1037–203	481–106	No data	Ektiren and Değirmenci (2018)
Adana	Drip	508–322	615–435	422–96	No data	Ertek and Kanber (2003)
Şanlıurfa	Drip	951–475	1033–588	616–285	0.48–0.61	Coşkun (2015)

Note: the values given in each row show values obtained in different years.

regions causes evapotranspiration to reach approximate values. In the Çukurova region, where cotton farming is most intense, it is predicted that the amount of rainfall will decrease by 29.4–34.7% until 2070, whereas the amount of agricultural water use will increase by 11% in cotton (Demir *et al.*, 2008).

Increasing temperatures alone reduced the 160-day growth and development time to 140 days from 1972 to 1992. Although this decrease is expected to lead to a decrease in evapotranspiration, it is clear that increases in average temperatures in the same period created significant increases in evapotranspiration.

Methods of irrigation

Public and private irrigated fields in Turkey use methods of surface irrigation (81.7%), sprinkler (16.6%) and drip irrigation (1.7%) (Anon, 2018). The cost of production habits and irrigation systems play an important role in the widespread use of irrigation methods. Although traditional irrigation methods are still widely used in regions with low income, fragmented land and pressurized irrigation methods are also becoming more and more common. Especially in places where water resources are limited and unit water cost and labour costs are high, the use of drip irrigation is increasing.

In addition, to expand the use of pressurized irrigation methods, the government incentivizes with policies of low interest rates and long payback periods. The most common irrigation methods in cotton plants in Turkey are surface irrigation and furrow irrigation. However, especially in the GAP region, farmers continue their initial irrigation with sprinkler irrigation and subsequent irrigation with furrow method. The most important reason why pressurized irrigation methods cannot be used continuously is that the energy cost (especially electrical energy) has increased in recent years. This situation causes producers in some regions to abandon drip-irrigation systems and switch back to furrow-irrigation methods.

Management of irrigation

Irrigation in cotton immediately after planting causes an increase in vegetative parts but a decrease in the number of fruits. Therefore, the first irrigation is done 30–40 days after planting. Even though the rate of water use is low in the pre-flowering period, farmers give first irrigation (10–20 mm) to wet the dry topsoil. The sprinkler method is generally used in first irrigation application. Almost half of seasonal evapotranspiration occurs during flowering. In all cotton-growing regions, there is almost no rainfall during this period.

Since water given more than the plant needs causes a continuous moistness in the soil, an increase in diseases and pests caused by moisture can be seen in the vegetative part. In addition to them, in areas with poor drainage, increases in the shallow ground-water level are generally observed during this period. In areas where the water-distribution network is not sufficient, farmers benefit from ground water. Soil salinity is observed especially in lands where shallow ground water is used.

In conditions of limited water and high water requirement, farmers turn to alternative water resources. For example, in the south-eastern Mediterranean region (Hatay province), where cotton farming is intensive, some farmers sometimes use drainage water or water collected at the end of the furrow for irrigation, especially at the end of the flowering period and near harvest. In cases of using furrow-irrigation method in cotton irrigation, one of the irrigations coincides with the beginning of the flowering period, two of them coincide with the period of maximum flowering and the other during the boll-formation period. In case of drip irrigation, it was seen from farmers' practices that two irrigations were made during this period.

In general, irrigation is stopped in the middle or third week of August, as after this date there is no contribution on yield. First water given to the plant, which uses the moisture remaining from the winter after planting, can meet the moisture needed by the plant about 10 days later. However, not too much water is given about 30–40 days until the first flowering. The length of the irrigation interval after the first irrigation depends mostly on the irrigation method used.

Since the average temperature and relative humidity values are approximately the same in the Aegean and Mediterranean regions, the same irrigation intervals are used in similar irrigation methods. In furrow-irrigation application, irrigation done approximately every 15 days will completely saturate the profile with water. In some areas of the Aegean, after the land is divided into large basins, the water is ponded in these basins and the excess water is drained from the pans by waiting for a while. In irrigation in this way, only two irrigations per season may be sufficient for producers. However, this application causes excessive softening of the root area and results in the above-ground parts of the plant lying on their side, resulting in yield losses.

In drip-irrigation applications, irrigation intervals vary between five and eight days. As the irrigation interval increases, the amount of irrigation water applied and deep infiltration increase, and sometimes an effective irrigation cannot be achieved. Since furrow irrigation is the common irrigation method in the south-eastern Anatolia region, irrigation intervals vary between 15 and 20 days.

Cotton Pests and Their Management

There is a diverse range of pests of cotton in Turkey (Table 13.2). *Liriomyza trifolii* (Burgess), *Phenacoccus solenopsi* Tinsleys (Gençsoylu, 2007; Kaydan *et al.*, 2013; Basal *et al.*, 2020) and *Leptodemus minutus* Jakovlev 1876 (Yazıcı and Sertkaya, 2020) have recently been added to the damage-causing insects in Turkey.

As elsewhere in the world, chemical control is prominent in combatting pests and mite species in cotton fields in Turkey as well. Ground equipment (field sprayers) is used in chemical controls (Fig. 13.4). The rate of using high-clearance sprayers is low. Some farmers try to modify their tractors so that the sprayer is raised and does not damage the cotton.

Aerial spraying, used on cotton fields in Turkey until 2001, has been prohibited completely since 2006. Although the prohibition of airborne pesticides contributed to the reduction of the negative effects of these pesticides on the environment, it has caused problems in combatting some pests. For example, in combatting pests such as *H. armigera*, the timing of which is important, farmers are obliged to devise a spraying schedule based not on the pest's biological periods but their irrigation dates. This leads to a decrease in the biological effects of the pesticides and an increase in the number of sprayings since they do not coincide with the appropriate biological period of the target pest. Chemical control is carried out in line with the Integrated Control Technical Instructions prepared by the Ministry of Agriculture and Forestry. The fact that the Economic Damage Thresholds specified in the instructions cannot be meticulously applied by the farmer on some harmful species

Table 13.2. The insects and mites of cotton in Turkey.

Scientific name	Order/family
Major pests	
Bemisia tabaci Genn.	(Hemiptera: Aleyrodidae)
Aphis gossypii Glov.	(Hemiptera: Aphididae)
Creontiades pallidus (Rumb.)	(Hemiptera: Lygaeidae)
Lygus spp.	(Hemiptera: Miridae)
Thrips tabaci Lind.	(Thysanoptera: Thripidae)
Helicoverpa armigera (Hb.)	(Lepidoptera: Noctuidae)
Tetranychus urticae Koch.; *T. cinnabarinus* (Boisd.)	(Acarina: Tetranychidae)
Minor pests	
Asymmetrasca decedens (Paoli)	(Hemiptera: Cicadellidae)
Empoasca decipiens Paoli	(Hemiptera: Cicadellidae)
Pectinophora gossypiella (Saund.)	(Lepidoptera: Gelechidae)
Agrotis ipsilon (Hufn.); *A. segetum* (D&S)	Lepidoptera: Noctuidae)
Spodoptera exigua (Hb.); *S. littoralis* (Boisd.)	(Lepidoptera: Noctuidae)
Agrotis segetum (D&S)	(Lepidoptera: Noctuidae)
Earias insulana Boisd.	(Lepidoptera: Nolidae)
Frankliniella spp.	(Thysanoptera: Thripidae)

Fig. 13.4. Tractor boom sprayer. (Photo: Cafer Mart)

such as cotton bollworm is one of the factors that increase pesticide consumption.

Turkey prohibits the use of pesticides that are prohibited by the EU. For this reason, many effective pesticides previously included in the technical instructions have been removed from the recommended pesticide lists against cotton pests. This can engender problems in combatting some pests due to the inadequacy of pesticides to be recommended. Examples of pesticides used most commonly in combatting cotton pests include the following: l-cyhalothrin + emamectin benzoate, chlorantraniliprole, chlorantraniliprole + l-cyhalothrin, spinosad, indoxacarb for *H. armigera*; acetamiprid, pyriproxyfen, sulfoxaflor, buprofezin for *B. tabaci*; acetamiprid, thiamethoxam, sulfoxaflor for *A. gossypii*. Due to the decrease in the biological effects of the pesticides, some of these are not used, although they are included in the technical instructions. It can be said that the biological effect decreases are due to the resistance problem. It is difficult to say that in using pesticides farmers fully meet the requirements of resistance management.

The average applications of pesticides per the pest are as follows: once on average for *Thrips* during the early growth period of cotton; once or twice for cotton bollworm, twice or three times for aphid, once or twice for whitefly during the boll development period; once or twice for aphid, once for cotton bollworm, once or twice for red spider and once or twice for *C. pallidus* and *Lygus* during the boll maturation period. In the last three years, aphids have been epidemic in almost all cotton-cultivation areas and in some regions the number of sprayings per season has reached 6–7, targetting only aphids. Due to the insufficient effect of the pesticides recommended for aphids, farmers have sought solutions by creating mixtures that are not included in recommendations. Depending on the high number of pesticide applications, agricultural control activities against pests and weeds reach 20% of the total cost.

It is observed that in Turkey, an important organic cotton producer in previous years, the production of organic cotton has decreased according to the data from 2019. Countries producing the highest amount of organic cotton in the world are as follows: India (47%), China (21%), Kyrgyzstan (12%), Turkey (6%), Tajikistan (5%), the USA (3%) and Tanzania (3%). According to the statistical data from 2019, in the organic cotton sector of Turkey, 266 farmers produced 11,652 bales of fibre in 5,418 ha of land. At the same period, Turkey has produced 6.4% of organic fibres worldwide and 1% of the national cotton production was organic (Anon, 2020e).

By years, the production of organic cotton in Turkey has been decreasing. Among the reasons for this are the traceability problem, the inability to sell the produced organic cotton for its real value, the decrease in cotton-cultivation areas in the country in general, the divergence of the producers from cotton agriculture due to the income parities between the products, and the difficulties in the supply of inputs suitable for organic agriculture, especially seeds.

In Turkish cotton fields, conventional cotton varieties are cultivated. It has been prohibited by law to cultivate transgenic cotton varieties. Domestic varieties developed by the private sector and public R&D units have a share of approximately 35% in the cottonseed market. In addition to the yield potential, newly developed varieties are aimed to increase the fibre quality as well. With the use of varieties with high yields, the use of certificated seeds exceeding 95% with the contribution of subsidy policies and with improvements in cultivation techniques, the producers have achieved today's yields.

Better Cotton

The production of 'better cotton', which is important for the sustainability of cotton cultivation, began in Turkey in 2013. Better Cotton (BCI), which is a multi-stakeholder and international initiative under which many stakeholders from the producer to the retailer unite and act together to reduce the negative environmental and social impacts of cotton production and to make the future of the sector safer, has begun production in Turkey following the agreement it signed with Better Cotton Applications Association (BCAA), founded in Turkey in 2013.

Better Cotton Applications Association has been undertaking training sessions, projects, supervision and certification activities for seven years in order for Turkish cotton farmers to produce cotton in accordance with Better Cotton standards. As of 2019, Better Cotton production has been realized in Aydın, Balıkesir, Manisa and İzmir in the Aegean region, in Adana and Kahramanmaraş in Çukurova, and in Diyarbakır and Şanlıurfa in south-eastern Anatolia.

As of 2019, the production area of Better Cotton has reached 53,400 ha and the number of farmers producing Better Cotton has reached 3299. The production area of Better Cotton amounts to 9.5% of the total area of cotton cultivation in Turkey. In 2020, affected by the decline in cotton-production areas in Turkey and the circumstances of the pandemic, the production area of Better Cotton fell to 34,353 hectares.

Cotton Diseases and Their Management

The main diseases encountered in cotton-production regions in Turkey are verticillium wilt, fusarium wilt, root rot and angular leaf spot disease. Verticillium wilt disease that is caused by *Verticillium dahliae* Kleb. leads to significant yield and quality losses made in the areas of cotton cultivation in Turkey. The rate at which cotton plants infected from *Verticillium* disease is approximately 27% in the Aegean region (İzmir, Aydın and Manisa), 25% in the Çukurova region (Adana), 16% in the south-eastern Anatolia region (Adıyaman, Batman, Diyarbakır, Mardin, Şanlıurfa and Siirt) and 14% in the Antalya region. Accordingly, it is reported that yield loss due to this disease is 12% in the Aegean and Çukurova regions and 4% in the Antalya region (Esentepe, 1979; Sezgin, 1985; Sağır *et al.*, 1995). Studies conducted to combat the disease indicated that the most effective way to reduce the loss of yield and quality in large areas where cotton is cultivated is to use resistant varieties (Wilhelm *et al.*, 1974a,b; ; Anon, 2000; Bell, 2001; Nemli, 2003; Agrios, 2005; Akışcan, 2011). Seedling root rot disease, *Rhizoctonia solani Kühn., Fusarium* spp., *Pythium* spp., *Alternaria* spp., *Verticillium* spp. *and Thielaviopsis basicola (Berk. and Broome)* are some of the soil-borne fungi. This disease, which is frequently seen in cotton-cultivated areas, negatively affects the yield by making damage at a level that requires replanting in rainy and cool years

(Mart, 2017). In a study conducted in the Aegean region, it was reported that the rate of seedling loss due to disease in different regions varied between 7.58% and 17.77% in 2000, and 13.56% and 25.06% in 2001 (Nemli and Sayar, 2002). In addition to applying chemicals to the seeds for the disease management, use of certified seeds, crop rotation, preventing the formation of cream after sowing, ridge planting and to avoid early and deep sowing is recommended. Cotton angular leaf spot disease caused by *Xanthomonas axonopodis* pv. Malvacearum (Atkinson, 1891) is observed in all growth stages of the cotton. However, it is observed infrequently at seedling stage in Turkey (Oğlakçı *et al.*, 2010). And, Alternaria leaf spot disease caused by *Alternaria* spp. occurs with low intensity in some years and in some regions.

Weed management

Cotton, which has an important place in the agriculture, industry and trade sectors in the world, with its use in different fields, is one of the plants sensitive to weed competition and the yield amount decreases with the effect of weeds.

While the yield losses caused by weeds in cotton reach approximately 36% worldwide, it is reported that these losses can be reduced to 8.6% with an appropriate weed control (Oerke, 2006; Doğan *et al.*, 2014; Jabran, 2016).

In addition to their direct damage to cotton, weeds host pests and diseases, increase the number of tillages, irrigation or fertilization, or make agronomic practices difficult. Most importantly, they significantly reduce productivity in mechanical harvesting. Weeds such as *Xanthium strumarium* (common cockleber), *Solanum nigrum* (black nightshade), *Datura stramonium* (jimsonweed) and *Setaria verticillata* (bristly foxtail), which occur in the late periods and especially after irrigation, make cotton harvest difficult in the short term and because of sticking to cotton fibres, they decrease its quality. Due to all these reasons, weeds can cause direct and indirect yield and quality losses in cotton farming.

When the results of studies carried out in Turkey to determine the weeds in cotton fields were evaluated, there were 105 species and 146 types of weeds associated with approximately 33 families (Kadıoğlu *et al.*, 1993; Uygur, 1997; Gönen, 1999; Uludağ and Üremiş, 2000; Yıldırım *et al.*, 2008; Arslan, 2018; Özkil *et al.*, 2019). These weeds are the main pests not only in cotton but also in many crops in the field/orchard. The ones that are considered to be very important according to their prevalence and density include: *Cyperus rotundus* (purple nutsedge), *Convolvulus arvensis* (field bindweed), *Sorghum halepense* (johnson grass), *Physalis* spp. (cutleaf ground cherry) and *Xanthium strumarium* (common cockleber). Those deemed important are *Amaranthus* spp. (pigweed), *Chenopodium album* (common lambsquarter's), *Cynodon dactylon* (Bermuda grass) and *Datura stramonium* (jimsonweed).

The important weeds are: morning glory, *Cucumis melo. agrestis* (field muskmelon), *Portulaca oleracea* (common purslane), *Setaria verticillata* (bristly foxtail) and *Solanum nigrum* (black nighshade) (Table 13.3). The others in the table are not as many as these two groups, but they are still weeds that should be considered and not neglected. In particular, there is *Ipomoea triloba* (three lobe morning glory), *Cucumis melo. agrestis Naudin* (field muskmelon) and *Amaranthus palmeri* (palmer amaranth) which are a very serious potential danger for all summer products for the coming years (Üremiş *et al.*, 2020).

For weed management, it is recommended to apply cultural measures to prevent the contamination of weeds on the cotton field, which is an anchor plant, and to ensure that cotton is grown in a healthy way. Thus, the healthy development of cotton can be increased and its competitive power against weeds can be increased (Tepe, 2014). Then, mechanical processes may be required for aquaculture. While this process is being done, it will naturally fight against weeds.

A method that can be used in biological control of weeds in cotton does not yet exist in Turkey. In the fight against weeds, the alternative of chemical control is not too much, it is easy to apply, it is effective in a short time, it is not so much affected by ecological conditions, and the cost of manpower is low compared to other methods, making herbicide use the most preferred method. Therefore, weed control is largely based on herbicide use (Jabran, 2016). In other words, the weed control in Turkish cotton production is mainly by means of pre-planting and post-emerging chemical treatments.

Table 13.3. Important weed species in cotton fields in Turkey and some characteristics of them.

Latin names	Common names	Families	Description	Plant type
Alhagi pseudalhagi (Bieb.) Desv.	Camelthorn	Leguminosae	Broadleaf	Perennial
Chrozophora tinctoria (L.) Rafin	Turnsoler weed	Euphorbiaceae	Broadleaf	Annual
Digitaria sanguinalis (L.) Scop.	Hair crabgrass	Poaceae	Grass weed	Annual
Echinochloa colonum (L.) Link.	Awnless barnyardgrass	Poaceae	Grass weed	Annual
Echinochloa crussgalli (L.) P. Beauv.	Barnyardgrass	Poaceae	Grass weed	Annual
Euphorbia spp.	Spurge	Euphorbiaceae	Broadleaf	Annual
Heliotropium spp.	Heliotrope	Boraginaceae	Broadleaf	Annual
Hibiscus trionum L.	Venice mallow	Malvaceae	Broadleaf	Annual
Prosopis farcta (Banks and Sol.) Macbride	Syrian mesquite	Fabaceae	Broadleaf	Perennial
Tribulus terrestris L.	Puncture vine	Zygophyllaceae	Broadleaf	Annual
Convolvulus arvensis L.	Field bindweed	Convolvulaceae	Broadleaf	Perennial
Cyperus rotundus L.	Purple nutsedge	Cyperaceae	Sedge	Perennial
Physalis spp.	Cutleaf ground cherry	Solanaceae	Broadleaf	Annual
Sorghum halepense (L.) Pers.	Johnsongrass	Poaceae	Grass weed	Perennial
Xanthium strumarium L.	Common cockleber	Asteraceae	Broadleaf	Annual
Amaranthus spp.	Pigweed	Amaranthaceae	Broadleaf	Annual
Chenopodium album L.	Common lambsquarter's	Chenopodiaceae	Broadleaf	Annual
Cynodon dactylon (L.) Pers.	Bermuda grass	Poaceae	Grass weed	Perennial
Datura stramonium L.	Jimsonweed	Solanaceae	Broadleaf	Annual
Ipomoea spp.	Morning glory	Convolvulaceae	Broadleaf	Annual
Cucumis melo var. *agrestis* Naudin	Field muskmelon	Cucurbitaceae	Broadleaf	Annual
Portulaca oleracea L.	Common purslane	Portulacaceae	Broadleaf	Annual
Setaria verticillata (L.) P. Beauv.	Bristly foxtail	Poaceae	Grass weed	Annual
Solanum nigrum L.	Black nightshade	Solanaceae	Broadleaf	Annual

Recent Developments and Future of Cotton in Turkey

Although cotton is considered as a strategic produce because of its importance in the textile industry, there are many obstacles challenging the sustainability of cotton production in Turkey. First, as stated above, is the fact that income parities between products are against cotton, leading the producers to alternative products, especially corn. The amount of support should be improved to eliminate the negativity caused by the income parity between products.

One of the issues that negatively affect cotton production in Turkey is the production costs. According to the 2019 data of the National Cotton Council, the cost per ha reaches $3268. The approximate cost of a kg of fibre is $1.86 (Anon, 2020e). These values are quite high and the high production cost causes producers to move away from cotton agriculture. Prominent factors in production costs are agricultural struggle, harvesting, irrigation and plant nutrition.

Fibre quality values derived from varieties of cotton grown in Turkey are at a satisfactory level, although the availability is problematic for various reasons preferred by industry. One of the main reasons for this is the contamination problem. Harvest and delaying applications will adversely affect the fibre quality ginning stage. Textile manufacturers will increase the demand for cotton produced in Turkey. Harvesting process in Turkey is largely done by machine. In particular, technical and economic insufficiencies in harvesting and post-harvest operations and storage and ginning processes cause losses in the added value of all segments from agriculture to industry. To prevent these losses, it is recommended to rehabilitate ginning operations in line with today's needs, to achieve economies of scale, to construct cleaning and drying units, and to improve the incentives provided. Their application will reduce the loss of quality in Turkish cotton and the demand is expected to increase. Widespread licensed warehousing operations in some regions in Turkey will be useful for the sustainability of cotton in cotton farming.

Greece

Greece is the EU's main cotton grower, accounting for more than 80% of total European production. Cotton is a crop of high importance for Greek agricultural production, with more than 8% of total agricultural output. Thessaly, Macedonia, Thrace and mainland Greece are the major cotton-producing areas. Cotton in Greece is grown mainly by small farmers with an average of 3–4 ha of cotton cultivation and is planted from 1 March to 15 April. The crop life cycle is usually 170–210 days, depending on the variety and weather conditions. The harvest normally occurs from 1 October to 30 November, and most of the cotton is machine-harvested. Almost all cotton is irrigated. In 2017/18, the cotton acreage had increased 9.5%, registering 230,000 ha, at the expense of durum wheat and corn acreage, with an estimated production of 1.24 million 480-lb bales, up 20% from the previous season due to favourable weather conditions during harvest; and good yields, which are expected to be higher than the previous year in the major cotton-growing districts of Thessaly and Macedonia (Fig. 13.5).

The cultivation of cotton in Greece dates back to the 2nd century BC described by Pausanias. Apart from fibre production, cottonseed, which remains after processing, is used to produce oil and oilseed cake for animal feed. Many producers seek ISCC certification to prove their products are sustainable. With ISCC certification, producers can address requirements for all markets as biofuels, food, feed and fibre. Both environmental and social sustainability criteria are set at a high level, with traceability in the supply chain and a greenhouse gas calculation also covered. EUROCERT is the market leader for ISCC audits of cotton farmers and processors in Greece.

The presence of *Helicoverpa armigera* was monitored over four years from 2002 to 2005 at four locations with three funnel traps in each location. The pheromone, z-11-hexadecenyl aldehyde [0.36% w/w] was used. An insecticide, dichlorvos, was included to kill the moths attracted into the trap, although in the last two years, additional traps without insecticide were also used. The latter traps had significantly fewer insects than those with insecticide. The number of insects trapped varied from year to year, with more adult male insects captured in pheromone traps when cotton production was low (Stavridis *et al.*, 2008).

Helicoverpa armigera had been controlled effectively with chemical insecticides in the major cotton crop production areas of northern Greece for many years. However, a resurgence of the pest was observed in 2010, which significantly affected crop production. A survey from 2007 to

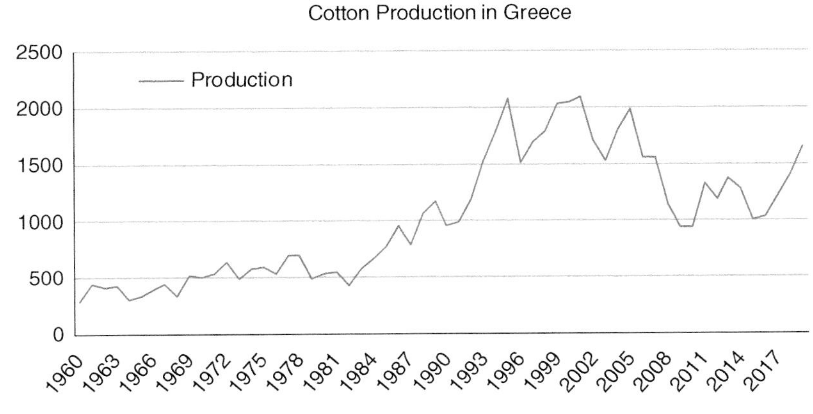

Fig. 13.5. Production changes since 1960.

2010 examined the insecticide resistance status of *H. armigera* populations from two major and representative cotton-production areas in northern Greece against seven insecticides (chlorpyrifos, diazinon, methomyl, alpha-cypermethrin, cypermethrin, gamma-cyhalothrin and endosulfan). Bioassays with topical application of a full dose on third instar larvae showed that resistance levels were relatively moderate until 2009, with resistance ratios below ten-fold for organophosphates and carbamates and up to 16-fold for the pyrethroid alpha-cypermethrin. Resistance increased to 46- and 81-fold for chlorpyrifos and alpha-cypermethrin, respectively, in 2010, when the resurgence of the pest was observed (Mironidis *et al.*, 2013). Cotton fields were sprayed using tractor-mounted sprayers or motorized ground equipment on the larger farms.

Spain

Cotton production in Spain is concentrated in the Guadalquivir Valley, Andalucía, Spain's southernmost region. It is grown with irrigation under salty and warm conditions, where few alternative crops can be grown profitably. The area of cotton grown has varied depending on water supplies and price relations and until recently the subsidies from the EU. In this area of Spain, it is a critical crop from the environmental, social and economic perspective.

A major cotton pest in Spain is *Helicoverpa armigera*, which required extensive use of insecticides. Some resistance was detected (Torres-Vila *et al.*, 2002), but Avila and Gonzalez-Zamora (2010) pointed out that of four insecticides evaluated in 2004, only endosulfan was at a moderate resistance level, whereas for methomyl, chlorpyrifos and lambda-cyhalothrin, it was at a low level. Endosulfan and methomyl have been used 2–9 times against the bollworm in the Saville area, where control failures had been reported during 2003. Farmers generally use tractor sprayers as iillustrated in Fig. 13.4, although boom width varied. Aerial spraying of cotton was not permitted within the EU.

The initial evaluation of the hand carried Electrodyn sprayer for the control of insect pests on cotton was investigated in Spain. Spray deposition was evaluated to determine how the height of nozzle above the crop, the speed of travel, wind speed and position of the sprayer nozzle relative to the row define the optimum method of use. Field trials of the ultra-low-volume, low-volatile formulations measured spray deposition and distribution at two heights on the plants using a fluorescent tracer. Plants were sampled from one row upwind and three rows downwind from each single spray run. More than 80% of the spray was deposited on 'top' leaves. A smaller proportion of the fully charged spray cloud penetrated to the lower half of the crop than an uncharged spray. The proportion deposited on the main stem and branches in the top of the plant was four times greater with charged droplets, which also provided significant under-leaf cover, with a concentration of spray at the tips and edges of leaves and bracts. Overall, a charged spray deposited 2.5 times more spray compared to an uncharged spray. Morton (1982) concluded that the Electrodyn sprayer should be held downwind of the operator walking at 1.0 m/s with the nozzle held 0.2 m directly above the row and all rows are sprayed. This sprayer was subsequently used in several countries with the nozzle provided as part of the insecticide container, so the operator did not have to prepare the spray as the container was fitted directly onto the sprayer.

In 2006, Spain suffered a significant decline in the area planted to cotton as a result of the implementation of the EU cotton reform, reaching a record low in 2008/09 (Fig. 13.6). In 2009/10, the Regulation (EC) 637/2008 was amended and the national guaranteed area was reduced from 70,000 ha to 48,000 ha with a total budget of €67.2 million. Since 2009/10, the cotton aid increased in value/ha, but less acreage can benefit from this payment. Between 2005 and 2014, the yield of cotton dropped by 36%. Since 2014/15, the reference amount for the area payment has been reduced to €1267.53/ha with specific conditions to be eligible to receive this support, defined annually in Spain's *National Gazette*. These include the requirement that only agricultural plots that were not planted to cotton in the previous season were planted to cotton at least once in the marketing years 2000–2003. The cotton varieties sown should be those listed in the *EU Plant Varieties Common Catalogue*. The regulations also state that the seeding should result in over 100,000 plants/ha in irrigated plots, although interspecific

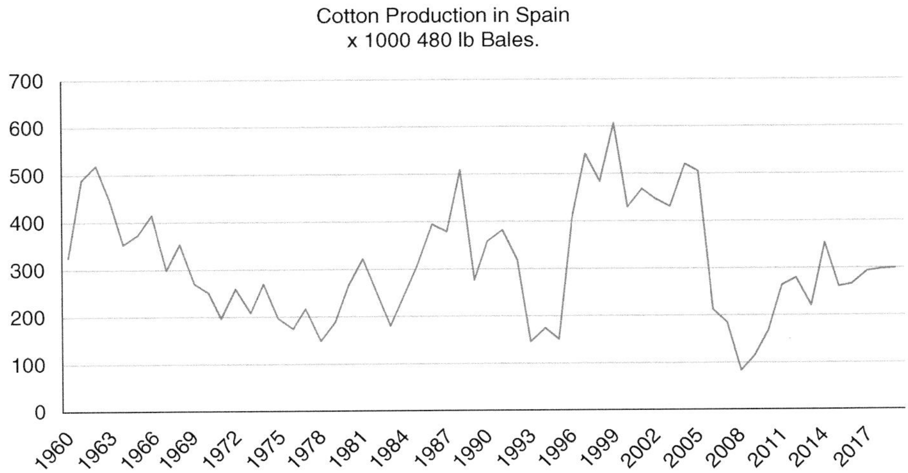

Fig. 13.6. Variation in production of cotton in Spain since 1960.

hybrid varieties can be grown at 75,000 plants/ha. The production obtained must meet minimum quality requirements.

In 2017/18, pest incidence was low, which favoured good crop development. A study indicated significant economic benefits of growing Bt cotton in Spain (Ceddia *et al.*, 2008), but as genetically modified cotton varieties were not allowed for planting in the EU, farmers have relied exclusively on the use of pesticides to reduce pest incidence. However, Ceddia *et al.* (2008) did point out that cotton farming is still less profitable than alternative crops grown for oil or protein production and estimated that the cotton area would decline.

Spain has eight ginning plants in Andalucía, of which only seven are currently operational, whereas the textile industry is concentrated in Catalonia.

Israel

Farmers started to grow cotton in Israel in the 1950s. Among the insect pests entering the crop, the spiny bollworm, *Earias insulana*, was first to cause problems, which led to the development of a pest-monitoring system based on economic thresholds and to formulate pest-control recommendations. This required two applied entomology research teams, one to study the biology and seasonal history of the insect pests and another to screen insecticides and their efficiency as pesticides and the development of pest resistance.

A pest-control policy was developed which resulted in routine field-level checking of all pesticides registered and recommended for the control of cotton pests, while there was a search for non-chemical alternatives and selective pesticides. Hoping to adopt biological control, experiments were carried out with *Trichogramma* spp. egg parasites for control of *Helicoverpa armigera*. Until the late 1980s, most cotton pests were controlled with conventional pesticides such as organophosphates, carbamates and pyrethroids. Since 1999, production of cotton has decreased (Fig. 13.7).

Prior to 1980, cypermethrin was considered effective in controlling *Bemisia tabaci* but then lost its efficacy in the early 1980s, and subsequent pyrethroids, fenpropathrin and bifenthrin, also lost their potency against larval stages of *B. tabaci* within two years. To delay the onset of resistance in *B. tabaci*, an IRM strategy was implemented in 1987 (Horowitz *et al.*, 1994). The main principles of the strategy were: restricting the use of insecticides to one pest-generation period and alternating compounds with different modes of action. The number of treatments was minimized according to action thresholds and by monitoring the development of resistance. The use of more selective insecticides, such as buprofezin, pyriproxifen and diafenthiuron, were introduced. Imidacloprid was also applied as a drench treatment. Detergents

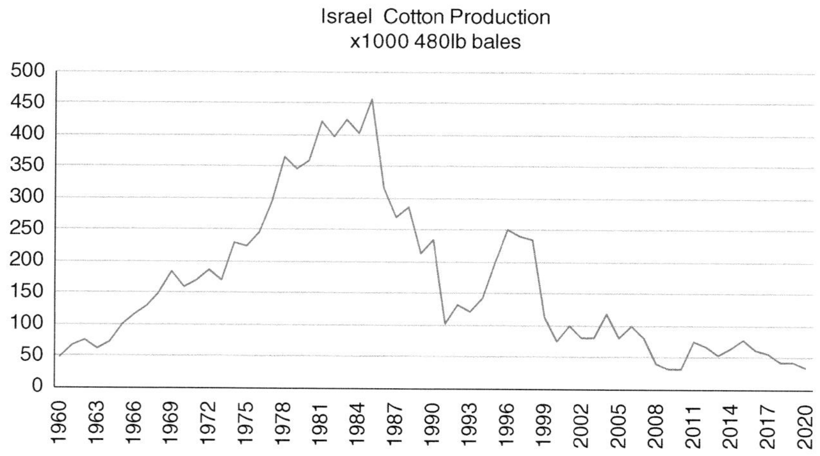

Fig. 13.7. Increase and then decline in cotton production in Israel since 1960.

and mineral oils were also tried on an experimental scale mainly to control of *B. tabaci* (Horowitz *et al.*, 1995).

The growing season was divided into four phases, with a specific insecticide group allocated to each phase. Resistance monitoring was focused on *B. tabaci*, with dosage-mortality curves for different insecticides compared between a susceptible reference strain and field-collected *B. tabaci* adults. Specific bioassays have been worked out for the various IGRs, taking into account their particular mode of action. The aim was to minimize insecticide applications during April–July in order to prevent any unwanted destruction of natural enemies capable of suppressing *B. tabaci* and *H. armigera* populations. One application of pyriproxifen at the end of July was sufficient to control the whitefly populations through the end of the cotton season, but if an additional treatment was required, buprofezin was recommended (Horowitz *et al.*, 1995).

In addition to whitefly, other sucking pests affecting cotton are *Aphis gossypii*, species of Thrips – *Thrips cinnabarinus*, *Thrips tabaci* – and leafhoppers, especially the cotton jassid, *Empoasca lybica*. The main chewing insects affecting cotton are *H. armigera* and *P. Gossypiella*, *S. littoralis* and *E. insulana*, populations of which increased after a period of relatively low activity. Pheromones are used mainly for monitoring and disruption of *P. gossypiella* populations.

Due to the availability of potent pesticides from both the chitin biosynthesis inhibitors group as well as juvenile hormone mimics, and adherence to the insecticide resistance management programme, a reduction in insecticide applications against the entire range of cotton pests, and especially *B. tabaci*, was observed, but this reduction in the amount of pesticide applications on cotton should be attributed first of all to the introduction of insecticides of high potency, mainly of the IGR group, and a reduction in the number of pest-control applications during one season. The IPM programme was the responsibility of the Cotton Board, with the support of both extension and research staff. The Board centralized the procurement of pesticides, supplying them to producers at a reasonable price. During the cropping season, a state-level pest-control committee comprising representatives of the board, crop husbandry and crop-protection extension, research and growers met weekly to evaluate reports on crop-growth development, pest occurrence and trends in resistance build-up. The weekly recommendations of the committee were then sent by fax to all pest scouts and reached all growers on the day they were drawn up.

Cotton crops were sprayed using aircraft with applications confined to the period from sunrise to 08.30, although in some areas with more temperate climates it was considered that the spraying period might be safely and economically extended, which would materially lower the cost of aerial spraying by increasing the utilization of each aircraft and thus reducing the total number of aircraft required (Lomas *et al.*, 1964). However, concern about spray drift from aircraft indicated a need to develop a technique using tractor-mounted sprayers. This resulted in the addition of an inflatable PVC sleeve, mounted along the spray boom, through which air was forced through a series of holes across the boom at up to 40 m/s to project the spray downwards into the crop canopy.

References

Agrios, G.N. (2005) *Plant Pathology*, 5th edn. Department of Plant Pathology, University of Florida, Elsevier Academic Press.

Akışcan, Y. (2011) Pamukta (*Gossypium hirsutum* L.) *Verticillium* Solgunluğu (*Verticillium dahliae* Kleb.) Hastalığına Dayanıklılık, Erkencilik, Verim ve Kalite Özelliklerinin Kalıtımı. Çukurova Üniversitesi, Fen Bilimleri Enstitüsü, Tarla Bitkileri ABD, Doktora Tezi.

Anon (2000) Pamukta Entegre Mücadele Teknik Talimatı, Tarım ve Köy İşleri Bakanlığı, TAGEM, Bitki Sağlığı Arş. Daire Bşk., Ankara.

Anon (2012) Durum ve Tahmin: Pamuk. Tarımsal Ekonomi ve Politika Geliştirme Enstitüsü. Available at: http://www.tepge.gov.tr (accessed 20 June 2013).

Anon (2018) Yılı Genel Tarım Sayımı sonuçları (www.die.gov.tr 16092020).

Anon (2020a) *Türkiye Genel, Hazır Giyim ve Tekstil Dış Ticareti*. Web sayfası: www.ihkib.org.tr (accessed 17 June 2020).

Anon (2020b) Yılı Pamuk Raporu. T.C. Ticaret Bakanlığı, Esn., San. ve Koop. Gen. Müd. Web sayfası: www.ticaret.gov.tr (accessed 17 June 2020).

Anon (2020c) Türkiye İstatistik Kurumu, Bitkisel Üretim İstatistikleri. Web sayfası: www.tuik.gov.tr (accessed 17 June 2020).

Anon (2020d) Milli Çeşit Listesi: Tarım ve Orman Bakanlığı, TTSM. Web sayfası: www.tarimorman.gov.tr/BUGEM/TTSM (accessed 24 June 2020).

Anon (2020e) *Textile Exchange Organic Cotton*. Web sayfası: store.textileexchange.org (accessed 2020).

Arslan, Z.F. (2018) Şanlıurfa ili pamuk tarlalarında sulama sonrası yabancı otlar ile ilgili yaşanan değişimler, sorunlar ve çözüm önerileri. *Harran Tarım ve Gıda Bilimleri Dergisi* 22, 109–125.

Atkinson G.F. (1891) The black rust of cotton. *The Alabama Agricultural Experimental Station Bulletin* 27, 1–16.

Avila, C. and Gonzalez-Zamora, J.E. (2010) Monitoring resistance of *Helicoverpa armigera* to different insecticides used in cotton in Spain. *Crop Protection* 29, 100–103.

Basal, H., Karademir, E., Goren, H.K., Sezener, V., Doğan, M.N., Gencsoylu, İ. and Erdoğan, O. (2020) Cotton production Turkey and Europe. In: Khawar, J. and Chauhan, B.S. (eds) *Cotton Production*. John Wiley & Sons, pp. 297–315.

Bell, A.A. (2001) *Verticillium* Wilt, In: Kirkpatrick, T.L. and Rothrock, C.S. (eds) *Compendium of Cotton Disease*, 2nd edn. APS Press.

Can, D. and Ödemiş, B. (2017) *Pan ve toprak nem açığına bağlı sulama suyu gereksinimini kurağa dayanıklı ve duyarlı pamuk çeşitlerinde kullanım olanaklarının araştırılması*, Yüksek Lisans Tezi, Mustafa Kemal Üniversitesi, Fen Bilimleri Enstitüsü.

Ceddia, M.G., Gomez-Barbero, M. and Rodriguez-Cerezo, E. (2008) An ex-ante evaluation of the economic impact of Bt cotton adoption by Spanish farmers facing the EU Cotton Sector Reform. *AgBioForum* 11, 82–92.

Coşkun, Z. (2015) Harran ovasında damla sulamanın pamuk verimine etkisi. Harran üniversitesi Fen Bilimleri Enstitüsü Tarımsal Yapılar Ve Sulama Anabilim Dalı (Yüksek Lisans Tezi).

Crozier, C.R. and Cleveland, B.R. (2014) *Fertilization. Cotton Information*. North Carolina Cooperative Extension Service, College of Agriculture and Life Sciences, North Carolina State University.

Demir, İ., Kılıç, G.and Coşkun, M. (2008) Precıs Bölgesel İklim Modeli ile Türkiye için İklim Öngörüleri: Hadamp3 Sres A2 Senaryosu, Iv. Atmosfer Bilimleri Sempozyumu, Bildiriler Kitabı, 365–373.

Doğan, M.N., Jabran, K. and Unay, A. (2014) Integrated weed management in cotton. In: Chauhan, B.S. and Mahajan, G. (eds) *Recent Advances in Weed Management*. Springer, The Netherlands, pp. 197–222. DOI: 10.1007/978-1-4939-1019-9_9

Ektiren, Y. and Değirmenci, H. (2018) Kısıntılı sulama uygulamalarının pamukta (*Gossypium hirsutum* L.) yaprak bitki besin elementlerine etkisi. *KSÜ Tarım ve Doğa Derg* 21, 691–698.
Ertek, A. and Kanber, R. (2003) Effects of different irrigation programs on the lint out-turn of cotton under drip irrigation. *Ksu Journal of Science and Engineering* 6, 106–116.
Esentepe, M., (1979) Adana ve Antalya İllerinde Pamuklarda Görülen Solgunluk Hastalığının Etmeni, Yayılışı, Kesafeti ve Zarar Derecesi ile Ekolojisi Üzerine Araştırmalar. Bölge Zirai Mücadele Araştırma Enstitüsü, Araştırma Eserleri Seri no:32, İzmir.
Gençsoylu, I. (2007) Evaluation of yellow sticky traps on populations of some cotton pests. *American-Eurasian Journal of Agricultural and Environmental Science* 1, 62–67.
Gönen, O. (1999) Çukurova bölgesi yazlık yabancıot türlerinin çimlenme biyolojileri ve bilgisayar ile teşhise yönelik morfolojik karakterlerin saptanması. Doktora Tezi, Çukurova Üniversitesi, 233 s, Adana.
Görmüş, Ö. (2014) *Lif Bitkileri Pamuk*. Çukurova Üniversitesi Ders Kitapları Yayın No: A-93, Adana/ Türkiye.
Haliloğlu, H. and Oğlakçı, M. (2000) Effects of Different Nitrogen Rates on Earliness, Yield and Yield Distribution of Cotton. The Interregional Cooperative Research Network on Cotton. A Joint Workshop and Meeting of the All Working Groups 20–24 September, Adana, Turkey.
Horowitz, A.R., Forer, G. and Ishaaya, I. (1994) Managing resistance in *Bemisia tabaci* in Israel with emphasis on cotton. *Pesticide Science* 42, 113–122.
Horowitz, A.R., Forer, G. and Ishaaya, I. (1995) Insecticide resistance management as a part of an IPM strategy in Israeli cotton fields. *Challenging the Future: Proceedings of the World Cotton Research Conference* CSIRO, Melbourne, Australia, pp. 537–544.
Jabran, K. (2016) Weed flora, yield losses and weed control in cotton crop. 27. Deutsche Arbeitsbesprechung uber Fragen der Unkrautbiologie und -Bekampfung, 23–25 February 2016 in Braunschweig.
Kadıoğlu, İ., Uluğ, E. and Üremiş, İ. (1993) Akdeniz bölgesi pamuk ekim alanlarında görülen yabancıotlar üzerinde araştırmalar. Türkiye I. Herboloji Kongresi (3–5 şubat 1993, Adana) Bildiriler, 151–156.
Kanber, R. (ed.) (1982) Türkiye'de Sulanan bitkilerin Su tüketim Rehberi. TOPRAKSU Genel Müdürlüğü Yay., No. 718, sayfa: 630. Ankara.
Karademir, Ç., Karademir, E., Doran, I., ve Altıkat, A. (2005) Gaziosmanpaşa Üniversitesi, *Ziraat Fakültesi Dergisi, cilt* 22, 55–61.
Kaydan, M.B., Calıskan, A.F. and Ulusoy, M.R. (2013) New record of invasive mealybug *Phenacoccus solenopsis* Tinsley (Hemiptera: Pseudococcidae) in Turkey. *Bulletin OEPP/EPPO Bulletin* 43, 169–171.
Kazgöz Candemir, D. and Ödemiş, B. (2018) Yapraktan uygulanan farklı kükürt dozlarının pamuk bitkisinin (*Gossypium hirsutum* L.) değişik gelişme dönemlerindeki su stresinin azaltılması üzerine etkileri. *Derim* 35, 161–172.
Lomas, J., Frankel, H. and Hirsch, I. (1964) Meteorological considerations in determining the permissible time for cotton spraying from the air in Israel. *Agricultural Meteorology* 1, 225–234.
Mart, C. (2017) Pamukta entegre üretim. Ankamat matbacılık, ISBN: 975-98356-0-6, Hatay.
Mironidis, G.K., Kapaantaidaki, D., Bentila, M., Morou, E., Savopoulou-Soulyani, M. and Vontas, J. (2013) Resurgence of the cotton bollworm *Helicoverpa armigera* in northern Greece associated with insecticide resistance. *Insect Science* 20, 505–512.
Morton, N. (1982) The 'Electrodyn' sprayer: first studies of spray coverage in cotton. *Crop Protection* 1, 27–54.
Nemli, T. (2003) 'Pamuk Hastalıkları ve Savaşım Yöntemleri', Pamukta Eğitim Semineri, 14–17 Ekim, 103–111, Bornova/zmir.
Nemli, T. and Sayar, İ. (2002) Aydın Söke yöresinde pamuk çökerten hastalığının yaygınlığı, etmenlerinin ve önlenme olanaklarının araştırılması, Türkiye Bilimsel ve Teknik Araştırma Kurumu (TÜBİTAK), Tarım Orman ve Gıda Teknolojileri Araştırma Grubu (TOGTAG), Türkiye Tarımsal Araştırma Projesi (TARP), Proje No: TARP-2535.
Ödemiş, B., Akışcan, Y., Akgöl, B. and Can, D. (2017) Kısıtlı su koşullarında yapraktan uygulanan kükürt dozlarının pamuk bitkisinin kuraklık toleransına etkileri. 214–254 numaralı Tübitak projesi.
Oerke, E.C. (2006) Crop losses to pests: centenary review. *The Journal of Agricultural Science* 144, 31–43.
Oğlakçı, M., Bölek, Y. ve Çopur, O, (2010) *Pamuk bitkisinde zararlanma, ürün kayıpları ve dayanıklılık*. Ticaret borsası yayınları-4, şanlıurfa, Turkey.
Özbek, N. (2017) Türk pamuklarında standardizasyonun gelişimi ve Türk pamuklarının durumu. *TÜRKTOB Dergisi* 21, 45–48.
Özer, M.S. and Dağdeviren, İ. (1986) Harran Ovası Koşullarında Pamuğun Azotlu Gübre İsteği. Köy Hizmetleri Araştırma Enstitüsü Müdürlüğü Yayınları No:25, Şanlıurfa, Türkiye.

Özer, S.M. (1992) Harran Ovası Koşullarında Pamuğun Fosforlu Gübre İsteği, Köy Hizm. Araştırma Ens. Müd., Yayın No: 25, Rapor Serisi No:17, Şanlıurfa, Türkiye.
Özkil, M., Serim, A.T., Torun, H. and Üremiş, İ. (2019) Determination of weed species, distributions and frequency in cotton (*Gossypium hirsutum* L.) fields of Antalya Province. *Turkish Journal of Weed Science* 22, 185–191.
Sağır, A., Tatlı, F. ve Gürkan, B. (1995) Güneydoğu Anadolu Bölgesinde Pamuk Ekim Alanlarında Görülen Hastalıklar Üzerine Çalışmalar, GAP Bölgesi Bitki Koruma Sorunları ve Çözüm Önerileri Sempozyumu, 27–29 Nisan, sayfa 5–9, *Şanlıurfa*.
Sarı, Ö. ve Dağdelen, N. (2010) Damla sulama yöntemiyle sulanan pamukta farklı lateral aralıklarının pamuk su-verim ilişkileri üzerine etkileri. *ADÜ Ziraat Fakültesi Dergisi* 7(1), 41–48.
Sezgin, E. (1985) Pamuk Solgunluk Hastalığı ile Savaşımda Kültürel İşlemlerin Önemi. *Yıllık* 3, 23–31.
Sönmez, B., Özbahçe, A., Akgül, S. ve Keçeci, M. (2018) Türkiye topraklarının bazı verimlilik ve organik karbon (TOK) içeriğinin coğrafi veritabanının oluşturulması. T.C. Tarım ve Orman Bakanlığı, Tarımsal Araştırmalar ve Politikalar Genel Müdürlüğü, Toprak Gübre ve Su Kaynakları Merkez Araştırma Enstitüsü, Proje no: TAGEM/TSKAD/11/A13/P03, Ankara.
Stavridis, D.G., Gliatis, A., Deligeorgidis,P.N., Giatropoulos, C., Giatropoulos, A., Deligeorgidis, N.P. and Ipsilandis, C.G. (2008) Cotton production in the presence of *Helicoverpa armigera* (Hb.) in central Greece. *Pakistan Journal of Biological Sciences* 11, 2490–2494.
Tepe, I. (2014) Yabancı Otlarla Mücadele. Sidas Medya Ziraat Yayın No:031, İzmir, 292s.
Torres-Vila, L.M., Rodriguez-Molina, M.C., Lacasa-Plasencia, A. and Bielza-Lino, P. (2002) Insecticide resistance of *Helicoverpa armigera* to endosulfan carbamates and organophosphates: the Spanish case. *Crop Protection* 21, 1003–1013.
Tozan, Ş., (1990) Büyük Menderes Havzası Topraklarında Azot, Fosfor ve Potasyum Gübrelerinin Pamuğun Topraktan Kaldırdığı Besin Maddesi Miktarları ve Bazı Lif Kalitesi Üzerine Etkiler*i*. Ege Üniv. Fen Bilimleri Ens, Toprak Anabilim Dalı, Doktora Tezi, İzmir, Türkiye.
Uludağ, A. and Üremiş, İ. (2000) A perspective on weed problems of cotton in Turkey: the inter-regional cooperative research network on cotton. Proceedings of a Joint Workshop and Meeting of the All Working Groups, 20–24 September, Cukurova University Press, Adana, Turkey, pp. 194–199.
Üremiş, İ., Soylu, S., Kurt, Ş., Soylu, E.M. ve Sertkaya, E. (2020) Hatay ili havuç ekim alanlarında bulunan yabancı ot türleri, yaygınlıkları, yoğunlukları ve durumlarının değerlendirilmesi. *Tekirdağ Ziraat Fakültesi Dergisi* 17, 211–228.
Uygur, S., (1997) Çukurova Bölgesi yabancı ot türleri, bu türlerin konukçuluk ettiği hastalık etmenleri ve dağılımları ile hastalık etmenlerinin biyolojik mücadelede kullanılma olanaklarının araştırılması. Doktora Tezi, Çukurova Üniversitesi, 148s., Adana.
Wilhelm, S., Sagen, J.E. and Tietz, H. (1974a) Resistance to *Verticillium* wilt in cotton: sources, techniques of identification, inheritance trends, and the resistance potential of multiline cultivars. *Phytopathology* 64, 924–931.
Wilhelm, S., Sagen, J.E. and Tietz, H., (1974b) *Gossypium hirsutum* subsp. mexicanum var. nervosum, Leningrad Strain, a source of resistance to *Verticillium* Wilt. *Phytopathology* 64, 924–931.
Yazıcı, G. and Sertkaya, E., (2020) A new pest of *Gossypium hirsutum* in Turkey: *Leptodemus minutus* Jakovlev 1876 (Hemiptera: Heteroptera: Lygaeidae: Oxycareninae). *Mustafa Kemal Üniversitesi Tarım Bilimleri Dergisi* 25, 256–261.
Yildirim, A., Işık, D., Bülbül, F. and Kaçan, K. (2008). Zirai Mücadele Teknik Talimatları, Cilt: VI. Gıda Tarım ve Hayvancılık Bakanlığı, Tarımsal Araştırmalar ve Politikalar Genel Müdürlüğü Yayınları, Ankara, 286s.

14 A Look Forward

Graham Matthews*

There have been many changes in the way cotton plants have been grown over the centuries, with mechanized harvesting and processing of the seed cotton significantly reducing the number of people needed to grow the crop. However, yields remained low as insect pests, diseases and weeds affected the plants, reducing the number of bolls that could be harvested. The period of relying on spraying pesticides undoubtedly made an enormous difference and yields have significantly risen, especially when crops could be irrigated. However, the management of the use of pesticides failed as although a particular product was effective and economical to use, the amount of pesticide applied in a season tended to be too much, allowing the pests to select a resistance mechanism which ultimately resulted in the pesticide being no longer effective, even when higher doses or more sprays were applied.

Genetically Modified Cotton Varieties

The ability to develop new plant-breeding technology and change the genetic make-up of the variety enabled plants to express the toxins of *Bacillus thuringiensis*. The new transgenic cotton varieties have been principally concerned with *Gossypium hirsutum*. However, with plants designed to produce fibres, the expression of the initial Bt genes used initially ceased when bolls were being produced and matured. Additional genetic manipulation soon overcame this initial set back, but it is likely that some resistance will develop. This seems inevitable as the toxins are insecticides incorporated within the plant.

Wild cottons and *G. barbadense* tend to have a higher content of gossypol, which resulted in fewer bollworms compared with *G. hirsutum* in the Sudan. Such differences were clearly shown when glandless varieties were compared to those with glands having a toxin, known as gossypol (Jenkins *et al.*, 1966). Insects were also shown to survive better on varieties with low gossypol content compared with conventional cotton (Hagenbucher *et al.*, 2019). However, many want less gossypol in cottonseeds so they can be used for feeding animals, as acute clinical signs of gossypol poisoning can occur where high concentrations of free gossypol are in the seeds (Gadelha *et al.*, 2014).

In 2019, the US Food and Drug Administration gave the green light to ultra-low gossypol cottonseed (ULGCS) to be utilized as human food and in animal feed. Dr Rathore (Texas A&M

*g.matthews@imperial.ac.uk

DOI: 10.1079/9781800620216.0014

Table 14.1. Global farm income benefits from growing GM crops 1996–2014 (millions US$).

Trait	Increase in farm income 2014	Increase in farm income 1996–2014	Farm income benefit in 2014 as % of total value of production of these crops in GM-adopting countries	Farm income benefit in 2014 as % of total value of global production of crop
GM insect-resistant cotton	3940.8	44,834.3	12.5	8.9
GM herbicide-tolerant cotton	146.5	1654.2	0.5	0.3

Note: All values are nominal. From Brookes and Barfoot (2016) *GM Crops: Global Socio-economic and Environmental Impacts* 1996–2014. PG Economics Ltd, Dorchester, UK.

Institute for Plant Genomics and Biotechnology) developed, tested and obtained deregulation for the transgenic cotton plant TAM66274. This unique cotton plant with ultra-low gossypol levels in the seed makes the protein from the seeds safe to consume, but also maintains normal plant-protecting gossypol levels in the rest of the plant, making it ideal for the traditional cotton farmer [https://today.tamu.edu/2019/10/14/fda-approves-ultra-low-gossypol-cottonseed – accessed 12 August 2021]

The present system of growing Bt cotton is to have an area of non-Bt cotton to slow down the selection of resistant bollworms or other pests. Wan *et al.* (2017) have shown that by crossing transgenic Bt plants with conventional non-Bt plants and then sowing the second-generation seeds provides a random mixture within fields of three quarters of plants that produce Bt protein and one quarter that do not. A study showed this strategy countered resistance to Bt cotton of pink bollworm, indicating that this seed mixture can improve the survival of susceptible insects and delay resistance. Similarly, by growing varieties designed to make the cotton plant tolerant of certain herbicides has failed when vast areas of cotton tolerant to glyphosate is grown alongside other crops such as soybean and maize, which also express tolerance to glyphosate. Inevitably, the weeds become resistant to the herbicide.

In Turkey, due to EU banning GM crops, plant breeders have, nevertheless, improved the yield potential of cotton varieties by more conventional means. Simply growing Bt cotton using imported seed in Burkina Faso was not the answer as lint quality was soon regarded as poor, so local plant selection was clearly still crucial.

These studies illustrate the importance of research in the different environments in which cotton is grown to enable farmers to benefit from new technology.

Plant breeding needs to be a key area of research to ensure the plants produce the quality of cotton to meet the requirements of the textile industry, while incorporating other important traits such as drought and heat-tolerance, especially with changes in climate. High yields need to provide the natural cotton fibres at a profitable price that can compete with synthetic fibres.

In Africa, since South Africa began to try GM technology, other countries have started since 2018. These include Sudan, Eswatini, Malawi, Nigeria and Ethiopia, with Mozambique, Niger, Ghana and Zambia expected to do so.

Resistance to Pesticides

The mistake with chemical pesticides is that the widespread use of relatively low-cost pesticide, being more economic to use than manual weeding or crop inspections to assess pest populations, inevitably results in the pest developing resistance. Many farmers have assumed that a higher dose of the pesticide is needed, or it should be mixed with another product, but this does not answer the increased resistance by the pest.

Resistance Management

One technique that can delay resistance depends on reducing the exposure of the pests to a

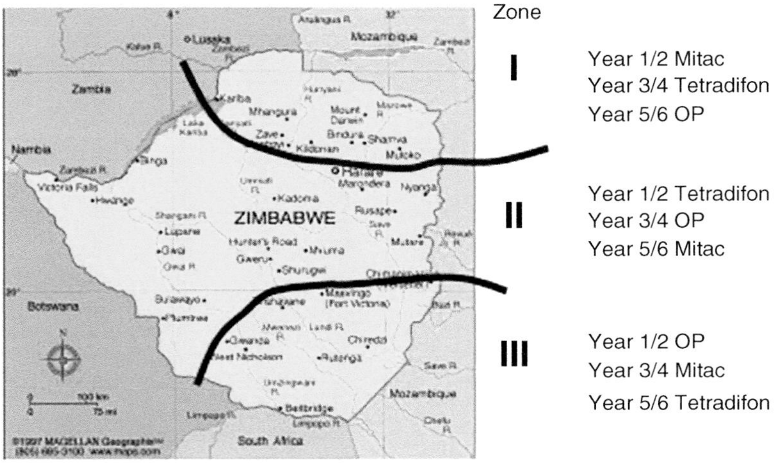

Fig. 14.1. Map of Zimbabwe showing zones for different acaricides.

particular pesticide by using a rotation of pesticides with different modes of action. Thus, from 1973, Zimbabwe was divided into three zones. The control of red spider mites was achieved with an acaricide rotation scheme in which red spider mites were exposed to one particular mode of action in zone 1 for no longer than two seasons, before moving to zone 2 for two years and then to zone 3. It could return to zone 1 after four years elsewhere in the country. The system needed acaricides with three different modes of action, managed on a national scale to provide distinct separation of the areas where different modes of action are used. Similarly, farmers need to rotate herbicide chemistries and include cultural and mechanical methods to avoid overusing some herbicides. Hopefully, with the development of new insecticides, such as spiropidion, regarded as kind to nature but hard on sucking pests, management of its use will minimize the risk of the targeted pests soon becoming resistant to it.

Pesticide Application

Whether GM cotton is sown or not, there may be occasions, when pest populations exceed the economic threshold. Development of new biopesticides or selective insecticides may fit into an IPM programme, but if a more conventional insecticide is applied, efforts should be made to minimize the dose applied. In Zimbabwe, the dose of insecticides was noted to be about a third of that recommended in the USA, as the spray distribution and especially the timing of sprays was aimed at controlling the first instar bollworm larvae before they entered a bud or boll.

Applying pesticides diluted with water inevitably meant spray deposits were prone to the impact of rain, thus moving the chemical into the soil or ultimately into a stream and then to a river. Perhaps in the future the importance of a formulation, applied at ultra-low volume, perhaps based on a vegetable oil, might be recognized. Development of formulations using nanoparticles may also allow lower dosages to be effective.

Use of biopesticides will increase, but their application will need careful research so that the spray reaches where the pests are located as their effect will usually require close contact with the pests.

Crop Rotation

Cotton, being a deep-rooted crop, is ideally sown in rotation with other crops, including maize and a legume, such as mung bean, pigeon pea, chickpea, faba bean, dolichos or vetch, depending on the local climate (Williams *et al.*, 2005). The maize yields are often better following cotton, while cotton crops benefit by the nitrogen fixation of legumes. Routine rotation from one crop to another is claimed to reduce disease pressure, reduce weeds, result in less insect damage, and improved nutrient levels in the soil. Many farmers are also exploring intercropping cotton and wheat, especially in China (Dai and Dong, 2016).

Plant Spacing

Rows have traditionally been sown 0.9–1.0 m apart, but in some situations a narrower plant spacing has been used. This may depend on equipment, such as a stripper harvester, if used. Intra-row spacings are not critical, although generally seed has been sown at 15–22 cm apart. With modern equipment, the optimum plant population on irrigated fields is 90,000–120,000 plants/ha, but on rainfed crops the population is lower at 45,000–60,000 plants/ha. There may be higher plant populations – 220,000 plants/ha – with a narrow row spacing, 0.52 m inter-row spacing, in conjunction with use of a stripper harvester. Future designs of mechanical harvesting equipment are likely to determine plant and row spacing.

Irrigation

Cotton plants are remarkable in surviving long dry periods once the deep root has been established, but yields have been significantly increased where the crop can be irrigated in areas where rainfall is not reliable and is essential, especially in very arid areas. Water is nevertheless not always available when needed so in some countries, provision of small reservoirs is needed. What is particularly important is a system that minimizes the amount of water used in the cotton fields but does not affect the yield (Pereira *et al.*, 2002). Drip irrigation is more efficient than furrow or sprinkler irrigation (Cetin and Bilgel, 2002), but is more expensive to install and maintain. In Uzbekistan, 18–42% of water was saved by using drip irrigation instead of furrow (Ibragimov *et al.*, 2007).

Drones

An unmanned aerial vehicle (UAV), usually referred to as a drone, fitted with a sprayer was developed in Japan and has been used predominately to apply pesticides over rice crops since 1990. The technology has developed significantly since then and drones have become very popular in China to spray cotton (He *et al.*, 2017). The farmers prefer the drone, which has sufficient downward airflow to provide good distribution of spray, so avoid carrying a heavy knapsack sprayer through the crop. The small drones have a limited capacity to carry a spray liquid and the operational duration is restricted, so these are used on relatively small farms. Larger drones fitted with a fixed wing and with sensors and AI-vision technology offer the benefit of speed and accuracy in application and, potentially, will replace the use of large tractor-mounted or trailed sprayers with very large spray tanks. This will reduce soil compaction in fields, while enabling large fields to be treated rapidly. The speed of these drones can be equivalent to tractor speeds in contrast to conventional aerial spraying and the nozzles will be closer to the crop canopy than with aircraft, so down-flow of air should improve spray coverage within the crop canopy. With rotary nozzles to select a narrower droplet spectrum and apply low or ultra-low volumes/ha, treatments can be applied much quicker when required in relation to economic thresholds. Drones have already been used to map fields and detect specific weeds and plant damage due to fungal diseases or nutrient deficiency, although for some pests, field inspection and use of pheromone traps may be still needed for certain pests. Utilization of modern devices, such as mobile phones, may also provide new ways of recording insect counts or assist in identification of pests. They have also been used in China to apply defoliants to facilitate harvesting and studies have indicated that drones could be used to assess crop nitrogen status to indicate whether fertilizer can be applied and still have an effect on yield (Ballester *et al.*, 2017).

Based on these results, when using UAV to assess cotton variability in crop nitrogen status, monitoring the simplified canopy chlorophyll content index (SCCCI) was the index that best explained the variation in N uptake and plant N% between treatments. An SCCCI at early stages of the crop was recommended, from first flower to peak bloom, when the variability can be detected, and fertilizer applications can still have an effect on lint yield. Moreover, there is the possibility of using the normalized difference red edge (NDRE) index measurements at maturity as a potential tool to identify areas with higher or lower defoliant application needs to make prescriptive applications to increase harvest efficiency.

Weed Management/No-till

Farmers have used ploughs to remove weeds and prepare the surface prior to sowing, but this adversely affects many organisms in the soil. Instead of ploughing, 'no-till' has been adopted in some areas to decrease soil erosion and improve water infiltration. Without ploughing, fields need some minimal preparation of the surface to sow the next crop. Such fields are usually also sown with a cover crop in the inter-row between the rows of cotton. On small farms, traditionally, weeds growing after sowing a crop were removed manually by hand or with hoes, but the introduction of herbicides avoided loosening the soil surface and damage to cotton plants. The genetic manipulation of crop varieties to achieve herbicide tolerance has resulted in simplifying the selection of herbicides, but weeds can become resistant to a herbicide, if it is used too often, notably when a particular herbicide is used with genetically modified, herbicide-tolerant cotton varieties. The development of mapping and spot treatment could reduce the selection pressure resulting in resistant weeds. This could be achieved either with drones or robotic ground equipment. Similarly, with identification of weeds from crop plants, mechanical removal of weeds has been achieved with certain crops (McAllister *et al.*, 2019). A study has already shown that early- to mid-season weeds can be mapped in cotton using a drone utilizing high spatial resolution true-colour imagery to determine weed density in cotton (Sapkota *et al.*, 2020).

Robots

The development of robots could provide small-scale farms with a means of mechanical harvesting using a smaller version of equipment currently used on large farms. Where sufficient labour is not available for harvesting or is becoming too costly, mechanical harvesting is likely to increase. Fue *et al.* (2020) examined current opportunities and challenges for robotic agricultural cotton harvesting research and commercial development and indicated that the robots will need effective sensing systems and manipulators to locate and pick cotton bolls, including those at the bottom of the plants. The power requirements, footprint and cost of such robots that pick one boll every three seconds will be critical as harvesting by hand has a distinct advantage of producing seed cotton with less trash and thus cleaner lint.

Integrated Pest Management (IPM)

The concept of IPM was advocated in the USA as restricting insecticide use by monitoring the crops and determining an economic threshold to determine when a spray should be applied. For many, it implied endeavouring to use biological controls, but in Africa, IPM began by the introduction of a pubescent variety resistant to jassids in the 1920s; later, the addition of the policy of a closed season combined with banning the ratooning of plants, as pests could build up on these plants before the cotton season. Later, a disease-resistant variety was added and the timing of insecticide applications following scouting the pests enabled the significant yield increases in the 1960s.

Growing a variety with pubescent leaves also had the advantage when insecticides were sprayed as the hairy leaf surfaces effectively collected the spray and the hairs slowed the movement of young larvae attempting to reach a bud or boll. The crucial factor was developing a variety suitable for the local environment. Looking ahead, improved plant-breeding technology will be vital to get varieties more resistant to pests, which can be integrated with crop rotations, together with conservation areas to encourage the survival of beneficial insects, including bees. Although cotton is considered to be self-pollinating, pollination by honey bees and wild bees assessed in Burkina Faso showed significantly increased yield quantity and quality, on average by up to 62%, while exclusion of

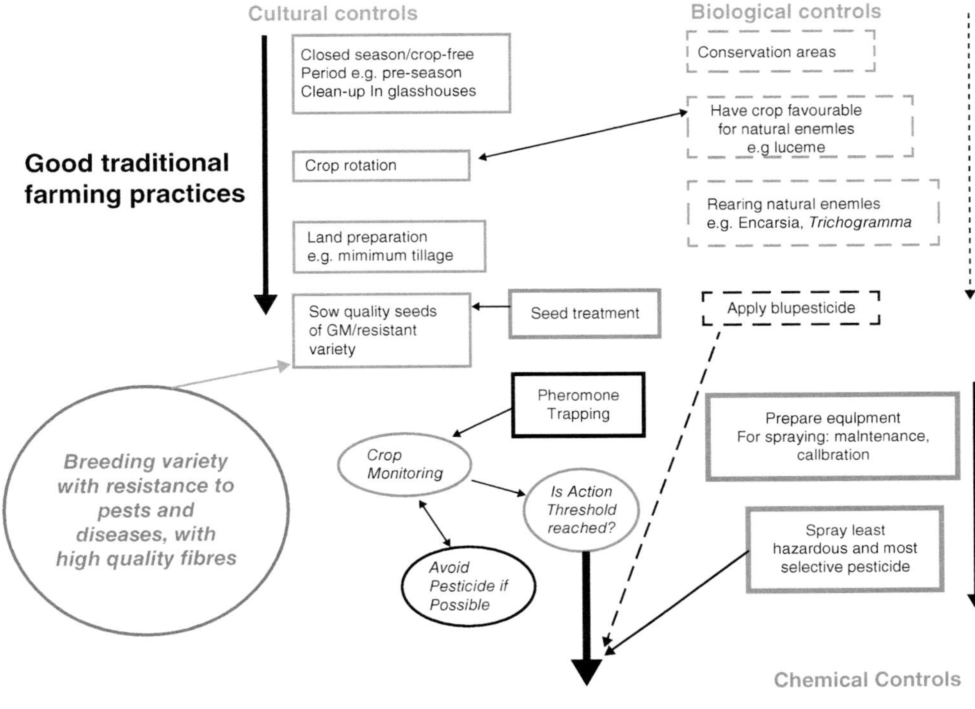

Fig. 14.2. Factors to be considered in integrated pest management.

pollinators caused an average yield gap of 37% (Stein *et al.*, 2017).

Organic Cotton

A number of countries have areas in which the aim has been to avoid using 'chemical' insecticides and protect their crops with predators and parasites and using natural products, such as neem as biopesticides. The most extensive system was having biofactories established to produce natural enemies, such as *Trichogramma* in Uzbekistan. Smaller attempts elsewhere have also examined the techniques used.

In Australia, trials investigated attraction of natural enemies into cotton fields, using food baits sprayed on the cotton crop and having lucerne or similar crop nearby to maintain an area in which natural enemies of cotton pests could develop. Some of the techniques used in Australia have been tried in other countries, such as Benin and Ethiopia, supported by the Pesticide Action Network (PAN) and the Better Cotton Initiative (BCI), which is a non-profit, multi-stakeholder governance group promoting better standards in cotton farming and practices across 21 countries. Similar schemes have operated on a limited area and efforts have switched to the adoption of growing genetically modified cotton varieties, despite many people opposing this technology.

An innovative way of growing cotton has been developed in the UK by growing cotton plants hydroponically within a controlled environment. The plants receive fertilizer via a drip-irrigation system and when pest insects are detected, predatory insects are deployed. This growing approach reduces the quantity of water and fertilizers needed by up to 80%, leads to 100% pesticide-free cultivation and can increase yield and fibre quality when compared to numerous cotton-growing regions. The venture is aimed to effectively address the UN's Sustainable Development Goals 6, 12, 15 and 17 and is currently expanding with a two-hectare pilot farm in India.

Fig. 14.3. Hydroponic pima cotton growing in a controlled environment in the UK.

The use of hydroponics might prove to be suitable in California where a shortage of water has predicted a decrease in acreage of cotton to be grown in 2021, which may represent a 34% decrease in pima acreage and a 14.3% decrease in upland acreage as compared to 2020. Drip-irrigation techniques are already in use in Asia to minimize water consumption.

Climate Change

Concern about changes in the climate has increased due to increasing emissions of greenhouse gas – CO_2, methane and nitrous oxide – levels in the atmosphere following the use of coal and oil in the industrial world over the last two centuries. Apart from the average increase in temperatures, some areas have experienced much hotter periods, while other areas have had higher rainfall. An elevated temperature may improve yields regardless of atmospheric CO_2 if sufficient water is available, but an elevated temperature will not mitigate any negative effects due to rising temperature on cotton growth (Broughton *et al.*, 2016). Time of sowing may need adjustment depending on how factors such as rainfall and temperature impact on the crop. Much will depend on breeding varieties adapted to the changes in climate and improvements in water management.

Knowledge Transfer

Historically, farmers planning to grow cotton could rely on a government-supported research team which selected improved varieties and developed agronomic and crop-protection advice, which was passed on to the farmers by a national extension service. This provided leaflets and training courses plus advice at field days and other meetings. This independent advice linked the farmers with the research programme, which facilitated farm-scale trials.

Although some countries, such as the USA, still have an effective extension programme, in many countries the extension service has lacked government support, so farmers have had to seek advice more from commercial organizations. Modern social media have, in some cases, provided information. While the identification of pests and diseases might be made easier, the more practical aspects of how to manage a crop are not necessarily taught adequately for farmers.

With so many new developments and the changes in climate, it is evident that an International Cotton Centre, similar to the centres within CGIAR (Consultative Group for International Agricultural Research), should be established to facilitate the development of improved cotton varieties through gene editing and other new technology and make the outcome of research to be freely available to resource-poor farmers.

References

Ballester, C., Hornbuckle, J., Brinkhoff, J., Smith, J. and Quayle, W. (2017) Assessment of in-season cotton nitrogen status and lint yield prediction from unmanned aerial system imagery. *Remote Sensing* 9, *1149.*

Broughton, K.J., Smith, R.A., Duursma, R.A., Tan, D.K.Y., Payton, P. *et al.*, (2016) Warming alters the positive impact of elevated CO_2 concentration on cotton growth and physiology during soil water deficit. *Functional Plant Biology* 44, 267–278.

Cetin, O. and Bilgel, L. (2002) Effects of different irrigation methods on shedding and yield of Cotton. *Agricultural Water Management* 54, 1–15.

Dai, J. and Dong, H. (2016) Farming and cultivation technologies of cotton in China. *DOI*: 10.5772/64485

Fue, K.G., Porter, W.M., Barnes, E.M. and Rains, G.C. (2020) An extensive review of mobile agricultural robotics for field operations: focus on cotton harvesting. *AgriEngineering* 2, 150–174.

Gadelha, I.C.N., Fonseca, N.B.S., Catarina, S., Oloris, S., Melo, M.M. and Soto-Blanco, B. (2014) Gossypol toxicity from cottonseed products. *The Scientific World Journal*, article ID 231635.

Hagenbucher, S., Eisenring, M., Meissle, M., Rahore, K.S. and Romeis J. (2019) Constitutive and induced insect resistance in RNAi-mediated ultra-low gossypol cottonseed cotton. *BMC Plant Biology* 19, article ID 322.

He, X.K., Bonds, J., Herbst, A. and Langenakens, J. (2017) Recent development of unmanned aerial vehicle for plant protection in East Asia. *International Journal of Agricultural and Biological Engineering* 10, 18–30.

Ibragimov, N., Evett, S.R., Esanbekov, Y., Kamilov, B.S., Mirzaev, L. and Lamers, J.P.A. (2007) Water use efficiency of irrigated cotton in Uzbekistan under drip and furrow irrigation. *Agricultural Water Management* 90, 112–120.

Jenkins, J.N., Maxwell, F.G. and Lafever, H.N. (1966) The comparative preference of insects for glanded and glandless cotton. *Journal of Economic Entomology* 59, 352–356.

McAllister, A., Osipychev, D., Davis, A. and Chowdhary, G. (2019) Agbots: weeding a field with a team of autonomous robots. *Computers and Electronics in Agriculture* 163.

Pereira, L.S., Oweis, T. and Zairi, A. (2002) Irrigation management under water scarcity. *Agricultural Water Management* 57, 175–206.

Sapkota, B., Singh, V., Cope, D., Valasek, J. and Bagavathiannan, M. (2002) Mapping and estimating weeds in cotton using unmanned aerial systems-borne imagery. *AgriEngineering* 2, 350–366.

Stein, K., Coulibaly, D., Stenchly, K., Goetze, D., Porembski, S., Lindner, A., Konaté, S. and Linsenmair, E.K. (2017) Bee pollination increases yield quantity and quality of cash crops in Burkina Faso, West Africa. *Scientific Reports* 7, 17691. DOI: 10.1038/s41598-017-17970-2

Wan, P., Xu, D., Cong, S., Jiang, Y., Huang, Y. *et al.* (2017) Hybridizing transgenic Bt cotton with non-Bt cotton counters resistance in pink bollworm. *Proceedings of the National Academy of Sciences of the United States of America* 114, 5413–5541.

Williams, E., Rochester, I. and Constable G. (2005) Using legumes to maximise profits in cotton systems. *The Australian Cottongrower* 26, 43–46.

Index

Note: Page numbers in *italics* denote tables and figures